UNTOLD TALES FROM THE BOOK OF REVELATION

Society of Biblical Literature

Resources for Biblical Study

Tom Thatcher, New Testament Editor

Number 79

UNTOLD TALES FROM THE BOOK OF REVELATION

SEX AND GENDER, EMPIRE AND ECOLOGY

By

Stephen D. Moore

SBL Press
Atlanta

Copyright © 2014 by SBL Press

All rights reserved. No part of this work may be reproduced or transmitted in any form or by any means, electronic or mechanical, including photocopying and recording, or by means of any information storage or retrieval system, except as may be expressly permitted by the 1976 Copyright Act or in writing from the publisher. Requests for permission should be addressed in writing to the Rights and Permissions Office, SBL Press, 825 Houston Mill Road, Atlanta, GA 30329 USA.

Library of Congress Cataloging-in-Publication Data

Moore, Stephen D., 1954– author.
  Untold tales from the Book of Revelation : sex and gender, empire and ecology / by Stephen D. Moore.
      p. cm. — (Society of Biblical Literature Resources for Biblical Study ; number 79)
  ISBN 978-1-58983-990-8 (paper binding : alk. paper) — ISBN 978-1-58983-992-2 (electronic format) — ISBN 978-1-58983-991-5 (hardcover binding : alk. paper)
  1. Bible. Revelation—Criticism, interpretation, etc. 2. Sex in the Bible. 3. Politics in the Bible. 4. Ecology—Biblical teaching. I. Title. II. Series: Resources for biblical study ; no. 79.
  BS2825.52.M66 2014
  228'.06—dc23   2014027587

Cover art by Robert Roberg; The Whore of Babylon, c. 1991, acrylic and glitter on canvas. From the collection of Albert Keiser Jr., Hickory, NC. Promised gift to the Hickory Museum of Art. Photo by Margaret Day Allen. Cover design by Kathie Klein.

Printed on acid-free, recycled paper conforming to
ANSI/NISO Z39.48-1992 (R1997) and ISO 9706:1994
standards for paper permanence.

Contents

Abbreviations vii

1. What Is, What Was, and What May Yet Be 1

2. Mimicry and Monstrosity 13

3. Revolting Revelations 39

4. Hypermasculinity and Divinity 75

5. The Empress and the Brothel Slave (co-authored with
   Jennifer A. Glancy) 103

6. Raping Rome 125

7. Retching on Rome 155

8. Derridapocalypse (co-authored with Catherine Keller) 179

9. Quadrupedal Christ 201

10. Ecotherology 225

Bibliography 245

Index of Modern Authors 285

Abbreviations

AB	Anchor Bible
ACCS NT	Ancient Christian Commentary on Scripture New Testament
ACNT	Augsburg Commentaries on the New Testament
ANRW	*Aufstieg und Niedergang der römischen Welt: Geschichte und Kultur Roms im Spiegel der neueren Forschung.* Edited by Hildegard Temporini and Wolfgang Haase. Berlin: de Gruyter, 1972–.
AOTC	Abingdon Old Testament Commentaries
BADG	W. Bauer, W. F. Arndt, F. W. Gingrich, and F. W. Danker. *Greek-English Lexicon of the New Testament and Other Early Christian Literature.* 2nd ed. Chicago: University of Chicago Press, 1979.
BibInt	*Biblical Interpretation*
BNTC	Black's New Testament Commentaries
BTB	*Biblical Theology Bulletin*
BWANT	Beiträge zur Wissenschaft vom Alten und Neuen Testament
EPRO	Études préliminaires aux religions orientales dans l'Empire romain
FRLANT	Forschungen zur Religion und Literatur des Alten und Neuen Testaments
ExpTim	*Expository Times*
HDR	Harvard Dissertations in Religion
HNT	Handbuch zum Neuen Testament
HTR	*Harvard Theological Review*
HTS	Harvard Theological Studies
IBC	Interpretation: A Bible Commentary for Teaching and Preaching
ICC	International Critical Commentary

Int	*Interpretation: A Journal of Bible and Theology*
JAC	Jahrbuch für Antike und Christentum
JAOS	*Journal of the American Oriental Society*
JBL	*Journal of Biblical Literature*
JECS	*Journal of Early Christian Studies*
JÖAI	*Jahreshefte des Österreichischen archäologischen Institutes*
JRS	*Journal of Roman Studies*
JSNT	*Journal for the Study of the New Testament*
JSNTSup	Journal for the Study of the New Testament Supplement Series
JSOTSup	Journal for the Study of the Old Testament Supplement Series
JTS	*Journal of Theological Studies*
KEK	Kritisch-Exegetischer Kommentar über das Neue Testament
LCL	Loeb Classical Library
LNTS	Library of New Testament Studies
LSJ	H. G. Liddell, R. Scott, H. S. Jones. *A Greek-English Lexicon*. 9th ed. with revised supplement. Oxford: Oxford University Press, 1996.
NCB	New Century Bible
NIBCNT	New International Biblical Commentary on the New Testament
NICNT	New International Commentary on the New Testament
NIGTC	New International Greek Testament Commentary
NovTSup	Supplements to Novum Testamentum
NPNF 1	*The Nicene and Post-Nicene Fathers*, Series 1. Edited by Philip Schaff. 1886–1889. 14 vols. Repr. Peabody, Mass.: Hendrickson, 1994.
NTD	Das Neue Testament Deutsch
NTS	*New Testament Studies*
ÖTK	Ökumenischer Taschenbuch-kommentar
OTL	Old Testament Library
OTP	*Old Testament Pseudepigrapha*. Edited by James H. Charlesworth. 2 vols. New York: Doubleday, 1983.
PMLA	*Publications of the Modern Language Association*
RB	*Revue biblique*
SBLEJL	Society of Biblical Literature Early Judaism and Its Literature

SBLRBS	Society of Biblical Literature Resources for Biblical Study
SBLSP	*Society of Biblical Literature Seminar Papers*
SBLSymS	Society of Biblical Literature Symposium Series
SEJC	Studies in Scripture in Early Judaism and Christianity
SemeiaSt	Semeia Studies
SNTSMS	Society for New Testament Studies Monograph Series
SP	Sacra pagina
TDNT	*Theological Dictionary of the New Testament*. Edited by Gerhard Kittel and Gerhard Friedrich. Translated by Geoffrey W. Bromiley. 10 vols. Grand Rapids: Eerdmans, 1964–76.
TLG	*Thesaurus linguae graecae: Canon of Greek Authors and Works*. Edited by L. Berkowitz and K. A. Squitier. 3rd ed. Oxford: Oxford University Press, 1990.
USQR	*Union Seminary Quarterly Review*
VCSup	Supplements to Vigiliae Christianae
WBC	Word Biblical Commentary
WUNT	Wissenschaftliche Untersuchungen zum Neuen Testament
ZBK	Zürcher Bibelkommentare

1
What Is, What Was, and What May Yet Be

> Do not seal up the words of the prophecy of this book, for the time is near. (Rev 22:10)

Revelation does not carry its warning on its label. It is only when we have devoured the book, avidly read it right through, that we learn that its seal was always already broken. Revelation is an unsealed book. Toxic poisons trickle from it. Consciousness-altering fumes waft out of it. Desperate hope and vindictive joy issue from it.

> Question: What kind of person spends innumerable hours poring obsessively over this unsafe apocalypse, breathing in its vapors and mulling over its mysteries?
>
> Answer: Either a member of an apocalyptic sect or a biblical scholar.

Both the apocalyptic believer and the apocalyptic specialist are consumed by the same desire. Affect theorist Lauren Berlant defines desire as "a state of attachment to something or someone, and the cloud of possibility that is generated by the gap between an object's specificity and the needs and promises projected onto it" (2012, 6). I myself have experienced intense, if ambivalent, attachment to the Apocalypse; the present book is testimony to that. And I have stumbled around in the Apocalypse's cloud of possibility ("Then I looked, and there was a white cloud," 14:14), at once toxic and euphoric, for more decades than I care to count, first as a member of an apocalypse-avid house church, then as a biblical critic. "Desire visits you as an impact from the outside," continues Berlant, "and yet, inducing an encounter with your affects, makes you feel as though it comes from within you" (2012, 6). I first encountered the Apocalypse in my late teens, flicking impatiently through the pages of the New Testament ("of Our Lord and Saviour Jesus Christ: Newly Translated out of the Original Greek; and with the Former Translations Diligently Compared and Revised, by His Majesty's Special Command"), hungry for palpable religious experience. My eye and drug-addled brain were caught and held by Rev 4:1: "After this I looked, and, behold, a door was opened in heaven: and the first voice which I heard was as it were of a trumpet talking with me; which said, Come up hither, and I will shew thee things which must be hereafter" (KJV). And all at once I was on the cloud elevator rising, if not to heaven, then to a transformed earth, now ubiquitously electric with cryptic code, but decipherable to those who knew how to deploy the (code)book of Revelation. I had become a member of the esoteric Church of the Apocalypse worldwide.

I have long since migrated, of course, to another equally far-flung, no less esoteric community, the guild of biblical scholars. "What does it mean about love," asks Berlant, "that its expressions tend to be so *conventional*, so bound up in institutions like marriage and family, property relations, and stock phrases and plots?" (2012, 7). By extension, what does it mean about the love, however ambivalent, that I feel for the Apocalypse that its expressions tend to be so tightly bound up in the austere, abstracted institution of biblical scholarship, despite Revelation's own extravagant imagistic excesses and urgent behavioral demands?

At least I will always know what it feels like to realize that the world will end at noon this Sunday.

I had stopped watching the clouds ("Look! He is coming with the clouds; every eye will see him," 1:7) by the time I started writing on Revelation. By then, too, certain unprecedented challenges to the monochrome model of Revelation scholarship had been voiced. That model had been trundling along for more than a century, pushed from behind and pulled from the front by the laboring horde of historian-philologists, their blinders set to screen out any context for the Apocalypse other than the ancient one, together with any awkward questions about its ethics or ideology. The fundamental premise of the historical-philological model was already in place by the eighteenth century; Johann David Michaelis articulated it concisely as follows: "The Apocalypse contain[s] prophecies, with which the very persons to whom it was sent, were immediately concerned" (1801, 4:504).[1]

---

1. The first German edition of Michaelis's work appeared in 1750.

1. WHAT IS, WHAT WAS, AND WHAT MAY YET BE

Far and away the most impertinent early challenge to the inherited model of Revelation scholarship was posed by Tina Pippin in her *Death and Desire: The Rhetoric of Gender in the Apocalypse of John* (1992a). Pippin's was not the first feminist reading of Revelation; it had been preceded by the feminist studies of Elisabeth Schüssler Fiorenza (1981; 1985; 1991) and Adela Yarbro Collins (1987; cf. 1993a). But Pippin's was the most scathing critique of Revelation—or, arguably, of any New Testament text—to have appeared up to that point. More than an instance of feminist criticism, it was an instance of ideological criticism, a development that had only recently coalesced in biblical studies (see Jobling and Pippin 1992).[2]

The liberatory scholarly agendas of Pippin, Yarbro Collins, and, most explicitly, Schüssler Fiorenza, expressed as feminist scholarship on Revelation, emerged out of the broad current of liberation hermeneutics that had been flowing around, and occasionally through, the field of biblical studies for decades. Liberation hermeneutics also found searing expression in Revelation scholarship in books by Allan Boesak, black South African anti-apartheid activist (Boesak 1987), and Pablo Richard, Chilean socialist and advocate for the poor (Richard 1995).

These various streams are swollen by now, and have overflowed in different directions. Feminist studies of Revelation[3] have spilled over into masculinity studies[4] and womanist studies,[5] and, through slightly more circuitous channels, have also flowed into queer studies.[6] Forms of liberationist exegesis other than the feminist forms, meanwhile,[7] have overflowed into empire-critical and postcolonial strategies of reading.[8] But empire-critical and postcolonial approaches have also mingled with feminist or

---

2. And which *The Postmodern Bible*, coauthored by a team of scholars that included Pippin, subsequently defined as a form of criticism designed to analyze biblical texts "for their ideological content and mode of production," and "to grasp the ideological character of contemporary reading strategies" (Bible and Culture Collective 1995, 277).

3. See also Garrett 1992; Keller 1996; Pippin 1992b; 1992c; 1994b; 1995; 1999; 2005; 2012; Rossing 1999a; Vander Stichele 2000b; Levine 2009; Carson 2011; Samuelsson 2012; Huber 2013.

4. See Moore 1996, 117–40; Frilingos 2004, 64–115; Huber 2008.

5. See Martin 2005; Smith 2012; 2014.

6. See Pippin and Clark 2006; Runions 2008a; Moore 2009; Huber 2011.

7. See also Míguez 1995; González 1999; Rowland 2004; Blount 2005; 2007; Rhoads 2005.

8. See Howard-Brook and Gwyther 1999; Ruiz 2003; Westhelle 2005; Moore

other gender-attuned approaches.⁹ Explorations of Revelation attuned to literary theory, or critical theory more broadly, have also appeared, some narrative-critical,¹⁰ others poststructuralist.¹¹ Many ecological readings have also taken root in Revelation.¹² Surprisingly, given the richness of the soil, studies of Revelation's reception in contemporary popular culture were late-blooming,¹³ but have now come into their own.¹⁴

Old-school historical-critical commentaries on Revelation, meanwhile, began to break the scales in the late 1990s, David Aune's monumental—and magnificent—three-volume commentary weighing in at around 1,500 pages (Aune 1997; 1998a; 1998b), soon followed by G. K. Beale's 1,200 page commentary (Beale 1999) and Grant Osborne's 900-page commentary (Osborne 2002). The advent of colossal commentaries in any subfield of biblical studies may be taken to signify either an unprecedented flowering of that subfield or terminal exhaustion of the critical paradigms in which the commentaries are rooted. What will feel like vitality to the scholars most invested in the paradigms will seem like fatigue—an exhaustive and exhausting recital of the all already said—to the scholars less invested in the paradigms. In Revelation scholarship, the former scholars overwhelmingly outnumber the latter scholars, a situation not likely to change in the foreseeable future—although the lines between the two groups should not be drawn too starkly. Work on Revelation like Steven Friesen's and especially Christopher Frilingos's (Friesen 2001; Frilingos 2004) showed how clunky historical criticism could be trans-

---

2006, 97–121; Carey 2006; 2008; Seesengood 2006, 66–84; Kang 2007; Sánchez 2008; Carter 2009; 2011; Darden 2011; Diehl 2013.

9. See Kim 1999; Moore 2001, 173–99; McKinley 2004; Keller 2005, 33–94; Schüssler Fiorenza 2007, 111–47; Marshall 2009; Nelavala 2009; Smith 2012.

10. See Barr 1998; 2001; 2003; Resseguie 1998; 2005, 213–40; 2009.

11. See Derrida 1992b; 2007; Quinby 1994; Price 1998; Keller 2002; Keller and Moore 2004; Royalty 2004; Chrulew 2008; Samuelsson 2012.

12. See Rossing 1999b; 2002; 2005a; 2008; Keller 2000; 2005, 67–94; Reid 2000; Maier 2002; Hawkin 2003; Martin 2009; Bauckham 2010, 174–78; Bredin 2010, 165–80; Cate 2010, 145–55; Horrell, 2010, 98–101; Sintado 2010, 271–334; cf. Adams 2007, 236–51.

13. For rare early examples, see Dellamora 1995; Brasher 1998; Vander Stichele 2000a.

14. See Rossing 2005b; Frykholm 2007; Lyons and Økland 2009; Walliss and Quinby 2010; Gribben and Sweetnam 2011; Howard 2011; Clanton 2012; Partridge 2012; Runions 2014; cf. Blount 2005, 91–118.

formed into elegant cultural history through a modest infusion of theory, postcolonial theory in Friesen's case and postcolonial and gender theory in Frilingos's case.

In the early to mid-1990s, my own interests as a New Testament scholar expanded from poststructuralism into cultural studies and gender studies, especially masculinity studies, and soon branched out additionally into queer theory and postcolonial studies. More recently, posthuman animality studies, a poststructuralist inflection of ecological studies, and affect theory, a post-poststructuralist reckoning with emotion and other associated states, have been my main intellectual preoccupations. For me, however, the passage from one passion to the next has never entailed the abandonment or renunciation of the previous passion. They all move eclectically in and out of focus as I read and write, as is perhaps apparent in certain of the later essays in this collection.

To my mind, Revelation irresistibly invites engagement from all the methodologies or reading strategies I have just named, which is why my own passage through these interlocking approaches has been tightly bound up with Revelation almost from the start (see Moore 1995a; 1998; 1999). Consider, for instance, what a powerful magnet Revelation is for gender studies. Revelation has provoked vigorous feminist engagement, as have certain other New Testament texts. What is distinctive, however, about Revelation is the degree of passion it arouses. No other New Testament text, arguably, has induced such deep divisions among feminist interpreters. These divisions have been epitomized by "the 'Great Whore' debate," with scholars such as Tina Pippin (1992a) and Caroline Vander Stichele (2000b) in one corner and scholars such as Elisabeth Schüssler Fiorenza (1998, 205–36) and Barbara Rossing (1999a) in the other. At issue is the question of whether Revelation's symbol-soaked female characters—Jezebel and the great whore, on the one hand, the woman clothed with the sun and the bride, on the other—are harmful to flesh-and-blood women.

Queer theory, meanwhile—that term classically naming poststructuralist analysis of sex and sexuality, particularly in their instability, fluidity, constructedness, and malleability—finds in the Apocalypse a more anomalously sexed and aberrantly gendered universe than any other in the New Testament.[15] To begin with, Revelation *has* characters who perform

---

15. Admittedly, Revelation pales in this regard relative to certain extracanonical early Christian texts, most especially *Odes of Solomon* 19:1-6.

sexual acts, which exceedingly few New Testament texts do: a "fornicating" female prophet (2:20-22; cf. 2:14), a prodigiously promiscuous prostitute (14:8; 17:1-2; 18:3; 19:2). More significantly for queer theory, however, Revelation also has a Jesus with female breasts ("girt about the paps [*tois mastois*] with a golden girdle" [1:13], as the King James translators matter-of-factly put it); a choir of 144,000 male virgins (14:1-4); a bride whose groom is a sheep (19:7-9; 21:9); and other arresting deviations from standard sex/gender scripts, whether ancient or modern.

Revelation's attraction for empire-critical and postcolonial studies is also immense. No other New Testament text thematizes "earthly" empire as single-mindedly as Revelation—and precisely in order to attack it with scathing intensity. What exactly the authors of the Gospels and Acts or the apostle Paul thought about Rome is a subject for nuanced scholarly deliberations and heated disagreements. Almost no critical interpreter of Revelation, however, doubts that it was intended as an all-out attack on imperial Rome. Revelation is the New Testament example par excellence of anti-imperial resistance literature (whether or not one sees that resistance as compromised by a compulsion to model God's empire on Rome's empire). As such Revelation invites, and has received, intense scrutiny both from scholars who wish to reconstruct Revelation's biting religio-cultural and socioeconomic critique of imperial Rome and from scholars who wish to turn that critique on contemporary neocolonialism or global capitalism (not that these are always two different groups of scholars).[16]

Revelation has also been a magnet for ecological work on the New Testament. Other New Testament authors predict a divinely ordained dissolution of the cosmos (see especially 2 Pet 3:7, 10, 12), but none describes it with such apparent relish as John of Patmos. Revelation's most spectacular ecocidal visions are concentrated in the seven trumpets and seven bowls sequences (see especially 8:7–12; 16:2–12). Eventually, "the first heaven and the first earth" are bulldozed away altogether to make room for "the new heaven and new earth" (21:1). As we shall see, a remarkable number of interpreters have managed nevertheless to wrest positive ecotheological significance from the jaws of ostensible ecocidal disaster in Revelation

---

16. On the different varieties of empire-attuned work in New Testament studies, see Moore 2006, 8-23; 2011a.

1. WHAT IS, WHAT WAS, AND WHAT MAY YET BE 7

by highlighting 22:1–2, the Edenic city park in the new Jerusalem with its healing tree and life-giving stream.

Sex, gender, empire, and ecology are the ingredients that slosh around singly or, more often, in combination in the interlinked essays that make up this volume. Chapter 2, "Mimicry and Monstrosity," begins with scene-setting sections that attempt, in somewhat traditional style, to resituate Revelation in its original imperial context: the Roman province of Asia under the principate, perhaps in the latter decades of the first century CE. The postcolonial theory of Homi Bhabha is then wheeled in, and Revelation's relations to Rome are reframed in terms of Bhabha's key analytic categories of colonial ambivalence, hybridity, and especially mimicry. Chapter 3, "Revolting Revelations," continues to reflect on Revelation and empire, now crossreading Revelation with two further intertexts, one proximate and the other distant. What links Revelation, the roughly contemporary Jewish text 4 Maccabees, and modern Irish nationalism is a shared preoccupation with blood sacrifice and martyrdom. Gender also looms large in this essay, particularly Revelation's construction of the masculinities of God and his Messiah through repeated acts of war: war making men making war making men making war…. Chapter 4, "Hypermasculinity and Divinity," analyzes the hegemonic yet curiously queer masculinity of Revelation's deity more fully. In Revelation, a numinous, aphasic, phallic male form is the object of unceasing adoration and the central fixture of the narrative's throne-room spectacle. This theme is explored in tandem with the contemporary cultural spectacle of male bodybuilding. As such, this essay is also an exercise in cultural studies. The final stretch of the essay attempts a defamiliarizing reframing of Revelation's climactic big reveal with a different, more mundane cultural spectacle: the TV reality show makeover.

Chapter 5, "The Empress and the Brothel Slave," was coauthored with Jennifer A. Glancy. The focus here shifts from Revelation's God and Christ to its "great whore," Babylon, a figure whom traditional scholarship has tended to construe as a courtesan or well-heeled prostitute. Glancy and I counterargue, through appeal to the now extensive body of classical scholarship on ancient Roman prostitution, that the *pornē* Babylon is better construed as a tattooed brothel slave, albeit one who, paradoxically, is also represented as an "empress." The essay then moves to a crossreading of Revelation's Babylon and another "whore-empress," Juvenal and Tacitus's Messalina. In chapter 6, "Raping Rome," Babylon remains the focus, but now as the goddess Roma, the (singularly queer) personification of Rome and its military might. The cult of Roma had particularly deep roots in Roman Asia. As Babylon, Roma is mercilessly parodied in Revelation. She is stripped of her habitual armor and decked out as a drunken prostitute—but only to be punitively stripped once more, violated, and annihilated. Judith Butler, equipped with her theory of gender performativity, is called in to decipher this multilayered scene of gender masquerade and sexual humiliation. Chapter 7, "Retching on Rome," marks the book's final return to Revelation's sexualized violence, but now through the medium of affect theory, the name for the post-poststructuralist analysis of emotions and still more elemental forces rooted in bodies and passing between bodies. Through Sara Ahmed's brand of affect theory in particular, Revelation's "whore" may be reconceived as a circulating object that is saturated or "sticky" with affect, and the complex dynamics of Revelation's affective economy may be teased out. That economy works by sticking "figures of hate" together: Jezebel, the whore, the beast(s), and the dragon. Affect theory also enables us to better understand why Rome is figured in intensely sexualized terms in Revelation: the intolerable cultural closeness of Rome requires representation that evokes intimate contact felt on the surface of the skin, contact at once alluring and repellent.

Chapter 8, "Derridapocalypse," was coauthored with Catherine Keller. We take turns deploying the later writings of Jacques Derrida to read Revelation in its context and ours. "Later Derrida" is the Derrida of the so-called turn to religion. The later writings are replete with concepts such as "the messianic," "faith," "the absolute secret," and "justice beyond the law"—all illuminatingly applicable to Revelation and its interpretation. Empire is again a unifying theme in this essay, whether as the protocolonial Roman Empire or the neocolonial American Empire, specifically in its post-9/11 incarnation. In chapters 9 and 10, "Quadrupedal Christ" and "Ecotherology," later Derrida remains an enabling resource. Now, however, it is Derrida's animality theory that is employed to reframe Revelation. These complementary chapters are applied exercises in posthuman animality studies, the name for theoretical analysis of the systemic othering of the animal by which the human is constituted. The chapters take as their point of departure the fact that Revelation is an animal book extraordinaire, a theological bestiary. They explore certain prominent aspects of Revelation that have been curiously underremarked by other ecological interpreters, such as that Revelation's Christ moves through most of the narrative not on two legs but on four, and they ponder at length the ecotheological implications of that oddity. Chapter 10 ends by taking the mea-

1. WHAT IS, WHAT WAS, AND WHAT MAY YET BE

sure of the immense megalopolis in which Revelation's paradisal stream and tree are situated, asking whether that über-urban space, or even its stream and tree, merit the treatment they have received from so many ecological interpreters of Revelation—that is, as symbols of release from our current environmental nightmare.

All of these essays have been published previously, some recently, others some time ago. In effect, this book is a freestanding companion to *The Bible in Theory: Critical and Postcritical Essays* (Moore 2010), which was a larger collection of my previously published articles and essays—excepting those on Revelation, which I was holding over for this volume. As with the earlier volume, each essay in this volume is prefaced with a specially composed headnote that contextualizes it. As with the earlier volume, too, I did not attempt, as I revised or retouched the older essays in this volume, to incorporate scholarship that appeared subsequent to the essay's original date of publication—mountains of scholarship that would have been exhausting to scale. But the main reason I decided not to take a time capsule into the past to rewrite surreptitiously and thoroughly the early essays of my younger self while he gazed out the window and daydreamed was that my mind has changed relatively little about Revelation since I first began to teach it and write about it. What has mainly changed is that I now see Revelation as a Jewish text through and through and all the way down, a realization reflected particularly in chapter 7 of this volume. I have also become more agnostic about the date of Revelation. In some of the earlier essays in the volume, I tend to side in the great dating debate with the late-in-the-reign-of-Domitian team over against the shortly-after-the-death-of-Nero team. But now I tend to see that entire debate as a textbook example of Stanley Fish's once famous pronouncement on interpretive disagreements. Revelation's dating clues—the cipher 666 (13:18); the code name "Babylon" (17:5; also 14:8; 16:19; 18:2, 10, 21); the five-have-fallen-one-is-living riddle (17:9–11); the mortal wound that has been healed (13:3, 12, 14); the measuring of the temple (11:1–2); and the handful of other lesser clues—"provid[e] just enough stability for the interpretive battles to go on, and just enough shift and slippage to assure that they will never be settled," allowing us to continue to debate the date earnestly and heatedly, "but with no hope or fear of ever being able to stop" (Fish 1980, 172).

The Bible in Theory would not have come about if Tom Thatcher had not had the idea for it, and since that volume is parent to this one, Tom's idea has borne double fruit. I am also doubly grateful to Tom for including this volume too in his Resources for Biblical Study series, to keep the first one company, and to Bob Buller, SBL editorial director, for demonstrating once again that his concept of "biblical study" is, like Tom's, a commendably capacious one. I am particularly grateful to Jennifer Glancy and Catherine Keller, first, for the exhilarating experience of being able to co-write on Revelation with each of them, and second, for permitting the results of those collaborations to be reprinted in this volume. Tina Pippin was the ultimate inspiration for the string of essays that make up this collection. Her *Death and Desire* (1992a) came out when she and I were comrades in the Bible and Culture Collective (see Bible and Culture Collective 1995), plotting the revolution that never quite came about, and she enabled me to see that there were problems in Revelation more profound than whether the temple was still standing or had fallen when it was composed. She impelled me to wrestle with those problems in my teaching and finally to write on them myself.

Most of the essays in this book began as SBL papers or invited lectures. For the latter I owe debts of gratitude to Joseph Bristow at UCLA; Jennifer Glancy at the University of Richmond; David Jasper at the University of Glasgow; Brigitte Kahl and Hal Taussig at Union Theological Seminary; Amy-Jill Levine at Vanderbilt University; Hugh Pyper at the University of Leeds; and Mark Vessey, Sharon Betcher, and Harry Maier at the University of British Columbia/Vancouver School of Theology. I was fortunate to have some wonderful respondents along the way, notably Randall Bailey, Kwok Pui-lan, and Erin Runions. Also, an invitation from Brigitte Kahl to teach a "minicourse" at Union on Revelation, empire, gender, and ecology challenged me to begin to put all of these elements together.

The students who have participated in my doctoral seminars on Revelation have been an ongoing source of inspiration to me, not least the four who, to date, have written, or are writing, dissertations on it: Lynne Darden, Shinwook Kang, Christy Riley, and Shanell Smith. I found the 2013 Revelation seminar especially energizing, and I feel compelled to issue a shout-out (modeled on the "great shout" of Rev 10:3) to all the doctoral students who took it: Perry Brock, Sarah Emanuel, Lindsey Guy, Midori Hartman, Jimmy Hoke, Jonathan Koscheski, Paige Rawson, and Karri Whipple. Maia Kotrosits visited the class and opened our eyes to the potential that affect theory represents for Revelation. She and Alexis Waller have been my guides as I have ventured into this area. A special word of gratitude is due to Karri Whipple, who served as my research assistant as I attempted to turn my Revelation essays into a book. This long-delayed collection would still be languishing in limbo were it not for her energy and efficiency.

2

Mimicry and Monstrosity*

Postcolonial studies began in the 1950s and 1960s with the searing writings of Frantz Fanon, Aimé Césaire, and other cultural critics and literary authors themselves living the tortuous transition from colonialism to postcolonialism; it began again in the late 1970s and 1980s with the poststructuralist-inflected colonial discourse analysis of Edward Said, Gayatri Chakravorty Spivak, and Homi Bhabha; and it began yet again in the 1990s with the eruption of postcolonial studies as a wildly proliferating, multidisciplinary field. It sucked me into its vortex along with many other biblical scholars on the various peripheries of the discipline. Admittedly, I was easy prey. I had been born into an Ireland that was still classically postcolonial and grown up in the shadow of the Northern Ireland conflict, that late colonial war.

All told, the ambitions of postcolonial theory were staggering. Its multifaceted object of investigation was nothing less than the tectonic forces that had shaped the geopolitical history of the planet: imperialism, colonialism, neocolonialism, globalization, and a plethora of other interrelated phenomena. The essay that follows was written in 2003 for *A Postcolonial Commentary on the New Testament Writings*, which, however, did not appear until 2007. It seemed to me long before 2003 that Bhabha's writings, most of all his analytic categories of colonial ambivalence, mimicry, and hybridity, admitted exegeti-

---

* First written for Fernando F. Segovia and R. S. Sugirtharajah, eds., *A Postcolonial Commentary on the New Testament Writings* (The Bible and Postcolonialism 13; New York: T&T Clark, 2007), 436–54, under the title "Revelation"; reprinted in lightly revised form with permission.

cal "application" to an extent that the writings of Said or Spivak did not.¹ And Revelation, whose theological centerpiece, the divine throne room, replicates even as it repudiates Roman imperial court ceremonial, seemed in particular to cry out for redescription in terms of these analytic categories. My postcolonial "commentary" on Revelation, however, itself mimicked Bhabha in a way I had not intended, namely, in his neglect of gender in his colonial discourse analysis. At least half of the remaining essays in this volume attempt to remedy that neglect.

Bhabha's influence on empire-attuned scholars of the New Testament has been considerable (if indirect, since few are avid readers of Bhabha). Gradually, the notion that no New Testament text exhibits unadulterated opposition to Rome, uncompromised by any reinscription of Roman imperial ideology, has become commonplace. Bhabha's concept of colonial ambivalence has leavened the lump, in other words, so much so that one now feels impelled to ask, what lies beyond ambivalence? That beyond is just beginning to be glimpsed (see Kotrosits 2014).

The Emperor himself invited and feasted with those ministers of God whom he had reconciled…. Not one of the bishops was wanting at the imperial banquet, the circumstances of which were splendid beyond description. Detachments of the bodyguard and other troops surrounded the entrance of the palace with drawn swords, and through the midst of these the men of God proceeded without fear into the innermost of the imperial apartments, in which some of the Emperor's own companions were at table, while others reclined on couches arranged on either side. One might have thought that a picture of Christ's kingdom was thus shadowed forth, a dream rather than a reality. (Eusebius, *Life of Constantine* 3.15, LCL)

We mimic men of the New World. (Naipaul 1967, 146)

---

1. An impression reinforced by some of the earliest biblical-critical engagements with postcolonial theory, which, when exegesis rather than reception history was their focus, tended to gravitate toward Bhabha (Runions 1998; 2002; Liew 1999; Samuel 2002; Thurman 2003).

2. MIMICRY AND MONSTROSITY

> Colonial mimicry is the desire for a reformed, recognizable Other, *as a subject of a difference that is almost the same, but not quite*. (Bhabha 1994d, 86)

Imperium Romanum

To ponder the book of Revelation's relations to empire is hardly a novel gesture. Critical scholars of Revelation have customarily read it as, perhaps, the most uncompromising attack on the Roman Empire, and on Christian collusion with the empire, to issue from early Christianity. To put it another way, and in terms that have become familiar following the recent "turn to empire" in New Testament and early Christian studies, Revelation is commonly read as a signal instance of ancient anti-imperial resistance literature. But what is "imperialism"?

In contemporary postcolonial studies, "imperialism" generally denotes the multifarious, mutually constitutive ideologies—political, economic, racial/ethnic, and religious—that impel a metropolitan center to annex more or less distant territories and determine its subsequent dealings with them (see Said 1993, 9). Although the English term *imperialism* did not emerge, apparently, until the late nineteenth century and was first used in connection with European expansionism and the ideologies that undergirded it, the term's etymological and conceptual roots lie in the Latin word *imperium*, which, under the Roman republic, designated the authority vested in consuls, magistrates, and other select officials to exercise command and exact obedience, and, under the principate, resided supremely in the person of the emperor. The latter's *imperium*, voted to him by the Roman senate at his accession, extended in principle to all peoples and territories under Rome's dominion (see Lintott 1993, 22, 41–42, 115–22 passim).

At first or even second glance, then, Revelation would appear to be an anti-imperial(istic) text that, in effect, announces the transfer of worldwide *imperium* from the Roman emperor to the heavenly emperor and his Son and co-regent, the "King of kings and Lord of lords" (19:16; cf. 17:14). As Revelation itself memorably phrases this transfer, "The world empire [*hē basileia tou kosmou*] has become the empire of our Lord and his Messiah" (11:15; cf. 14:6–8).[2] The paramount question the present essay will raise,

---

2. With Howard-Brook and Gwyther (1999, 115 n. 77; see also 224–25), I prefer "empire" to "kingdom" as a less anodyne translation of *basileia* in Revelation.

however—one prompted by the particular body of postcolonial theory it will be appropriating, that of Homi Bhabha—is whether or to what extent Revelation merely reinscribes, rather than effectively resists, Roman imperial ideology.

Coloniae Romanum

Revelation is explicitly addressed to seven urban churches in the Roman province of Asia (1:4, 11), the westernmost province of the larger geographical region known (somewhat confusingly) as Asia Minor, which extended from the Aegean to the western Euphrates, thus corresponding roughly to modern Turkey. The history of colonization in Asia Minor extended back to the hellenizing campaigns of Alexander the Great and his successors, who sowed Greek cities (*poleis*) throughout the region—although several of Revelation's seven cities, notably Ephesus and Smyrna, were Greek colonies well before the advent of Alexander. The extent, indeed, to which the multilayered hellenization of Asia Minor effected a cultural colonization that expedited its eventual absorption by the consummately hellenized Romans can hardly be exaggerated.

The English term *colony* derives from the Latin term *colonia* (the equivalent Greek term being *apoikia*). It would, however, be misleading to conceive of Roman *coloniae* purely on the model of the European colonies of the early modern period and its aftermath. The classic Roman *colonia* was a civic foundation, which is to say a city or town. Essentially, *coloniae* were civic communities of Roman citizens settled outside Italy and composed mainly of military veterans. The *colonia* was one of the three principal types of Roman provincial community, all of them urban; the others were the *municipia* (confined mainly to the Latin West, and of lesser status than the *colonia*), and the city or town that was neither an official *colonia* nor *municipia*, and as such less "Romanized" than either. The classic unit, then, of Roman colonization (in the contemporary sense of the term) was urban, and it was through an infrastructure of self-governing cities that Roman provinces were administered.

What of the province of Asia? Julius Caesar and especially Augustus had each engaged in the settlement of military veterans in various pockets of Asia Minor generally, which is to say that they "seeded" the region with *coloniae*, but the systematic introduction of new settlers became rare in the post-Augustan period. How might all or any of this be related to modern European colonial practices?

Contemporary postcolonial discourse frequently distinguishes between *settler colonies*, on the one hand, and *colonies of occupation*, on the other (while also acknowledging that many colonies fit neatly into neither category but straddle both at once). Settler colonies (also known as settler-invader colonies) are those in which the indigenous population is decimated and uprooted, eventually becoming a minority in relation to the majority settler-invader population; modern examples of such colonies would include Australia, Canada, and the United States (Johnston and Lawson 2000). In contrast, colonies of occupation are those in which the indigenous population remains in the majority numerically but is subjugated and governed by a foreign power; modern examples would include pre-independence India and Ireland.

Which of these two modes of colonization best approximates the situation of Roman Asia? As will be apparent from what was said earlier, Asia could in no wise be regarded as a settler-invader colony (using the term *colony* now in its modern sense); it better fits the colony-of-occupation model instead. Roman culture was concentrated in the (mainly coastal) cities of the province in contrast to the rural Anatolian interior, which managed to preserve its indigenous character, conspicuous especially in its native languages and religious cults, more or less intact until the third century CE. Even in the cities, however, the Roman presence would have been relatively slight. In general, the number of elite Roman officials allotted to any one Rome province was minuscule relative to the amount of territory to be administered. Asia was one of the "ungarrisoned" provinces of the empire, moreover, meaning that no full legion was stationed there; and what military presence there was tended to be concentrated in the interior. What, then, was the mechanism that enabled continuous Roman control of Asia?

Hegemony

At this point, another concept commonly invoked in contemporary postcolonial studies may usefully be invoked, that of *hegemony*, in the special sense accorded to the term many decades ago by the Italian Marxist intellectual Antonio Gramsci (1971, esp. 145, 571, 765–66, 810). Hegemony, in the Gramscian sense, means *domination by consent*—in effect, the active participation of a dominated group in its own subjugation, whether a social underclass, say (Gramsci's own principal focus), or a colonized people. The attraction of the concept for postcolonial studies is that it serves to account for the ability of an imperial power to govern a colonized territory whose indigenous population overwhelmingly outnumbers the army of occupation. In such cases (thinking alongside Gramsci), the indigene's desire for self-determination will have been displaced by a discursively inculcated notion of the greater good, couched in such terms as social stability (whether in the form of a *Pax Romana*, say, or a *Pax Brittanica*) and economic and cultural advancement. The more efficient an imperial administration, indeed, the more it will rely on hegemonic acquiescence and the less it will have recourse to military force in the retention and exploitation of its colonial possessions—in which case the neocolonial empires of contemporary global capitalism would represent a quantum leap in administrative efficiency when measured against the relatively unwieldy empires of the past.[3]

The concept of hegemony usefully illuminates the situation of Roman Asia. The province itself originated not in an invasion but in an invitation: Attalus III of Pergamum bequeathed his kingdom to the Romans. It became *provincia Asia* after Attalus's death in 133 BCE, and expanded in increments over the next half century or so, gradually assuming the form it would take under the principate. Like any Roman province, the routine governance of Asia depended on the active cooperation and participation of the local urban elites. The administrative infrastructure consisted of a loose coalition of self-governing cities, each having responsibility for the territorial hinterland attached to it. The mainspring of the complex hegemonic mechanism that enabled Roman governance of Asia, however—economically a jewel in the imperial crown, rich in natural resources, agriculture, and industry—was the intense competition for imperial favor and recognition in which the principal Asian cities were permanently embroiled (Ephesus, Pergamum, and Smyrna in particular, although the rivalry extended to many lesser cities as well). A vital expression of this competition was the city's public demonstration of the measure of its loyalty to the emperor, the ultimate patron or benefactor in relation to whom the city was a client or dependent, and as such in rivalry with the other client cities of the province for a limited quantity of goods and privileges. And the principal mechanism in turn (the wheel within the wheel) for formal demonstrations of such loyalty was the *imperial cult*: the rendering of divine honors to Roman emperors, living or dead.

---

3. A shift provocatively explored by Hardt and Negri 2000.

2. MIMICRY AND MONSTROSITY

Divus Caesar

Officially instituted in 42 BCE, when the Roman Senate posthumously recognized Julius Caesar as divine, the imperial cult—to the extent that it can be spoken of in the singular: it was profoundly marked by regional variation, as we shall see—infiltrated the religio-political life of every province in the empire during the Augustan and post-Augustan periods (with the hard-won exception of the province of Judea). Whereas in the western provinces the imperial cult tended to be imposed by Rome, in the eastern provinces it was a "voluntary" affair. It could well afford to be. Ruler worship in the east predated Roman expansion, having been catalyzed in particular by the spectacular conquests of Alexander the Great. Whereas in Rome itself divine honors were offered as a rule only to deceased emperors (impatient exceptions notwithstanding, notably Caligula, Nero, and Commodus), the worship of currently reigning emperors was tolerated and even encouraged in the provinces. What more reassuring token of an apparent willingness to be conquered could a conqueror possibly desire?—even if the provincial imperial cults may, in historical hindsight, also be construed as surreptitious determinations on the part of the emperor's subjects of who and what he was to be for them, thus paradoxically setting subtle limits on his autonomy in the very act of acknowledging his absolute authority.

From an extremely early stage, the local Asian elites enthusiastically embraced the Roman imperial cult, dedication to which became a major vehicle of competition between the leading cities of the province.[4] But it was a highly regulated competition. Delegates of the various civic communities met annually as the Council or Assembly (*koinon*) of Asia in one of the five official provincial cities (Ephesus, Pergamum, Smyrna, Sardis, or Cyzicus) in order to conduct the business of the province, a crucial element of which was the organization of the imperial cult. In 29 BCE, a mere two years after Octavian/Augustus's accession to supreme power, the Assembly of Asia had requested and was granted the honor of erecting a provincial temple to Roma and Augustus at Pergamum. The establishment of a cult of Roma and Augustus in Asia and in the neighboring province of Pontus-Bithynia became a model for other eastern provinces. The cult of *Dea Roma* or *Thea Rhōmē* ("goddess Rome," the divine personification of

---

4. Price 1984 remains the standard study of this phenomenon.

the city) is a particularly telling manifestation of hegemony (again, in the Gramscian sense), since no such cult existed in the capital itself. It was not imposed or even modeled by those at the apex of power, in other words, but was invented by Roman subjects instead (elite subjects, however, a point to which I shall return below). A temple to *Dea Roma* had existed at Smyrna since 193 BCE, the first such temple in Asia Minor.[5]

But the Assembly of Asia devised still more extravagant ways to acknowledge Rome's intimate and apparently irresistible hold on the destiny and daily life of the province. Early in the principate, the assembly, in consultation with the Roman proconsul of Asia, determined to honor *Divus Augustus* by creating a new calendar for the province that would begin, not on January 1 as in the standard Roman calendar, but on September 23, the emperor's birthday—again a signal instance of those nearer the base of the pyramid of power surpassing those nearer the apex (those elites, that is, in the capital itself with physical access to the emperor) in the symbolic performance of subjection—a performance all the more remarkable for the fact that prior to the principate of Augustus the province had suffered acutely under Roman rule, due to rapacious governors, crushing taxes, and a disastrously unsuccessful rebellion. The energy and rapidity with which the province of Asia subsequently set about deifying the conqueror and sweeping the sordid history of exploitation under the rug of myth testifies to the unprecedented efficiency of the Roman hegemonic apparatus under the principate—an efficiency that would be almost inexplicable were it not for the fact that the most extravagant expressions of consent to Roman domination of the region arose from the ranks of the local elites, who stood to gain infinitely more from ostentatious displays of acquiescence than the mainly impoverished urban and rural populations whom they purported to represent. Considerable prestige attached to the priesthoods and other offices of the provincial imperial cults—they could, indeed, form the pinnacle of a local political career. Major priesthoods in the imperial cults, moreover—most especially that of annual president or chief priest (*archiereus*) of the provincial assembly—could also form crucial stepping stones to a political career in Rome itself for the select few, or at least for their sons or grandsons.

---

5. Chapter 5 below deals with the relationship between the goddess Roma and another, less reverent female personification of Rome, the prostitute Babylon.

In due course, therefore, each of Revelation's seven cities, along with others in the province, erected temples or altars to Roman potentates living or dead: Julius Caesar (coupled with *Dea Roma*), Augustus (also with *Dea Roma*), Tiberius (with the Roman Senate), Vespasian, Domitian, and Hadrian. The leading cities competed for the coveted title of *neokoros*, "temple warden/caretaker," awarded at the discretion of the Senate and the emperor to cities containing an imperial temple with panprovincial status (see Friesen 1993). And elaborate imperial festivals became a prominent feature of the religious life of the province, enmeshing the populace in a communal symbolic articulation of the omnipresence and immanence of absolute power in the absent person of the Roman emperor, whose arms encircled the civilized world by virtue of the *imperium Romanum*.[6]

CATACHRESIS

How best to situate Revelation in relation to the complex matrix of power relations that determined the religio-political life of Roman Asia? Consummately counterhegemonic in thrust (in the special sense in which we have been using the term *hegemony*), Revelation indeed appears to represent a stunning early instance of an anti-imperial literature of resistance (to return to the theme with which we began). In shocking contrast to the official prayers offered to the Greek gods of the Olympian pantheon by priests of the local imperial cults for the health of the Roman emperor and the length of his reign (for prayers on behalf of the emperor were more common than prayers addressed to his image), Revelation gleefully predicts the imminent destruction of Rome instead, which it mockingly renames "Babylon" (14:8; 16:9; 17:5; 18:2, 10, 21), in answer to the counterprayers offered by Christians to their own god (6:9–11; 8:3–4; cf. 16:5–7; 19:1–2). In effect, faithful Christians constitute an imperial countercult in Revelation, a priesthood (1:6; 5:10; 20:6) dedicated to the Christian emperor and his co-regent, Jesus Christ, in relation to which the official cult is meant to be seen as a monstrous aberration: worship of a hideous beast that derives its ultimate authority from Satan (13:4, 8, 12, 14–15; cf. 14:9–11; 16:2; 19:20; 20:4).

---

6. See chapter 3 below for a rather different set of reflections on the Roman imperial cult as it relates to Revelation.

This cunning polemical strategy can be construed as a signal instance of *catachresis*. Originally an ancient Greek term denoting "misuse" or "misapplication," catachresis has been revived and adapted by postcolonial theorist Gayatri Spivak to designate a process whereby the victims of colonialism or imperialism strategically recycle and redeploy facets of colonial or imperial culture or propaganda. Catachresis, in this sense, is a practice of resistance through an act of creative appropriation, a retooling of the rhetorical or institutional instruments of imperial oppression that turns those instruments back against their official owners. Catachresis is thus also an act of counterappropriation: it counters the appropriative incursions of imperialist discourse—its institutional accouterments, its representational modes, its ideological forms, its propagandistic ploys— by redirecting and thereby deflecting them.[7] As a strategy of subversive adaptation, catachresis is related to parody, which can be defined in turn as an act or practice of strategic misrepresentation. In the context of imperialist and anti-imperialist discourse, indeed, parody is best regarded as a species of catachresis.

Parody of the Roman imperial order permeates Revelation, reaching a scurrilous climax in the depiction of the goddess Roma, austere and noble personification of the *urbs aeterna*, as a tawdry whore who has had a little too much to drink (17:1–6).[8] The most fundamental instance of catachresis in Revelation, however, is its redeployment of the term "empire" (*basileia*) itself. In Asia as in any Roman province, the most immediate and most encompassing referent of the term *basileia* would have been the *imperium Romanum* (see Howard-Brook and Gwyther 1999, 224). Revelation, in keeping with its catachrestic proclivities, far from dispensing with the category of empire in pronouncing upon the divine sphere, retains the imperial model instead (down to its details, as we shall see), but makes certain audacious adjustments to it—most significantly, switching the figure at its

---

7. References to catachresis are scattered throughout Spivak's work; see, e.g., 1990, 111; 1991, 70; 1996, 145–54 passim; 1999, 14. The definition of *catachresis* offered above represents, to a degree, my own appropriation of Spivak's definition of it. For a similar elaboration of catachresis, see Ashcroft, Griffiths, and Tiffin 2001, 34. Spivak herself, however, also employs the concept to characterize aspects of *colonizing* discourse, on which see Morton 2003, 33–35.

8. Greg Carey observes: "As resistance literature, Revelation twists the standards of propriety for its own ends, engaging the loftiest of subjects—emperor and empire— through imagery that is bizarre, even base" (2006, 178).

2. MIMICRY AND MONSTROSITY

center so that it is no longer the Roman emperor, an exchange that effects a retooling of the entire model, producing a catachrestic realignment of the whole.

A Roman Throne Room in Revelation's Heaven

Speculation with regard to the details of this realignment has long been a standard feature of critical scholarship on Revelation. Chapters 4–5, for example, which constitute a notable case in point, have elicited observations such as the following:

- The acclamation "Worthy art thou" (*axios ei*), addressed to God or the Lamb by those assembled around the heavenly throne (4:11; 5:9; cf. 5:12), was also employed in Roman imperial court ceremonial to greet the emperor.
- The title "our Lord and God" (*ho kyrios kai ho theos hēmōn*), likewise used in the heavenly court (4:11; cf. 4:8; 11:17; 15:3; 16:7; 19:6; John 20:28), was also applied to the emperor Domitian (whether or not he himself actually demanded it),[9] under whose reign Revelation achieved its final form, if the scholarly majority is to be believed.
- The twenty-four elders around the throne (4:4) correspond to, among other things, the twenty-four lictors who regularly accompanied Domitian (lictors being fasces-bearing bodyguards whose number symbolized—indeed, trumpeted forth—the degree of *imperium* conferred on a Roman potentate).
- The elders' gesture of casting their crowns or wreaths (*stephanoi*) before the throne (4:10) corresponds with a form of obeisance frequently offered to Roman emperors.[10]
- The reappearance of Jesus in the guise of a Lamb standing in the presence of the Divine Emperor "as though it had been slaughtered" (*hōs esphagmenon*, 5:6) acquires added semantic clout from the fact that the image of the Roman emperor officiating at sacrifice was a pious commonplace from the reign of

---

9. On which see 90–91 below.
10. Any good-sized commentary on Revelation is likely to propose some or all of the first four parallels. I return to and consider them at greater length below (84–86).

Augustus onward, almost no one other than the emperor (and his immediate family) being depicted thus in the imperial iconography (sculptures, friezes, coins, and imprinted sacrificial cakes) that proliferated throughout the empire;[11] and so on.

The multiplication of such parallels by critical scholars has by no means been confined to Revelation 4–5; to a lesser extent, it has extended to the book as a whole. The sheer number of these alleged parallels, taken collectively, probably prohibits their outright dismissal as a product of scholarly mass hallucination: even if any specific parallel can always be contested, the existence of the general authorial strategy to which they gesture collectively is probably as secure as most fixtures in the gently quaking quagmire of Revelation scholarship. I have relabeled that strategy catachresis here, and noted its intimate relationship to parody. That Revelation's representation of the Roman imperial order is essentially parodic, however, has long been a tenet of critical scholarship on the book. In order to disclose what is really at stake in that tenet, and to rethink Revelation's relationship to empire more generally through the conceptual resources afforded by postcolonial theory, I will need to turn to the work of Homi Bhabha.[12] But first a final stage set needs to be wheeled into place.

The New Metropolis

One signal advantage of Bhabha's conceptual categories for a reading of Revelation, as we shall see, is that they enable, indeed impel, us to interrogate the metaphysical and ethical dualism that the book attempts to foist on us as one of its primary rhetorical strategies: its construction of the Roman Empire as the absolute antithesis of "the Empire of God and his Messiah" (11:15). The success of the strategy is evident from the fact that this binary opposition has been endlessly (and unreflectively) replicated even in critical commentaries on Revelation.

Within the book itself, this dualism attains its apogee in the construction of the new Jerusalem, a scene in which Babylon/Rome is both absent

---

11. This last is my own contribution to this heady speculative exercise, inspired by Beard, North, and Price 1998, 350–51.

12. For a rather different treatment of parody in Revelation, see Maier 2002a, 164–97. Whereas I (prompted by Bhabha) argue that the parody topples over into mimicry, Maier argues that it slides over into irony.

(because already annihilated: see 18:1–19:5; cf. 19:17–21) and present (because still required, as we are about to see). The scene concludes with a blessing and a curse: "Blessed are those who wash their robes, so that they will have the right to the tree of life and may enter the city by the gates. Outside are the dogs and sorcerers and fornicators and murderers and idolaters, and everyone who loves and practices falsehood" (22:14-15; cf. 21:8, 27). Here, then, is the cartography of paradise (cf. 2:7), an attenuated, absolutely hierarchized geography of difference, designed to distinguish a (hyperidealized) "metropolis"—the New Jerusalem—from a (demonized) "periphery"—that which until recently was designated "Babylon" in this book. Revelation's vision of paradise restored (see 22:1-2; Gen. 2:10; Ezek. 47:1-12) is thus the logical culmination of the dualism that has characterized its rhetoric throughout. The cartographic self-representations of the Roman Empire itself, in which the imperial territories gradually shaded over into the barbaric, the chaotic, and the monstrous the farther one ventured outward from the metropolis, is here countered with what is, in effect, a catachrestic parody of imperial cartography: immediately beyond the walls of the Christian metropolis, absolute alterity begins, with no incremental passage from sameness to difference to act as conceptual buffer (a binary conceit all the more curious for the contradictory fact that out in the negative zone entire nations are apparently poised to pay homage to the new megalopolis: see 21:24, 26). In Revelation's hyperdualistic cosmos, then, Christian culture and Roman culture must be absolutely separate and separable (see 18:4: "Come out of her, my people"). But are they? This is where Bhabha's strategies of colonial discourse analysis come into their own.

Ambivalence, Mimicry, Hybridity

Homi Bhabha's demandingly dense and elliptically titled essay collection *The Location of Culture* (1994c) ranks with Edward Said's *Orientalism* (1978) and Gayatri Spivak's "Can the Subaltern Speak?" (1988) as among the most influential, and controversial, products of postcolonial theory. Much of *The Location of Culture* amounts to a critical interrogation of any conceptual dichotomization of metropolis and periphery, empire and indigene, colonizer and colonized. Bhabha's enabling assumption is that the relationship between colonizer and colonized is characterized by ambivalence instead, which is to say attraction and repulsion at one and the same time (Bhabha 1994d; 1994a, esp. 129–38). Basing himself ultimately on the psychoanalytic contention that ambivalence is ubiquitous in psychic processes (1994a, 132), Bhabha's presumption is that the stance of the colonized vis-à-vis the colonizer is rarely, if ever, one of pure unequivocal opposition—which, by extension, calls a second dualistic distinction into question, that between the resistant colonial subject, on the one hand, and the complicit colonial subject, on the other. For Bhabha, resistance and complicity coexist in different measures in each and every colonial subject. And the complex conjoining of resistance and complicity is nowhere more evident than in the phenomenon of colonial mimicry.[13]

Colonial mimicry results when the colonizer's culture is imposed on the colonized and the latter is lured or coerced into internalizing and replicating it. This replication is never perfect, however—the colonized is never simply an exact copy of the colonizer ("*Almost the same but not white*," is how Bhabha wittily phrases the matter [1994d, 89])—nor does the colonizer desire that this mimicry be absolutely accurate, in any case, for then the hierarchical distinction between original and copy, primary and secondary, colonizer and colonized would collapse, and with it the linchpin of imperial ideology. Hence the necessary ambivalence of the colonizer's injunction to the colonized to mimic him (or, less often, her): "Replicate me/do not replicate me." This injunction, moreover, is fraught with risk for the colonizer: mimicry can all too easily topple over into mockery or parody, thereby menacing the authority, even the identity, of the colonizer.

The third concept that, together with ambivalence and mimicry, captures the complex psychic interpenetration of colonizer and colonized, for Bhabha, is hybridity (1994e, 111–22). In its "weak" sense, the term *hybridity* as used in contemporary postcolonial studies means no more than that the contact between colonizer and colonized is constantly productive of hybrid cultural manifestations. Bhabha, however, gives the concept of hybridity a decidedly Derridean twist,[14] seeing it not as a simple

---

13. "Of Mimicry and Man" (Bhabha 1994d) is again the key text. Bhabha's concept of colonial mimicry can be traced to various influences, but prominent among them is V. S. Naipaul's postcolonial novel *The Mimic Men* (as Bhabha himself acknowledges: 1994d, 87). "We pretended to be real," Naipaul's protagonist Singh reminisces; "we mimic men of the New World" (1967, 146). Further on Bhabha's appropriation of Naipaul, see Huddart 2006, 71–75.

14. Bhabha's debt to Jacques Derrida and other French poststructuralist theorists for his analytic strategies and critical sensibilities is enormous. Consider the following statement (Bhabha 1992, 439) in particular: "My growing conviction has been that the encounters and negotiations of differential meanings and values within 'colonial'

synthesis or syncretic fusion of two originally discrete cultures but rather as an in-between space—or "Third Space," to use his own preferred term (1991; 1994b, 37–39)—in which cultures are themselves simultaneously constituted and deconstructed: the identity of any cultural system only emerges as an effect of its differences from other cultural systems, but the infinitely open-ended differential network within which any given culture is situated radically and necessarily destabilizes its identity even as it generates it. In consequence, no culture can be pure, prior, original, unified, or self-contained; it is always already infected by impurity, secondariness, mimicry, self-splitting, and alterity. In a word, it is always already infected by hybridity.

In order to outline Bhabha's theory in brief, I have had to abstract it from its embeddedness in the analysis of disparate colonial texts and histories—most especially those of nineteenth-century British India, the prime catalyst for much of Bhabha's conceptual innovation—and systematize it to an extent that Bhabha himself, in good deconstructive fashion, studiously avoided.[15] But he has not been able to avoid scathing criticism (e.g., Moore-Gilbert 1997, 140–51; Parry 2004, 13–36, 55–74). His theory has been prodded, probed, and repeatedly contested over such issues as its alleged universalism—its application of "first world" psychoanalytic categories to "third world" psychic processes—and its alleged diminution of agency—its neglect of overt and conscious forms of resistance on the part of the colonized, not least armed opposition, in favor of covert and unconscious forms of resistance. While these are serious criticisms and concerns, certain of the supple concepts proposed by Bhabha, used cautiously and creatively, can enable a reappraisal not only of the book of Revelation's relations to empire but also of Revelation's theology. In what follows, therefore, I will be less interested in proving the theory than in reopening the book.

---

textuality, its governmental discourses and cultural practices, have enacted, *avant la lettre*, many of the problematics of signification and judgment that have become current in contemporary theory—aporia, ambivalence, indeterminacy, the question of discursive closure, the threat to agency, the status of intentionality, the challenge to 'totalizing' concepts, to name but a few." See also Bhabha 1990, 2–4 passim; 2002, 21.

15. Anybody delving into *The Location of Culture* and expecting to find tidy, systematic expositions of ambivalence, mimicry, hybridity, and so on will be sorely disappointed.

The Book of Mimicry

The phenomenon of mimicry is endemic to Revelation. The book's representation of the Roman imperial order is essentially parodic, as we have noted, and parody is a species of mimicry: it mimics in order to mock. Do Bhabha's pronouncements on colonial mimicry apply, then, to Revelation's parodic strategy? Yes and no, it seems to me. In contrast to the scenario adduced by Bhabha in which systemic mimicry of the agents and institutions of imperialism perpetually threatens to teeter over into parody or mockery,[16] Revelation presents us with a reverse scenario in which parody or mockery of the imperial order constantly threatens to topple over into mimicry, imitation, and replication. Revelation's implicit claim, as commentators never tire of telling us, is that Roman imperial court ceremonial, together with the imperial court itself, are but pale imitations—diabolic imitations, indeed—of the heavenly throne room and the heavenly liturgy. But commentators also routinely note that the heavenly court and liturgy in Revelation are themselves modeled in no small part on the Roman imperial court and cult (recall our earlier ruminations on Revelation 4–5)—which means in effect that the "heavenly" order in Revelation is busily engaged in imitating or mimicking the "earthly" order, notwithstanding the book's own implicit charge that the earthly is merely a counterfeit copy of the heavenly (see Royalty 1998, 99 n. 57, 246).

The latter observation borders on the obvious, perhaps. Yet the obvious is not without interest in this instance. Revelation's attempted sleight of hand ensnares it in a debilitating contradiction. Christians are enjoined to mimic Jesus, who in turn mimics his Father ("To the one who conquers I will give a place with me on my throne, just as I myself conquered and sat down with my Father on his throne," 3:21; cf. 20:4), who, in effect, mimics the Roman emperor, who himself (at least as represented in the imperial cult) is a mimetic composite of assorted royal and divine stereotypes. In Revelation, Christian authority (*exousia*) inheres in imitation. "To everyone who conquers and continues to do my works to the end," the Son of

---

16. As, for example, in this complaint by a nineteenth-century British missionary to India, whom Bhabha (1994d, 92) quotes: "Still everyone would gladly receive a Bible. And why?—that he may lay it up as a curiosity … or use it for waste paper.… Some have been bartered in the markets, others have been thrown in snuff shops and used as wrapping paper." In effect, the colonized has mimicked the colonizer's fetish for this book of books, but the mimicry easily slides into mockery.

God (see 2:18) promises, "I will give authority over the nations, to rule them with an iron rod" (2:26), the same iron rod that he himself wields (12:5). Imitate me, replicate me. Do as I have done, and you will become what I have become. But there is a destabilizing contradiction built into this mimetic mechanism. If the Roman imperial order is the ultimate object of imitation in Revelation, then, in accordance with the book's own implicit logic, it remains the ultimate authority, despite the book's explicit attempts to unseat it.

On Revelation's own account, of course, it is Rome, the sea beast, that is the consummate mimic—the mimic monster—with its ten horns and seven heads (13:1; 17:3), in imitation of the great red dragon (12:3, explicitly identified as Satan in 12:9 and 20:2), whose own appearance is in turn an imitation of various ancient Near Eastern mythic prototypes.[17] Furthermore, the unholy trinity of Satan, sea beast, and land beast/"false prophet" (for the latter epithet, see 16:13; 19:20; 20:10) mimics the holy trinity (strictly lowercase, of course; we are not yet within spitting distance of Nicaea) of God, lamb, and prophetic spirit (for the latter, see 2:7, 11, 17, 29; 3:6, 13, 22; cf. 1:10; 4:2; 17:3; 21:10). In addition to the general structural parallel of two antithetical triads, certain characteristics ascribed to the sea beast in particular mirror those ascribed to Jesus or God. A Christlike "resurrection" is attributed to the sea beast in 13:3, 14. Also, the thrice-repeated declaration that the sea-beast "was and is not and is to come" (which crops up twice, in 17:8 and again in 17:11, in variant forms) parodies the thrice-repeated acclamation of God as he "who is and who was and who is to come" (1:4, 8; 4:8). Also notable is the depiction of the land beast as a lamb beast: "it had two horns like a lamb" (13:11). Revelation is engaging, it would seem, in subtle mockery of Satan and his elect agents, here implying that they are best seen as distorted reflections of God and his elect agents.

Yet, as we have just seen, Revelation's deity cannot function as anchor for this mimetic chain, but is merely another link in it, because he is modeled on the Roman emperor—and we have not even begun to consider the extent to which this deity is also a composite copy of Ezekiel's deity, Daniel's deity, and so on, themselves in turn ultimately constructed on the model of the ancient Near Eastern monarch. If the Roman imperial court is, in Revelation, merely a dim, distorted reflection of the heavenly court,

---

17. Collins 1976 remains the classic study of this and related themes.

the latter is itself merely a magnified reflection of the former and sundry other earthly courts, so that the seer's vision of heaven occurs in a mimetic hall of mirrors.

Again, this observation smacks of the obvious, and as such falls short of profundity. Yet the "obvious" does not always command acknowledgment. The difficulty of effectively exiting empire by attempting to turn imperial ideology against itself is regularly underestimated, it seems to me, by those who applaud Revelation for decisively breaking the self-perpetuating cycle of empire. To my mind, Revelation is emblematic of the difficulty of using the emperor's tools to dismantle the emperor's palace. The seer storms out of the main gates of the imperial palace, wrecking tools in hand, only to be surreptitiously swept back in through the rear entrance, having been deftly relieved of his tools at the threshold.

Becoming Rome

More than any other early Christian text, Revelation is replete with the language of war, conquest, and empire—so much so, indeed, as to beggar description.[18] Note in particular, however, that the promised reward for faithful Christian discipleship in Revelation is joint rulership of the empire of empires soon destined to succeed Rome (3:21; 5:10; 20:4–6; 22:5), a messianic empire established by means of mass slaughter on a surreal scale (6:4, 8; 8:11; 9:15, 18; 11:13; 14:20; 19:15, 17–21; 20:7–9, 15) calculated to make the combined military campaigns of Julius Caesar, Augustus, and all of their successors pale to insignificance by comparison. All of this suggests that Revelation's overt resistance to, and expressed revulsion toward, Roman imperial ideology is surreptitiously compromised and undercut by covert compliance and attraction. Not for nothing is Rome figured in Revelation as a prostitute—indeed, as "the mother of whores" (*hē mētēr tōn pornōn*, 17:5): what better embodiment, for the seer, of seductive repulsiveness, of repulsive seductiveness?[19] Empire is a site of immense ambivalence in this book.

Bhabha's controversial contention (more implicit than explicit, however, in his work) is that since colonial discourse is inherently ambivalent, and as such internally conflicted, it contains the seeds of its own

---

18. Further on this theme, see 51–57 below, and for a different take on it, see Seesengood 2006, 66–84.

19. Another theme further elaborated below (175–76).

2. MIMICRY AND MONSTROSITY

dissolution, independently of any overt act of resistance on the part of colonized subjects. With regard to Revelation, however, the scenario is again reversed. Because Revelation's *anti*colonial discourse, its resistance to Roman omnipotence, is infected with the imitation compulsion, and hence with ambivalence, it contains the seeds of its own eventual absorption by that which it ostensibly opposes. (Actually, this too is consonant with Bhabha's theory, since he ascribes ambivalence to the colonized no less than the colonizer. The logical collapse of counterimperial discourse, however, under the weight of its own internal contradictions, is not the kind of phenomenon that Bhabha tends to emphasize or examine.) In this regard, Revelation epitomizes, and encapsulates for analytical scrutiny, the larger and later process whereby Christianity, in the (post-)Constantinian epoch, paradoxically *becomes* Rome.

As numerous colonial discourse analysts from Albert Memmi (1957) to Homi Bhabha have argued, the relationship between colonizer and colonized is best conceived as a mutually constitutive interaction. In terms of identity construction, the flow of effects is not all in one direction, from colonial overlord, say, to native subject, but each is caught up instead in a complex circulation of reciprocal effects and influences. To cite an elementary example, the relationship of the metropolis (London, say, or Rome) to its colonies or annexed territories becomes a crucial element in its ideological self-representation, and hence the construction of its cultural identity. Look no further than the Julio-Claudian Sebasteion at Aphrodisias in the province of Asia, which includes sculpted reliefs representing Rome as hypermasculine conqueror of feminized subject peoples.[20] As an example of the flow of effects surging in the opposite direction, the colonized eliciting imitation from the colonizer and thereby remolding the colonizer—but also themselves being remolded in the process—consider the post-Constantinian Christianization of the Roman Empire, arguably the most spectacular historical example of the co-constitution and reciprocal recreation of colonizer and colonized. As a means through which to conceptualize its own unique identity and destiny, metropolitan Roman culture absorbed and internalized Christianity, originally a peripheral, provincial product (although one to whose emergence Rome had already contributed the crucial catalyst by publicly executing its "founder"), and Rome reinvented itself in the process. As though anticipating this astounding act of

---

20. See further 137–38 below.

co-option, Revelation resolutely targets hybridity, and holds up for emulation a Christian praxis that is at once peripheral and pure.

Hybrid Harlotry

The threat of the hybrid is embodied for Revelation in the "works" and teaching of "the Nicolaitans" (2:6, 15), the teaching of "Balaam" (2:14; cf. Num 22–24; 31:8, 16; Deut 23:4-5; Josh 24:9-10; 2 Pet 2:15–16; Jude 11), and the teaching of "that woman Jezebel" (2:20; cf. 1 Kgs 16:31; 18:1–19; 19:1–3; 21:23, 25; 2 Kgs 9:22, 30–37). The Nicolaitans are otherwise unknown; subsequent references to them in patristic literature seem to depend ultimately on Revelation. The names Balaam and Jezebel, presumably, are symbolic. The phrase "the teaching of Balaam" would appear to be a synonym for "the teaching of the Nicolaitans." The context further suggests that Balaam is not a code name for a Christian teacher at Pergamum, but Jezebel *is* a code name for a Christian prophet at Thyatira—a Nicolaitan prophet to be precise: the content of her teaching ("teaching and beguiling my slaves to practice fornication and to eat food sacrificed to idols," 2:20) is described in terms identical to that of the Nicolaitans (2:14–15). Like the names Balaam and Jezebel, the practice of fornication (*porneusai*, 2:14, 20) with which the Nicolaitans are charged is probably symbolic, fornication being a common figure for idolatry in the Jewish Scriptures.[21] The Nicolaitans are best seen as Christian "assimilationists" (see Duff 2001, 132), who, like their counterparts in the Corinthian church (see 1 Cor 8:1–13; 10:23–11:1), took a relaxed or pragmatic view of Christian accommodation to certain cultural norms, specifically (to cite the practice that elicits the seer's censure), eating meat in assorted socio-religious settings, whether public settings, such as regular calendric festivals, including those of the imperial cult; or (semi)private settings, such as banquets or other meals hosted by trade guilds or other voluntary associations or social clubs; or simply eating temple "leftovers"—meat that had been sold in the marketplace after having been sacrificed and partially consumed in the temple cults.

---

21. For more detailed presentations of the arguments advanced thus far in this section, see Thompson 1990, 121–24; Aune 1997, 148–49; and Duff 2001, 36–47, 55–58. For an attempt to read Jezebel against broader cultural tapestries, see Pippin 1999, 32–42.

Revelation's stance, then, with regard to Christian participation in the regular civic life of Roman Asia—exemplified by participation in the many cultic and semicultic meals that constituted an important ingredient of the social glue of the province—is strenuously antiassimilationist, which is to say, "separatist." But this is also to say that Revelation's stance on Roman civic culture is also counterhegemonic (using "hegemonic" once again in its Gramscian sense):[22] Christians must not enact, through symbolic means, their own subjection to the Roman *imperium* by participating in the social and religious rituals that collectively prop up the far-flung canopy of the empire and enable it to cast its shadow over the day-to-day lives of the diverse populations under its sway. Revelation enjoins a practice of nonviolent resistance to Rome, a symbolic "coming out" of its empire (see 18:4: "Come out of her, my people, so that you do not participate in her sins") while continuing to remain physically within it—although whether a coming out to form full-fledged countercommunities (systematic antitypes of standard Asian communities) or a more ad hoc, guerilla-style coming out is unclear.

As such, the main pillar of Asian collaboration with Roman domination, the Assembly of Asia, an important aspect of whose function was the organization and promotion of the imperial cult, as noted earlier, is singled out for special condemnation in Revelation—provided that the land beast of 13:11–18, assigned with the responsibility of "making the earth and its inhabitants worship the [sea] beast" (Rome and its emperors: see 13:1; 17:3, 9), is to be identified as the priesthood of the imperial cult, as has often been suggested.[23] The land beast derives its authority from the sea beast, but the latter derives its own authority from the dragon, who is Satan (13:4; cf. 12:9; 20:2).

Revelation's unequivocal condemnation of collaboration with Rome, however—even (or especially?) collaboration conducted through symbolic (that is, ritual) means—extends, by implication, to all strata of Asian society, as its denunciation of Christian assimilationism makes clear. But why? Is it because the mortar of empire is inevitably mixed with the blood of its victims (2:13; 6:9; 13:15; 16:6; 17:6; 18:24), so that those who reap the benefits of empire, even when the benefits are relatively meager, are by extension guilty of the blood that keeps the wheels of empire oiled? By

---

22. See 17–18 above.

23. The suggestion already appears in the seminal critical commentaries of Bousset (1906) and Swete (1906), and remains a live option for contemporary commentators.

this logic, only empire's excremental outcasts—the detritus of the imperial system—may be deemed innocent of its systemic injustices.[24] If this is indeed Revelation's central assertion regarding the mechanics and ethics of empire, it is an utterly uncompromising and unsettling one.

In light of such a stance, the consistent demonization of imperial authority in Revelation becomes yet more comprehensible, as does its denunciation of assimilationist Christ-following. In order that Revelation's blanket critique of empire acquire full rhetorical force, the distinction between the agents of empire, on the one hand, and the victims of empire, on the other, must be asserted at an absolute, and hence metaphysical, level, and such a distinction is necessarily menaced by any manifestation of Christian hybridity, however innocuous. In Revelation, it is the Nicolaitans, epitomized by "Jezebel," who most fully embody the threat of hybridity, as we have seen.

But what is the precise relationship between "Jezebel" and the "great whore" (17:1; 19:2), that other female incarnation of iniquity in Revelation?[25] In other words, what is the relationship between Christian assimilationism and imperial seduction (14:8; 17:1-2; 18:3, 9; 19:2) in this book? The whore, it may be argued, represents the threat from without to the Christian *ekklesiae*, whereas Jezebel represents the threat from within. The threat from within, however, represented by the spectacle and specter of Christian assimilation, is precisely that the threat from without is not simply external: the outside has infiltrated and infected the inside. Jezebel and the whore thus represent but two sides of the same (counterfeit) coin in Revelation: on the one hand, an inside that has somehow strayed outside; on the other hand, an outside that has somehow stolen inside. Either way, contamination has occurred.

---

24. This construal of the extra-imperial would correspond in part with Spivak's conception of the subaltern as, among other things, the "detritus" or "flotsam" of the system (1987b, 245–46).

25. Beyond the fact that each name evokes an especially unappetizing fate, that of ending up on the wrong end of the food chain: the original Jezebel is devoured by dogs (1 Kgs 21:23; 2 Kgs 9:30–37), whereas the whore is devoured by a far more fearsome beast (Rev 17:16). Further on this connection, see 173–74 below.

2. MIMICRY AND MONSTROSITY

The Book of Empire

In its concern to maintain intact the high-walled partition separating imperial metropolis and Christian periphery, Revelation, although passionately resistant to Roman imperial ideology, paradoxically and persistently reinscribes its terms, to the extent that Roman imperial ideology (like subsequent European imperial ideology) itself pivoted around an interrelated series of dualistic distinctions between metropolis and periphery, civilized and barbaric, and so on—that brand of imperialistic dualism that Frantz Fanon aptly dubbed "Manicheanism" (e.g., Fanon 1968, 41, 93; see further JanMohamed 1983; 1995; Gibson 2003, 113–17). Of course, Revelation maintains the metropolis/periphery binarism only in order to stand it on its head: the hierarchical power relations that currently obtain between metropolis and periphery, Rome and (nascent) Christianity, are soon destined for spectacular reversal: "The world empire has become the empire of our Lord and his Messiah" (*egeneto hē basileia tou kosmou tou kyriou hēmōn kai tou Christou autou,* 11:15). Were the tightly sealed elements on each side of the binary opposition allowed to leak into each other, conceptually speaking, a simple reversal of the hierarchy would not be possible. Activities or ideologies that do not conform to the binary separation, therefore (such as participation of Christians in the imperial cult), are subjected to censure or rendered taboo in Revelation. But the inherent instability and untenability of the binary division comes to displaced expression in the elaborate mimicry that, as we saw, characterizes Revelation's depiction of the "other" empire, that of God and the Lamb, a mimicry that persistently blurs the boundaries between the two empires until it becomes all but impossible to say where one leaves off and the other begins. The Divine Empire that Revelation proclaims is neither separate nor independent from the Roman Empire. Instead, the former is parasitic on the latter.

In due course, however, the host absorbed the parasite, precipitating the host's mutation into the one monstrosity that the seer of Revelation seems incapable of imagining: an empire that is Roman and Christian at one and the same time. But the curious phenomenon of Constantinian Christianity itself bears monumental testimony to the fatal flaw in Revelation's ostensibly anti-imperial theology: the fact that it counters empire with empire. To proclaim that "the world empire has become the empire of our Lord and his Messiah" is also to proclaim that "the empire of our Lord and his Messiah has become the world empire." More than any other early Christian text, arguably, Revelation epitomizes the imperial theology that enabled the Roman state effortlessly to absorb Christianity into itself, to turn Christianity into a version of itself, and to turn itself into a version of Christianity—notwithstanding the paradox that Revelation is also ostensibly more hostile to Rome than any other early Christian text. The flaw in Revelation's theology inheres in three mutually reinforcing—and inescapably obvious?—features of the text (although the obvious is always hedged about with obliviousness, and hence never as inescapable as one would like). First of all, the throne is the paramount metonym for God in this book.[26] Second, the principal attributes of "the one seated on the throne" are stereotypically imperial attributes: incomparable glory and authority, absolute power, and punitive wrath. And third, the principal activities of the one seated on the throne and those of his elite agents are quintessentially imperial activities: the conduct of war and the enlargement of empire.[27]

To construct God or Christ, together with their putatively salvific activities, from the raw materials of imperial ideology is not to shatter the cycle of empire but merely to transfer empire to a transcendental realm, thereby reinscribing and reifying it. The dearth of nonimperial synonyms for the Christian theological commonplace(s), "the kingdom [or reign, or rule] of God [or Christ]," even in contemporary theological and pastoral discourse,[28] is symptomatic of the extent to which imperial metaphors have maintained, and continue to maintain, a virtual monopoly and stranglehold on the Christian theological imagination—one ultimately unchecked by the cross, I would venture to add, which all too easily folds up to form a throne. The product is an imperial divine amalgam or "essence" that is extremely difficult to dismantle or dislodge.

---

26. God all but vanishes behind the throne and into the throne in Revelation. "The one seated on the throne" is the preeminent title for God in this book (4:2, 9–10; 5:1, 7, 13; 6:16; 7:15; 21:5; cf. 7:10; 19:4).

27. Further on these activities, see 43–65 (left-hand column) below.

28. One such synonym, however, would be the neologism "kindom of God," associated with Cuban American liberation theologian Ada María Isasi-Díaz, who writes: "Two reasons compel me not to use the usual word employed by English Bibles, *kingdom*. First, it is obviously a sexist word that presumes that God is male. Second, the concept of kingdom in our world today is both hierarchical and elitist. The same reasons hold for not using *reign*. The word *kin-dom* makes it clear that when the fullness of God becomes a day-to-day reality in the world at large, we will be sisters and brothers—kin to each other" (2004, 213 n. 1).

And yet there is undoubtedly a place for what Gayatri Spivak, in a related context, has termed "strategic essentialism."[29] The envisioning of a cosmic counterempire presided over by a divine emperor may serve an important strategic function in struggles for liberation from situations of desperate oppression, as work on Revelation such as that of Allan Boesak (1987) or Pablo Richard (1995) eloquently testifies. Revelation is eminently well-equipped to speak to such situations (see also Ruiz 2003, esp. 121–23); to a greater or lesser extent, it was in such a crucible that Revelation itself was forged (not yet a situation of systemic state-sponsored persecution, apparently—but the seer's intuition that such oppression lay over the horizon was entirely accurate). Ultimately, however, if Christian theology is to be intellectually as well as ethically adequate, and as such less luridly anthropomorphic and less patently projectionist, might it not require what Revelation, locked as it is in visions of empires and counterempires, emperors and counteremperors, is singularly powerless to provide: a conception of the divine sphere as other than empire writ large?[30]

---

29. And which she glosses as "a strategic use of positivist essentialism in a scrupulously visible political interest" (Spivak 1987c, 205).

30. For a full-scale attempt to develop such a conception of the divine, see John D. Caputo's *The Weakness of God* (2006)—to my mind, an extraordinarily original and profound attempt despite my quibble with it below. "My idea is to stop thinking about God as a massive ontological power line that provides power to the world," Caputo explains, "instead thinking of something that short-circuits such power and proves a provocation to the world that is otherwise than power" (13). He finds a model for such thinking in the New Testament "under the name 'kingdom of God' ... filled in or fleshed out ... in soaring parables and mind-bending paradoxes.... You see the weak force that stirs within the name of God only when someone casts it in the form of a narrative, tells mad stories and perplexing parables about it, which is what Jesus did when he called for the kingdom of God.... The kingdom of God that is called for in the New Testament is an anarchized field.... In the kingdom, weak forces play themselves out in paradoxical effects that confound the powers that be, displaying the unsettling shock delivered to the reigning order by the name of God" (13–14). Caputo's Jesus, then (although this is not a book about Jesus; Paul plays a more prominent role in it, and Derrida a more prominent role than Paul), is the world-subverting sage familiar from certain strands of historical Jesus scholarship. We never discover what Caputo would do with the other kingdom of God in the Synoptic tradition, the one ushered in with irresistible force by an imperial Christ enthroned on the clouds and attended by angelic courtiers—which, however, is still a gentler return than that of the imperial Christ in Rev 19:11–21 (see 53–55, left-hand column, below).

3
Revolting Revelations*

This essay unpacks some underanalyzed assertions of the previous essay (despite having been written several years prior to it), such as that Revelation is a war book and that Revelation's deity is too anthropomorphically imperial to function effectively as a counterimperial icon. The present essay analyzes those entangled themes in gendered terms: traditionally, men make war, which, quintessentially, makes traditional men. The essay is about gendered violence, wars of extermination, and the construction of hegemonic masculinity. It attempts to unpick the dense knots that tie these themes together in Revelation.

The term *empire*, as employed in postcolonial studies, often has an everywhere-yet-nowhere quality to it. This essay also attempts to break with the abstraction of "empire" in its analysis of Revelation in favor of the affective specificity of nationalism with its visceral calculus of heroic martyrs, blood sacrifice, and ultimate glory. Here the essay takes a long sideways leap, for it invokes modern Irish nationalism to illuminate the protomartyrology of Revelation. This essay on Revelation is also about mythical Ireland, nationalist Ireland, and postcolonial Ireland.

Hence the essay is also about me. Its immediate inspiration was the giddy upsurge of autobiographical criticism in the 1990s (Anderson and Staley 1995; Moore 1995b; Staley 1995b; Kitzberger 1998). It was written as my first flush of infatuation

---

* First written for Ingrid Rosa Kitzberger, ed., *The Personal Voice in Biblical Interpretation* (New York: Routledge, 1998), and reprinted in revised form with permission.

with autobiographical criticism was cooling. The title and opening stretch of the essay parodies that critical genre. As the essay unfolds, however, I'm seduced by the genre all over again, even if several of the voices in the essay are not quite my own. When the essay bifurcates into two columns, the right-hand column is a dialogue between two essentially fictional characters. This device allowed me to explore positions on Revelation with which I did not entirely identify. The essay attempts to pull more disparate elements together than any other essay in the volume and so, potentially, was always the one most at risk of failing spectacularly.

This essay was originally commissioned for a collection titled *The Personal Voice in Biblical Interpretation*. Where I come from, however, the third word of this title could only be pronounced as *vice*.[1] But it is not the personal vice in biblical interpretation that I wish to ponder here, nor even my own personal vice (although I shall hardly be able to resist the temptation), so much as that of the only New Testament narrator to employ the personal voice throughout his narrative. I speak of the narrator of Revelation, whose unblinking "I" first transfixes us in 1:9—"I, John, your brother who shares with you in Jesus the persecution and the kingdom and the patient endurance, was on the island called Patmos"—not releasing its hypnotic hold on us until we pass 22:8: "I, John, am the one who heard and saw these things."

In any case, there is by now relatively little of my own vices and vicissitudes left to reveal. Buried in books that, if my royalty checks are anything to go by, are seldom exhumed and opened up, my published secrets are slowly decomposing, thin nourishment for the occasional prurient bookworm. For these secrets are a sadly unsensational lot. Still, had your own life been so desperately dreary during the past decade or so as to cause you to dig out and devour my every published word, you would have read, with ever-mounting ennui, of

1. my roots in the soggy soil of rural Ireland. My father was a butcher; you'd be weary of hearing that.[2] Later on he became a farmer; you might also have caught that. But you wouldn't yet

3. REVOLTING REVELATIONS

know that my mother was a hairdresser. That she was, however, is the real subject of an entire monograph (Moore 2001);

2. my LSD-induced psychosis in the summer of 1974 (well, DMT actually; is there anybody else out there who still remembers what that was?), which led, simultaneously, to

3. my conversion to Christianity; and

4. my incarceration in St. Joseph's Mental Hospital, Limerick;

5. my subsequent incarceration (voluntary, this time) in Mount Melleray Cistercian Abbey on the slopes of the Knockmealdown Mountains in County Waterford;

6. my current ... sorry, it's caught in my throat. I probably shouldn't have confessed it in the first place. And now it's out there, observing the progress of my career with malevolent interest, cleaning its weapons compulsively as it plans its next move. (I should have taken a leaf from John's book: "And when the seven thunders had sounded, I was about to write, but I heard a voice from heaven saying, 'Seal up what the seven thunders have said, and do not write it down'" [Rev 10:4].)

Anyway, that's six. Now, let's see. I need a seventh if I'm to keep up with John and his amazing four-legged friend ("When the Lamb opened the seventh seal, there was silence in heaven for about half an hour" [Rev 8:1]).[3] Well, there is one other thing, actually:

7. My relationships with other men, one ten-year affair in particular, which began in the spring of 1972, and persisted even through my period in the monastery (we entered the novitiate together). Unlike my first six revelations, I haven't written about this one before. The other party has, however; see Anon., *The Boys*, unpublished MS concealed in the bottom drawer of the gray file cabinet in my office, beneath my (equally unpublished) doctoral dissertation.[4] It appears that the latter favors the missionary position. This improbable couple eventually conceived and gave birth to *God's*

Gym: Divine Male Bodies of the Bible (Moore 1996), which ends with a preliminary exploration of the book of Revelation.

And it is to Revelation that I wish to return here. For my essay title is not intended to be a reference to my own revelations, insufficiently revolting as they are, as I had started to say, so much as those of this exquisitely bizarre book—the book of Revelations, as it is most often called. Even among Bible-reading Christians, surprisingly few refer to the book by its actual (short) title, "Revelation,"[5] much preferring the plural, "Revelations." The latter has a titillating ring to it, I suppose, that the more theological "Revelation" cannot match. "Revelations" doesn't conjure up the tablets of the law so much as the law of the tabloids. And what is the law of the tabloids—the tabloid press, and also tabloid TV, epitomized by the talk show? It is simply that secrets sell. Revelations is the grand dénouement of the biggest best seller of all time. So whose sordid secrets is it supposed to be peddling? God's or merely John's? Let us see.

* * * * *

"So what is Revelation actually about?" I ask myself.

"Easy," the answer comes back. "It's about the establishment of God's kingdom on earth."

"God's 'kingdom,' eh? Not his dukedom or fiefdom, then? I marvel at your propensity to shroud your theological thought in archaic political metaphors."

"Oh, the shroud isn't mine. I borrowed it from John."

"You look so comfortable in it. So just how is the kingdom of God to be established on earth, according to John?"

"How are kingdoms or empires ever established?"

"Through military conquest, no doubt."

"Well, there you have it."

"You mean Revelation is all about war?"

War Book I	War Book II
Assemble them for battle on the great day of God the Almighty. (Rev 16:14)	Bloodshed is a cleansing and sanctifying thing, and the nation which regards it as the final

3. REVOLTING REVELATIONS

"Yes, messianic war. Richard Bauckham is excellent on this. Now, if I can only find his discussion of it.... Ah yes, here we are.

> The prominence of Davidic messianism in Revelation can be gauged from Jesus' self-designation, "I am the root and the descendant of David, the bright morning star" (22:16). The first of these two titles comes from Isaiah 11:10 ("the root of Jesse") and is used of the Davidic Messiah.... The second refers to the star of Numbers 24:17, which (in the context of 24:17–19) was commonly understood to be a symbol of the Messiah of David who would conquer the enemies of Israel. "The root of David" is found also in Revelation 5:5, alongside another title evoking the image of the royal Messiah who will defeat the nations by military violence: "the Lion of Judah" (cf. Gen. 49:9; 4 Ezra 12:31–32). Further allusions to the Messiah of Isaiah 11, a favourite passage for Davidic messianism, are the sword that comes from Christ's mouth (1:16: 2:12, 16;

horror has lost its manhood. (Pearse 1924a, 99)

Revelation *is* a book of war, a book of warriors, but not an especially vivid one, at least to my mind. My own internal standard for an ancient war book is the *Táin Bó Cúailnge* ("The cattle raid of Cooley"), the oldest vernacular epic in Western literature, ancient Ireland's answer to the *Iliad*.[34] Between the ages of seven and twelve (numbers to which John of Patmos accords sublime significance), I had but a single teacher in the tiny school that I attended in the village of Adare in County Limerick, and he happened to be an ardent nationalist. Ancient Irish mythology took precedence over biblical mythology in my early formation consequently, even though the school was run by a religious order. The *Táin* in particular made an indelible impression on my fledgling imagination years before I even knew that John's war book existed.

Every page of the *Táin* is a paean to war. Even its accounts of warriors mentally preparing themselves for battle are marked by an exuberance and excess that I have yet to encounter in any other ancient literature. Judge for yourself. In the following excerpt, the hero of the *Táin*, Cúchulainn, prepares to take on an entire army single-handedly.

> The first warp-spasm seized Cúchulainn, and made him into

19:21) with which he strikes down the nations (19:15; cf. Isa. 11:4; 49:2) and the statement that he judges with righteousness (19:11; cf. Isa. 11:4).⁶"

"Stirring stuff. But does all this mayhem have a purpose—other than the establishment of 'God's kingdom,' that is (a phrase which, as you know, never fails to send a shiver down my spine)? What if its real purpose were to engender masculinity, to make men? War making men making war making men...."

"Ah, so we're to obsess about deessentialized manhood again, are we? The gender construct blues? Muddy Waters had it all wrong, no doubt. Instead of bellowing 'I'm a man!' he should have sobbed 'I'm a subject whose gender identity is purely performative, the product of a compulsory set of rituals and conventions, which conspire to engender retroactively the illusion that my masculinity is natural and innate, merely "expressed" by the actions, gestures, and speech that in fact produce it—' "⁷

"Sorry to interrupt your own performance and what—

a monstrous thing, hideous and shapeless, unheard of. His shanks and his joints, every knuckle and angle and organ from head to foot, shook like a tree in the flood or a reed in the stream. His body made a furious twist inside his skin, so that his feet and shins and knees switched to the rear and his heels and calves switched to the front. The balled sinews of his calves switched to the front of his shins, each big knot the size of a warrior's bunched fist. On his head the temple-sinews stretched to the nape of his neck, each mighty, immense, measureless knob as big as the head of a month-old child. His face and features became a red bowl: he sucked one eye so deep into his head that a wild crane couldn't probe it onto his cheek out of the depths of his skull; the other eye fell out along his cheek. His mouth weirdly distorted: his cheek peeled back from his jaws until the gullet appeared, his lungs and liver flapped in his mouth and throat, his lower jaw struck the upper a lion-killing blow, and fiery flakes large as a ram's fleece reached his mouth from his throat. His heart boomed loud in his breast like the baying of a watchdog at its feed or the sound of a lion among bears. Malignant mists and spurts of fire—the torches of the Badb[35]—flickered red in the vaporous clouds that rose boiling above his head, so fierce was his fury. The hair of his head twisted

ever peculiar gender configuration it's constructing even as you speak. But the following observation by Harold Washington suggests that gender construction is indeed germane to our topic: 'Until recently, critical discussion of violence, warfare, and the sacred in the Hebrew Bible has failed to consider the constitutive role of gender categories for these texts. This, I think, is remarkable, for what could be more acutely gendered than war, an activity historically described as performed by men only, in a space containing nothing but men?'[8] Now, given that it's the *Davidic* Messiah we've been discussing, let me cut to another essay, 'David the Man' by David Clines."

"The 'man' of the title isn't Clines himself, then?"

"No, or at least not entirely. He notes: 'The essential male characteristic in the David story is to be a warrior, a man of war … or a mighty man of valour…. It is essential for a man in the David story that he be strong—which means to say, capable of violence against other men and active in killing other men.' Later he adds:

> Or, to take another example, from a little

like the tangle of a red thornbush stuck in a gap; if a royal apple tree with all its kingly fruit were shaken above him, scarce an apple would reach the ground but each would be spiked on a bristle of his hair as it stood up on his scalp with rage. The hero-halo rose out of his brow, long and broad as a warrior's whetstone, long as a snout, and he went mad rattling his shields, urging on his charioteer and harassing the hosts. Then, tall and thick, steady and strong, high as the mast of a noble ship, rose up from the dead centre of his skull a straight spout of black blood darkly and magically smoking like the smoke from a royal hostel when the king is coming to be cared for at the close of a winter day.[36]

When that spasm had run through the high hero Cúchulainn he stepped into his sickle war-chariot that bristled with points of iron and narrow blades, with hooks and hard prongs and heroic frontal spikes, with ripping instruments and tearing nails on its shafts and straps and loops and cords. The body of the chariot was spare and slight and erect, fitted for the feats of a champion, with space for a lordly warrior's eight weapons, speedy as the wind or as a swallow or a deer darting over the level plain. The chariot was settled down on two fast steeds, wild and wicked, neat-headed and narrow bodied, with slender quarters and roan breast, firm in hoof

outside the David story itself, in 1 Sam. 4.9 the Philistines say to one another, having learned that the ark of Yahweh has come into their camp: 'Take courage (lit. be strong), and acquit yourselves like men, O Philistines, lest you become slaves to the Hebrews as they have been to you; acquit yourselves like men and fight.' This phrase 'acquit yourselves like men,' literally 'become men' … , means, very simply, that to be a man is to fight. The whole ideology surrounding this utterance is a little more complex than that, no doubt.… But as far as the gender issue is concerned, it is simple: men fight."[9]

"Enough! My testosterone level is shooting off the scale. But how does it all connect with the Davidic Messiah?"

"Don't pretend you don't see it. The Davidic Messiah, as the supreme warrior, would also have been the ultimate icon of masculinity."

"A male fantasy of phallic proportions?"

"Funny you should mention the phallus—"

and harness—a notable sight in the trim chariot-shafts. One horse was lithe and swift-leaping, high-arched and powerful, long-bodied and with great hooves. The other flowing-maned and shining, slight and slender in hoof and heel.

The battle itself is limned with lines no less hyperbolic.

In that style, then, he drove out to find his enemies and did his thunder-feat and killed a hundred, then two hundred, then three hundred, then four hundred, then five hundred, where he stopped—he didn't think it too many to kill in that first attack, his first full battle with the provinces of Ireland. Then he circled the outer lines of the four great provinces of Ireland in his chariot and he attacked them in hatred. He had the chariot driven so heavily that its iron wheels sank in the earth. So deeply the chariot-wheels sank in the earth that clods and boulders were torn up, with rocks and flagstones and the gravel of the ground, in a dyke as high as the iron wheels, enough for a fortress-wall. He threw up this circle of the Badb round about the four great provinces of Ireland to stop them fleeing and scattering from him, and corner them where he could wreak vengeance for the boy-troop.[37] He went into the middle of them and beyond, and mowed

"Well, I was missing it, rather. You usually insist on beating me over the head with it. You were saying?"

"You've made me forget. Oh, yes. I was reminded that Bauckham himself makes a rather phallic point about John's preoccupation with Psalm 2. Yes, here it is:

> One of John's key Old Testament texts, allusions to which run throughout Revelation, is Psalm 2, which depicts 'the nations' and 'the kings of the earth' conspiring to rebel against 'the LORD and his Messiah' (verses 1-2).... God promises to give this royal Messiah the nations for his inheritance (verse 8) and that he will violently subdue them with a rod of iron (verse 9). Allusions to this account of the Messiah's victory over the nations are found in Revelation 2:18, 26-8; 11:15, 18; 12:5, 10; 14:1; 16:14, 16; 19:15."[10]

"And your own point is what, precisely?"

"Oh, come on! Rod of iron?"

down great ramparts of his enemies' corpses, circling completely around the armies three times, attacking them in hatred. They fell sole to sole and neck to headless neck, so dense was that destruction. He circled them three times more in the same way, and left a bed of them six deep in a great circuit, the soles of three to the necks of three in a ring around the camp. This slaughter on the Táin was given the name Seisrech Bresligi, the Sixfold Slaughter. It is one of the three uncountable slaughters on the Táin: Seisrech Bresligi, Imslige Glennamnach—the mutual slaughter at Glenn Domain—and the Great Battle at Gáirech and Irgáirech (though this time it was horses and dogs as well as men). Any count or estimate of the number of the rabble who fell there is unknown, and unknowable. Only the chiefs have been counted. The following are the names of these nobles and chiefs: two called Cruaid, two named Calad, two named Cír, two named Cíar, two named Ecell, three named Crom, three named Caur, three named Combirge, four named Feochar, four named Furechar, four named Cass, four named Fota, five named Aurith, five named Cerman, five named Cobthach, six named Saxan, six named Dach, six named Dáire, seven named Rochad, seven named Ronan, seven named Rurtech, eight named Rochlad, eight named Rochtad, eight

"Well, you know what Freud said: 'Sometimes a rod of iron is just a rod of iron.'"

"Freud aside, I submit that what John is really saying is that the Messiah, when he comes, will fuck the nations into submission."

"You have such an exquisitely delicate way of putting things. Any nation in particular, though?"

"I think we both know the answer to that: 'Babylon the great, mother of whores and of earth's abominations.'[11] Which brings me back to Harold Washington: 'The language of war in the Hebrew Bible and other ancient Near Eastern literatures is acutely masculinist. Warfare is emblematically male and the discourse of violence is closely imbricated with that of masculine sexuality.'[12] He goes on to quote Harry Hoffner: 'The masculinity of the ancient was measured by two criteria: (1) his prowess in battle, and (2) his ability to sire children.... These two aspects of masculinity were frequently associated with each other.... Those symbols which primarily referred to his military exploits often served to remind him of his sexual ability as well.'[13] Here I'm reminded of an instruc-named Rinnach, eight named Coirpre, eight named Mulach, nine named Daithi, nine named Dáire, nine named Damach, ten named Fiac, ten named Fiacha and ten named Feidlimid. In this great carnage on Murtheimne Plain Cúchulainn slew one hundred and thirty kings, as well as an uncountable horde of dogs and horses, women and boys and children and rabble of all kinds.[38] Not one man in three escaped without his thighbone or his head or his eye being smashed, or without some blemish for the rest of his life. And when the battle was over Cúchulainn was left without a scratch or a stain on himself, his charioteer or either of his horses. (Kinsella 1970, 150–56)[39]

I must confess to finding John's parallel account of Jesus battling the beast, "the kings of the earth," and their armies single-handedly (for although he commands an army, apparently he doesn't need one) a tad insipid by comparison.

> Then I saw the beast and the kings of the earth with their armies gathered to make war against the rider on the horse and against his army. And the beast was captured, and with it the false prophet who had performed in its presence the signs by which he deceived those who had received the mark of the beast and those who worshiped its image. These

tive scene in Stanley Kubrick's Vietnam satire, *Full Metal Jacket*. You know the scene I mean? The Marine recruits are required to clutch their M16s firmly in one hand and their crotches just as firmly in the other, all the while chanting, 'This is my rifle, this is my gun; this is for fighting, this is for fun!'"

"You can let go of your crotch now. I wouldn't want you to injure yourself."

"Sorry, I got a little carried away. Washington also notes:

> The male is by definition the subject of warfare's violence and the female its victim. For example, the language of the siege instructions of Deut. 20.10–20 is densely supplied with syntactical groups joining a masculine singular verbal subject with a city as (feminine) object of attack.... Given a linguistic milieu where cities are so often portrayed in the figure of a woman—either mother (Isa. 66.8-13), queen (Isa. 62.3), or virgin daughter (Isa. 37.22), a woman married (Isa. 62.5), widowed (Isa. 47.8, 9; 54.4; Lam. 1.1), or raped (Jer. 6.1-8;

two were thrown alive into the lake of fire that burns with sulfur. And the rest were killed by the sword of the rider on the horse, the sword that came from his mouth; and all the birds of the air were gorged with their flesh. (Rev 19:19-21; cf. 20:7-10)

Whereas the *Táin* is a garish celebration of war, Revelation is a muted celebration of war. Superimposed on Revelation, the *Táin* colors in its blanks with lurid hues. Revelation is a war scroll (see Bauckham 1993a, 210-37).[40] But the *Táin* is what this war scroll would look like fully unfurled.

(Fearful of exposure, Revelation resorts to threats: "I warn everyone who hears the words of the prophecy of this book: if anyone adds to them, God will add to that person the plagues described in this book" [22:18]. The *Táin*, for its part, feeling its colors bleed into Revelation, ends on an equally nervous note: "A blessing on everyone who will memorise the Táin faithfully in this form, and not put any other form on it" [Kinsella 1970, 283].)

As will by now be readily apparent, perhaps, my own attitude toward violence is somewhat ambivalent. There is probably little of which the God of Revelation is guilty of which I myself would not also be capable, given certain extreme environmental stimuli—and a dash of omnipotence, needless to say. Which is why I fear this God as

13.22; Isa. 47.1–4; Nah. 3.5–6)—the concentration of feminine forms in Deut. 20.10–20 inescapably evokes the figuration of the city as an assaulted woman. In issuing the command to draw near to a city 'in order to attack it,' this text effectively enjoins the soldiers 'to attack her' [20.10].... The description of the submissive city 'opening' to the warrior [20.11] ... evokes an image of male penetration. Similarly, the law uses [the same verb] to describe the military seizure of a city [20.19] ... [as] for the forcible seizure of a woman in sexual assault [22.28]."[14]

"And all of this is relevant to Revelation, I take it?"

"Quite possibly. Compare, for instance, Tina Pippin's reading of Babylon in Revelation as a sexually assaulted woman. Her proof text, as you may recall, is Rev 17:16, 'they will make her desolate and naked; they will devour her flesh and burn her up with fire.'"[15]

"I also recall that the perpetrators of these dire deeds are the 'ten horns,' together much as I do, and resist him for all I am worth.

And yet, despite myself, I love Revelation for its beauty. Its intricate lacework of lurid images never fails to thrill me. Of course, my impression that Revelation is an exquisite work of language is willful illusion on my part. The English translations in which I ordinarily read it are cosmetic coverings concealing from view all of Revelation's unsightly grammatical blemishes, its "barbarous idioms," as Dionysius of Alexandria long ago dubbed them (so Eusebius, *Church History* 7.25.26–27). The translators take John's broken Greek and beautify it (as did numerous copyists before them), excising all its startling irregularities, a nip here, a tuck there. I know enough Greek to spot John's stunning solecisms, but not enough to *hear* them. I wish I could hear John's exotic Aramaic(?) intonations, listen as he fumbles in the warm Aegean night for the correct grammatical boxes in which to lock his glittering visions, occasionally picking up the wrong one in the deep, velvety darkness, as so many scholars have imagined him doing—although others have argued that John is wearing night vision goggles instead and verbalizing his visions with painstaking precision, freely opting, for whatever reason, not to express himself in standard Greek.[41]

Revelation seduces me no matter how much I resist it. There is another kind of resistance that I wish to bring

with the beast, not Christ, your phallic warrior."

"Admittedly, but ultimately it's the commander in chief, the one seated on the throne, who is responsible for Babylon's rape: 'For God has put it into their hearts to carry out his purpose' (17:17)."

"I'm nervous that you'll soon have him leaping naked from his throne to join in the action, so let's move on. So far you've been talking as though John's Jesus were a one-man army—"

"Or a one-*lamb* army, at least."

"—but he doesn't wage war alone; instead he leads an army against the enemies of God. Here's Bauckham again:

> Also derived from this militant messianism is Revelation's key concept of conquering. It is applied both to the Messiah himself (3:21; 5:5; 17:14) and to his people, who share his victory (2:7, 11, 17, 28: 3:5, 12, 21; 12:11; 15:2; 21:7). Once again we note the importance in Revelation of the Messiah's army. That the image of conquering is a militaristic one should be unmistakable, although

into the discussion, however, and also another kind of seduction. On the annual school excursion from Adare to Dublin, our Ireland-for-the-Irish schoolmaster would lead us in solemn procession around the cavernous interior of the General Post Office,[42] where the Irish rebels set up their headquarters during the armed revolt of 1916 against British colonial rule. He would deliver a hushed but heated soliloquy by the statue of Cúchulainn enshrined in the building to commemorate the uprising, a speech sodden with intoxicating excerpts from the Proclamation of Irish Independence. The act that ignited the 1916 conflict was the public reading of this proclamation from the steps of the building. We were obliged to memorize the proclamation at school, in common with most Irish schoolchildren of that era. As though fearful that the Irish public had forgotten the symbolism of the Cúchulainn statue, the authorities had it framed with a lengthy quotation from the proclamation long after our school pilgrimages to it.

> We declare the right of the people of Ireland to the ownership of Ireland, and to the unfettered control of Irish destinies, to be sovereign and indefeasible. The long usurpation of that right by a foreign people and government has not extinguished the right, nor can it ever be extinguished except by the destruction of the Irish people. In every generation

interpreters of Revelation do not always do justice to this. It is closely connected with language of battle (11:7; 12:7–8, 17; 13:7; 16:14; 17:14; 19:11, 19), and it is notable that not only do Christ's followers defeat the beast (15:2), but also the beast defeats them (11:7; 13:7), so that this is evidently a war in which Christ's enemies have their victories, though the final victory is his. We should note also that the language of conquering is used of all the three stages of Christ's work: he conquered in his death and resurrection (3:21; 5:5), his followers conquer in the time before the end (12:11; 15:2), and he will conquer at the parousia (17:14). Thus it is clear that the image of the messianic war describes the whole process of the establishment of God's kingdom as Revelation depicts it—"[16]

"Permit me to interrupt. May I hazard a précis of the plot thus far?"

"The plot? I was unaware that there was one."

"Here goes. Revelation can plausibly be said to be about the Irish people have asserted their right to national freedom and sovereignty; six times during the past three hundred years they have asserted it in arms. Standing on that fundamental right and again asserting it in arms in the face of the world, we hereby proclaim the Irish Republic as a sovereign independent state, and we pledge our lives and the lives of our comrades-in-arms to the cause of its freedom, of its welfare, and of its exaltation among the nations.

The Cúchulainn statue represents a calculated combination of vulnerability and indomitability. Too weak to stand through loss of blood, the hero has strapped himself to a stone pillar, sword still gripped in his now lifeless hand, defiant to the end and beyond, even in the face of unimaginable odds. A raven has settled on his shoulder, a signal to the armies assembled around him that it is now safe to approach him.

But whence this fatal loss of blood, you might ask, given that Cúchulainn earlier took on these same armies single-handedly without incurring a single scratch? Well, the blood loss was inflicted by Cúchulainn's "adored foster-brother" Ferdia, "the horn-skinned warrior from Irrus Domnann," "the burden unbearable and the rock fatal in the fray," who had been shamed into challenging Cúchulainn to single combat (Kinsella 1970, 168). Ferdia

the establishment of God's kingdom on earth. How is this kingdom to be established? Through the messianic war. And what is the messianic war? An activity that, on the symbolic level, is conducted exclusively by male subjects (note the notorious 14:4),[17] and is constitutive of the masculinity of those subjects, since it is ultimately directed against the feminine (note, again, the no less infamous 17:3–6)."[18]

"Stunning. But there's one small matter you've overlooked. This is an army that does not kill; on the contrary, it allows itself to be killed."

"How noble. Your friend Bauckham, too, makes much of the fact that—let me see—'just as 5:5-6 depicts Jesus Christ as the Messiah who has won a victory, but has done so by sacrificial death, not by military might, so 7:4-14 depicts his followers as the people of the Messiah who share in his victory, but do so similarly, by sacrificial death rather than by military violence.'[19] Well, what else should we expect John to say? A military campaign against 'the enemies of God' is hardly a viable option for Christians at the time in which he is writing. That will have to await the Parousia, he

emerged the loser, eventually finding himself at the wrong end of the *gae bolga*, the belly-barb, Cúchulainn's ultimate weapon, which, cast from the fork of his foot across the waters of the ford in which they were fighting, sliced through Ferdia's "deep and sturdy apron of twice-smelted iron, and shattered in three parts the stout strong stone the size of a millstone," which he had earlier stuffed inside the apron for good measure, "for fear and dread of the *gae bolga*," "and went coursing through the highways and byways of his body so that every single joint filled with barbs" (1970, 193, 196–97). Thereupon the taciturn Ferdia is moved to remark, "That is enough now; I'll die of that," Cúchulainn's javelin, which he had already thrust through Ferdia's heart so that "half its length showed out through his back," having failed to impress Ferdia sufficiently (1970, 196–97). But Ferdia got his licks in too. At the height of the contest, which raged for four days, the heroes were "piercing and drilling each other" with their "big burdensome stabbing-spears" (the *Táin* is not without its queer conceits), and a bewildering assortment of other heavy weaponry. "If even birds in flight could pass through men's bodies they could have passed through those bodies that day and brought bits of blood and meat with them out into the thickening air through the wounds and gashes," insists the narrator (1970, 189). And

supposes, when Christians will have Christ, as invincible divine warrior—king of kings and warlord of warlords (see 17:14; 19:16)—present in person to lead them forth into battle. (As you may have surmised, I understand the 'armies of heaven' in 19:14—'And the armies of heaven, arrayed in fine linen … followed him on white horses'—to be the Christian elect,[20] especially in light of 17:14, which tells us that those who stand with the Lamb in the final battle 'are called and chosen and faithful.') What hasn't even occurred to John, of course, is the possibility that Christians might be in a position to triumph over their enemies—not symbolically, though sacrificial martyrdom, but literally, through military might—long before the divine warrior gallops into view. And if the slaughter of the 'ungodly' should be permissible at the Parousia, then why not before? (Whether or not the rider on the white horse is *literally* whacking off heads with his sword in 19:21 is a moot point, it seems to me, given the subsequent fate of the owners of those heads: eternal death, or worse, in the lake of fire that burns with sulfur. The latter that was only the third day; the fourth was even worse.

Limerickman Frank McCourt, in his autobiographical Pulitzer Prize-winning novel *Angela's Ashes*, describes his first encounter with the dead hero Cúchulainn. Frank's penniless ne'er-do-well émigré father ("He fought with the Old IRA and for some desperate act he wound up as a fugitive with a price on his head" [McCourt 1997, 2]) has brought four-year-old Frank, his mother, and his brothers back from New York to Dublin, en route to Limerick, in the late 1930s. Now the family is being driven across Dublin to the train station:

> Dad asks the driver if he'd mind going by way of the G.P.O. and the driver says, Is it a stamp you want or what? No, says Dad. I hear they put up a new statue of Cuchulain to honor the men who died in 1916 and I'd like to show it to my son here who has great admiration for Cuchulain.
>
> The driver says he has no notion of who this Cuchulain was but he wouldn't mind stopping one bit. He might come in himself and see what the commotion is all about for he hasn't been in the G.P.O. since he was a boy and the English nearly wrecked it with their big guns firing up from the Liffey River. He says you'll see the bullet holes all over the front and they should be left there to remind the Irish of English perfidy. I ask the man what's perfidy

spectacle is elaborated in 14:9–11: 'Those who worship the beast and its image, and receive a mark on their foreheads or on their hands, they will also drink the wine of God's wrath, poured unmixed into the cup of his anger, and they will be tortured [*basanisthēsetai*] with fire and sulfur in the presence of the holy angels and in the presence of the Lamb. And the smoke of their torture goes up forever and ever.')[21] The Crusades, the Inquisition, even the Holocaust itself (the smoke rising day and night from the ovens of Auschwitz), are but some of the more notable manifestations of the mass-death ethic that animates Revelation. Indeed, any one of these campaigns might have claimed a warrant for its genocidal fantasies in the sinister logic of this most dangerous of biblical books."

"Sorry to disappoint you, but the title 'Most Dangerous Biblical Book' has already been awarded to the Gospel According to Luke."

"By Jane Schaberg, you mean?[22] Well, Luke *is* subtler, and to that extent more deadly, but he doesn't have nearly as many notches on his gun. Even Bauckham concedes that

and he says ask your father and I would but we're stopping outside a big building with columns and that's the G.P.O.

Mam stays in the motor car while we follow the driver into the G.P.O. There he is, he says, there's your man Cuchulain.

And I feel tears coming because I'm looking at him at last, Cuchulain, there on his pedestal in the G.P.O. He's golden and he has long hair, his head is hanging and there's a big bird perched on his shoulder.

The driver says, Now what in God's name is this all about? What's this fellow doin' with the long hair and the bird on his shoulder? And will you kindly tell me, mister, what this has to do with the men of 1916?

Dad says, Cuchulain fought to the end like the men of Easter Week. His enemies were afraid to go near him till they were sure he was dead and when the bird landed on him and drank his blood they knew.

Well, says the driver, 'tis a sad day for the men of Ireland when they need a bird to tell them a man is dead. I think we better go now or we'll be missing that train to Limerick. (1997, 54–55)

Padraic Pearse, poet and pillar of the 1916 rebellion, argued that the spilling of Irish blood was at least as important to the cause of Irish freedom as the spilling of English blood. "Bloodshed is a cleansing and sancti-

the body count in Revelation is astronomical:

> So the series of judgments affecting a quarter of the earth (6:8) and the series affecting a third of the earth (8:7–12; 9:15, 18) are not, as we might expect, followed by a series affecting half the earth.... But there is now to be only the final judgment, the sixth trumpet (10:7). When the content of the seventh trumpet is spelled out in detail as the seven bowls (15:1), they are total, not limited, judgments (16:2-21), accomplishing the final annihilation of the unrepentant."[23]

"Yes, arguably the unrepentant are to be annihilated *in toto*, according to John. But by whom? Not by the Lamb, apparently—"

"Or not yet, at any rate."

"—nor by the army he commands, but by the one seated on the throne and those whom he commands, his heavenly host. But these spectacular military strikes cannot induce repentance in the enemy (9:20-21; 16:9, 11). For that an altogether different kind of army is required—one

fying thing," he wrote, "and the nation which regards it as the final horror has lost its manhood. There are many things more horrible than bloodshed; and slavery is one of them" (Pearse 1924a, 99). Feminization is another, apparently, as Pearse's evocation of the specter of emasculation suggests, against the backdrop of the English colonial construal of Ireland as "not-England," an indispensable other (one of several) against which English identity could consolidate its (imaginary) contours. "If John Bull was industrious and reliable," the Irish postcolonial critic Declan Kiberd observes, "Paddy was held to be indolent and contrary; if the former was mature and rational, the latter must be unstable and emotional; if the English were adult and manly, the Irish must be childish and feminine" (1995, 30). The hypermasculine high hero Cúchulainn, resurrected in Standish O'Grady's *History of Ireland* (1878–1880) after centuries of decomposition and neglect, and enfleshed in English words for the first time (the better to meet the invader on his own terms and beat him back to the sea?), provided a ready model of masculinity for militant Irish nationalists such as Pearse. "What was in Patrick Pearse's soul when he fought in Easter Week but an imagination," the Irish nationalist and mystic George William Russell would later declaim,

> and the chief imagination which inspired him was that of a hero

designed not to kill but to be killed, as I remarked earlier, whose general is the Lamb who was slain."

"What's interesting to me is the way in which military metaphors are withheld from the one seated on the throne and his angelic agents—their most qualified recipients, one might suppose—and lavished instead on the Lamb-in-a-body-bag and the walking dead who accompany him. In Revelation, Jesus is not so much God become *man* as God become *masculine*—although it doesn't look like that at first. In chapter 1 Jesus is an angel,[24] and in chapters 5 and following he is a lamb, but in chapter 19 he is a superwarrior. As angel he is barely human, as lamb he is barely a man, but as warrior he is hegemonically hypermasculine. In the final analysis, John presents Christ, together with his Christians, as icons of masculinity, reserving feminine imagery for the enemy. Smells suspiciously like a smokescreen to me, suggesting—yes, I see from your supercilious smirk that you've anticipated what I'm about to say—a certain anxiety around the issue of masculinity."

"The dimensions of the seer's sacred member was

who stood against a host.... I who knew how deep was Pearse's love for the Cuchulain whom O'Grady discovered or invented, remembered after Easter Week that he had been solitary against a great host in imagination with Cuchulain, long before circumstances permitted him to stand for his nation with so few companions against so great a power. (Quoted in Kiberd 1995, 196–97)

Kiberd's own take on the Irish nationalist infatuation with Cúchulainn is a little less romantic.

So the ancient hero Cuchulain died strapped to a rock, single-handedly defending the gap of the north ... ; and as his life ebbed away, a raven alighted and drank his blood. This combination of pagan energy and Christ-like suffering was of just the kind recommended for the production of muscular Christians at Rugby, suggesting that the revivalist Cuchulain was little more than a British public schoolboy in drag. (1995, 31).

Pearse, then, would provide a still queerer spectacle, that of a Gaelic-speaking British public schoolboy, draped in the battle dress of an Irish Volunteer,[43] which has been sexily unbuttoned so as to reveal his underwear: the costume of a mythical Celtic superhero. His right hand holds a rifle aloft, while his left brandishes a sword.

hardly the issue. In his own mind, its measurement was likely a multiple of seven."

"No doubt. But domination versus submission might well have been the issue, or, more precisely, the cultural proclivity to construe the former as masculine and the latter as feminine in the ancient Mediterranean world.[25] I mean, consider the fate that John is convinced awaits him and his fellow martyrs-in-the-making, the prospect of having to submit themselves passively to being fucked with physically, unto death if necessary...."[26]

"Not a very manly way to go out of *that* world, admittedly, at least not without a modicum of rationalization. Actually, you find the same gendered rationalization of martyrdom in another, roughly contemporary text, 4 Maccabees.[27] There the atrociously abused martyrs are lauded as true men in the most explicit terms—even, or especially, the female martyr, the mother of the seven brothers—and at the expense of their tormentors, especially Antiochus Epiphanes, whose own masculinity is subtly but effectively called into question."[28]

The latter accessory is no fantasy, actually: Pearse did wear an ancient sword strapped to his waist in the GPO through much of Easter Week 1916, as artillery shells rained down on the building and sniper bullets whistled through it.[44]

When Pearse declared bloodshed to be "a cleansing and sanctifying thing," therefore, he was speaking not only in the shadow of the cross of Christ but also of the pillar of Cúchulainn, that other son of a god. (Have I mentioned that Cúchulainn, too, was born of a human mother and a divine father?—not the god of Israel, however, but rather a prince of the *síde*, that ancient race, now invisible, from whom the fairies are descended.) To abhor blood sacrifice is to accept emasculation, Pearse implies, to remain dependent and enslaved: to remain a woman (1924a, 99).[45] Independence, nationhood, manhood requires that one be willing to sever the bond of servility, to turn one's weapon upon oneself as well as upon the emasculating imperial overlord to whom one is in thrall. "We must accustom ourselves to the thought of arms, to the sight of arms, to the use of arms," insists Pierce, polishing and repolishing his ancient sword (1924a, 98). Ultimately it is our own shackled wrists, our own fettered ankles, that we will be hacking through, and we shall bleed to death as a result. But we shall just as surely rise again, proclaims Pearse, his patriotism taking on a characteristic

* * * * *

When I first penned the above, suggesting that the martial imagery applied to the (proto-) martyrs in Revelation was an apologia for passive resistance as a legitimate masculine stance, and appealing to 4 Maccabees for support, I was groping for closure and feared I was overreaching. Shortly afterward, however, I stumbled on some parallel claims that served to persuade me of my own suggestion. First, classicist Brent Shaw's "Body/Power/Identity: Passions of the Martyrs," which asserts of 4 Maccabees:

> Praises of active and aggressive values entailed in manliness (*andreia*) by almost all other writers in the world of [4 Maccabees] could easily fill books. The elevation to prominence of the passive value of merely being able to endure would have struck most persons ... as contradictory and, indeed, rather immoral. A value like that cut right across the great divide that marked elite free-status male values and that informed everything about bodily

Pauline inflection. For Pearse did not hesitate to cast his own "blood sacrifice" in explicitly christological terms, brazenly picking up Christ's discarded purple robe and draping it over his green Irish Volunteer uniform with the ancient Celtic battle dress underneath. The following poem/prayer, composed by Pearse while in prison, was delivered to his mother on the day of his execution.

A Mother Speaks

Dear Mary, that didst see thy
  first-born Son
Go forth to die amid the scorn of
  men
For whom He died,
Receive my first-born son into
  thy arms,
Who also hath gone out to die for
  men,
And keep him by thee till I come
  to him.
Dear Mary, I have shared thy
  sorrow,
And soon shall share thy joy.
(Quoted in Edwards 1977, 315–
  16)

"If you strike us down now, we shall rise again and renew the fight," Pearse proclaimed from the dock in which he was sentenced to death (Edwards 1977, 318). Nine months earlier he had delivered a rousing panegyric at the grave of fellow rebel O'Donovan Rossa.

behaviour. (Shaw 1996, 278–79)

Shaw finds in 4 Maccabees (and not only in 4 Maccabees: he also appeals extensively to Seneca, for instance) "the conscious production of a rather elaborate conception of passive resistance," or, more precisely, "the explicit cooptation of passivity in resistance as a fully legitimized male quality—a choice that could be made by thinking, reasoning and logical men" (1996, 280).

Second, a paper by another classicist, Tessa Rajak, which I chanced to hear in August 1997 at the annual meeting of the Studiorum Novi Testamenti Societas in Birmingham. (The SNTS seminar, Early Jewish Writings and the New Testament, had devoted its entire program that year to 4 Maccabees. For three days we circled this torturous text, our genteel scholarspeak enabling us to say anything and everything about it but the thing that most rises in my own throat as I read it: *This is a text that makes me want to vomit.*) Independently of Shaw, Rajak argued in her (since published) paper, "Dying for the Law: The Martyr's Portrait in Jewish-Greek Literature," that in the ancient

Life springs from death; and from the graves of patriot men and women spring living nations. The Defenders of this Realm have worked well in secret and in the open. They think that they have pacified Ireland. They think that they have purchased half of us and intimidated the other half. They think that they have foreseen everything, think that they have provided against everything; but the fools, the fools, the fools!—they have left us our Fenian dead,[46] and while Ireland holds these graves, Ireland unfree shall never be at peace. (Pearse 1924b, 136–37)

The British authorities proceeded to demonstrate the truth of Pearse's soteriological assertions by summarily executing him, along with the other captured leaders of the rebellion, thereby unwittingly creating a cadre of martyrs and transforming an unpopular and apparently unsuccessful uprising into a popular and ultimately successful one, in that it eventually led to Irish independence.[47] As if in acknowledgment of the christological contours of this saga of execution and vindication, the standard term in Ireland for the 1916 revolt has long been the Easter Rising.[48] And as though to drive the message home, the 1998 peace accord that promised to close the horrific chapter in Northern Irish history ushered in by the events of 1916 and its

Mediterranean world "the inflexible and obdurate mindset of the martyr was perceived by some logical spirits as the epitome of unreason" (1997, 51). Philo, for example, "says that opponents might consider the Jews' readiness to die for their laws as 'barbaric,' while in reality it was an expression of freedom and nobility (*Leg.* 215). Later, Marcus Aurelius, in expounding the Stoic way to die, was to observe that this should be with considered judgment [*lelogismenō*] and not, in the Christian manner, obstinately and showily [11.3]" (1997, 51–52).[29] Concerning "the terminology of male heroism" in 4 Maccabees, therefore, with its elaborate evocations of warfare and athletic prowess (as, for example, in 17:11–17), Rajak claims: "These images are intrinsic to martyrology, as the agents which effect the transmutation of shaming passivity into the highest of masculine virtues. What we are offered is a concentrated inversion of the competitive, physical values which constructed masculinity for Graeco-Roman society, a triumphant reversal of the power-structure, with the victim as the winner" (1997, 55).

aftermath happened to be called the Good Friday Agreement.

Nationalism has long been a religion in Ireland, a sacrificial rite and a cult of martyrs. Indeed, the latter motifs are explicitly inscribed in the Proclamation of Independence itself (hardly surprising since Pearse was its principal author):[49] "In this supreme hour, the Irish nation must, by its valour and discipline, and by the readiness of its children to sacrifice themselves for the common good, prove itself worthy of the august destiny to which it is called." All of which brings us back to *Angela's Ashes*.

> Come on, boys. Sing.
> Because he loved the motherland,
> Because he loved the green
> He goes to meet a martyr's fate
> With proud and joyous mien....
> You'll die for Ireland, won't you, boys?
> We will, Dad.
> And we'll all meet your little sister in heaven, won't we, boys?
> We will, Dad. (McCourt 1997, 36–37)

But if nationalism has long been a religion in Ireland, religion has long been a form of nationalism.

> Talking about First Communion makes the [school]master all excited. He paces back and forth, waves his stick, tells us we must never forget that the moment the Holy Communion is placed on

Third, classicist Carlin Barton's "Savage Miracles: The Redemption of Lost Honor in Roman Society and the Sacrament of the Gladiator and the Martyr," which includes a reflection on "the ambiguous reception met by Christians in the arena" (1994, 56):

> The laughing, joyous submission to the rigmarole of the arena, the tranquil accommodation to brutality, and the apology that publicized their motivations were meant to forestall the perception of them as mere *ludibria*, ridiculous, weak, and humiliated.... The condemned Christian, like the gladiator, needed to establish to the audience that he or she was redeemed and a redeemer, not insulted and insulting.... But there can be no doubt that the voluntary suicide had difficulty in sacralizing himself or herself to a Roman audience. Most often they remained objects of scorn. (1994, 57)

But back to the conversation I have so rudely interrupted.

our tongues we become members of that most glorious congregation, the One, Holy, Roman, Catholic and Apostolic Church, that for two thousand years men, women and children have died for the Faith, that the Irish have nothing to be ashamed of in the martyr department. Haven't we provided martyrs galore? Haven't we bared our necks to the Protestant ax? Haven't we mounted the scaffold, singing, as if embarking on a picnic, haven't we, boys?
  We have, sir.
  What have we done, boys?
  Bared our necks to the Protestant ax, sir.
  And?
  Mounted the scaffold singing, sir.
  As if?
  Embarking on a picnic, sir.
  (McCourt 1997, 135)

It feels strange, sacrilegious even, to be conjuring up these ghostly voices (which echo eerily in my own indoctrinated head no less than Frank McCourt's) in the wake of the Good Friday Peace Agreement. Like McCourt, I'm a bit of a dinosaur (although not a tIRAnnosaur).

"And do they still have to memorize the Proclamation of Independence?" I ask my sister in tones of mild apprehension, although I'm not quite sure which answer I'm most in dread of hearing. We're sitting at her kitchen table discussing Irish education in general and that of her own children

* * * * *

"You're saying that 4 Maccabees inhabits Revelation's textual unconscious, then, just as texts such as Daniel and Ezekiel inhabit its textual consciousness?"

"Is that what I said? I wonder what I meant by it. Actually, Revelation goes well beyond 4 Maccabees in its defensive feminization of the foe.[30] Fourth Maccabees never goes so far as to spit the epithet 'whore' at Antiochus, after all, to deck him out in drag,[31] then strip him naked and 'devour his flesh' (see Rev 17:16), whatever that might mean."

"Whatever, indeed. The phrase 'they will devour her flesh' trips a tad too lightly off the tongues of most commentators on Revelation. There's a chapter in *American Psycho*, simply titled 'Girl,' which opens with the said psycho slapping the eponymous girl around—and spitting the word 'whore' at her, as it happens—and, when that fails to arouse him sufficiently, forcing her to watch a home video—"

"That *can* be torture, I'll admit."

"—in which he literally devours the brain of another nameless woman, with great in particular. My sister still lives in County Limerick ("between Patrickswell and Crecora," I always write anxiously on the envelope, unable after so many years in the United States to believe that my letter will ever find its way otherwise to her unnumbered, unstreet-named, unzipcoded domicile), and her children are educated locally. She looks at me in wonder, or perhaps it is pity. "The proclamation of what? Sure, I doubt they'd ever even have heard of it."

The author of Revelation, too, attached immense importance to martyrs and martyrdom.[50] Indeed, his calculation of the symbolic value of martyrdom was to prove uncannily accurate. Christian martyrdom would eventually purchase a vast empire "for God" on earth, the Great Persecution under Diocletian ushering in imperial Christendom under Constantine. Understandably enough, John failed to foresee Constantinian Christianity, precisely.[51] But the motif of the millennium ("I also saw the souls of those who had been beheaded for their testimony to Jesus.... They came to life and reigned with Christ a thousand years," 20:4-6), which he smuggles into the climactic sequence of his eschatological scenario, neatly anticipates its advent. For the symbolic economy of martyrdom in Revelation can be reduced to a simple exchange. In order to be deemed worthy to dominate others, Christ and his followers first have to show, in good Greco-Roman

relish, slathering it with Grey Poupon—"³²

"Thanks for that, but I think I'll stick with Ezek 23:25–30 as my stomach-churning backdrop of choice for the sickening tableau in Rev 17:16."³³

"But what if *American Psycho* were merely making explicit what is already implicit in Ezekiel 23 and Revelation 17? What if—"

"Hang on a minute. Why are we discussing *American Psycho*?"

"Because I see it as a highly illuminating instance of menaced masculinity at the margin—or, rather, beyond every margin, every limit, which is what makes it so chilling. Anyway, what if *American Psycho* were the apocalypse of Revelation, the uncoverer of the uncoverer, which, juxtaposed with Revelation, exposes the gendered savagery that seethes beneath the latter's surface?"

"Nothing like a bit of good clean misogyny to make a man feel good about his own masculinity. Is that the common thread, then, winding its dismal way through Ezekiel and Revelation and tying them, across space and time, to *American Psycho* and

fashion, that they are able to dominate themselves.⁵²

And so we arrive back at 4 Maccabees, which, like the *Táin*, is a paean to war, but this time to war with oneself. For 4 Maccabees is a paean to martyrdom, understood as ultimate self-mastery (Moore and Anderson 1998). The Roman Catholic children's Bible used in Christian doctrine class in that small school in Adare did not contain the book of Revelation, even in abridged form. But it did contain something still more disturbing, the story of the scalping, dismemberment, and roasting of seven brothers, watched by their mother, from 2 Maccabees 7 ("Then the king fell into a rage, and gave orders to have pans and caldrons heated. These were heated immediately, and he commanded that the tongue of their spokesman be cut out and that they scalp him and cut off his hands and feet, while the rest of the brothers and the mother looked on. When he was utterly helpless, the king ordered them to take him to the fire, still breathing, and to fry him in a pan"—vv. 3–5), which is to say, 4 Maccabees in miniature.⁵³ And so my childhood was innocent of the book of Revelation but haunted by the *Táin Bó Cúailnge* and the tale of the Maccabean mother and her butchered, pan-fried sons. But perhaps providence was merely preparing me for Revelation. For it strikes me now that what I earlier said of the *Táin* in relation to Revelation would apply muta-

other leading snuff books of our day?"

"I'm afraid it might well be. But there's one other thing that troubles me."

"And what might that be?"

"Surely you're not persuaded by my reading?"

"Of course not. I'm merely patronizing you."

"Thank goodness. You had me worried."

tis mutandis to 4 Maccabees: Whereas 4 Maccabees is a garish celebration of martyrdom, Revelation is a muted celebration of martyrdom. Superimposed on Revelation, 4 Maccabees colors in its blanks with lurid hues.

Picking my way nervously through 4 Maccabees, I plunge through its blood-besotted pages and fall screaming into Revelation's upper atmosphere.

At which point I wake up in a cold sweat.

If you too feel Revelation to be a place of peril, perhaps it is because you, like me, cannot crawl beneath the altar with the souls of those "who ha[ve] been slaughtered for the word of God and the testimony they have borne" (6:9), those who have suffered such atrocious injustices that they feel justified in crying out not for justice but for vengeance—"Sovereign Lord, holy and true, how long will it be before you judge and avenge [*ekdikeis*] our blood on the inhabitants of the earth?" (6:10; cf. 16:4–7)—those for whom the book was written. Perhaps it is not addressed to you. Certainly it is not addressed to me.

"There was a time when it *might* have been addressed to you," a remonstrative voice whispers in my head, "during the centuries when the colonial overlord oppressed your people and tried to crush the Catholicism out of their bodies. They bled, they prayed, they died."

"But what does the cry for vengeance from under the altar, heard and heeded by the one seated on the throne, actually effect?" a second voice whispers in response. "An eye for an eye? No, not an eye for an eye. What Revelation seems to be saying is this: If you gouge out the eye of one of God's witnesses, or even refuse to heed them, God will gouge out both of your eyes in return. And not only that but he will puncture your eardrums as well, and tear out your tongue, and sever your spine, and plunge you into a timeless torment. Or, what amounts to much the same thing, he will have you tortured for all eternity, the smoke of your torment ascending like incense to his throne (14:9–11; 20:15; 21:8; cf. 2 Macc. 7:3: 'The king fell into a rage and gave orders to have pans and caldrons heated'). It's the 'forever and ever' that seems to make the punishment spectacularly incommensurate with the crime," the whispering voice continues, feebly belaboring the obvious, "however horrific the latter may be. In this respect, too, Revelation reveals that it is of a piece with 2 and 4 Maccabees (and 4 Macc. 12:11–12 in particular, in which the tyrant is told that in exchange for the temporary tortures inflicted on the martyrs, God has 'laid up for [him] intense and eternal fire and tortures [*aiōniō pyri kai basanois*], and these throughout all time will never let [him] go')."

But the real source of my unease with Revelation's God lies elsewhere, I suspect. As I intimated earlier, I find him disturbingly like myself. God knows, if anyone were to inflict grievous injury on one of my own loved ones, I myself would not be content merely to repay injury with comparable injury, given half a chance. I would seek more, much more. Because that is how I am. Had I set out to create a god in my own image and likeness, then, I could hardly have done better than the one who confronts me in Revelation. Reading Revelation is, for me, uncannily like looking in a mirror—while having a psychotic episode.

The Gospel of John is generally regarded as the New Testament witness par excellence to the Christian doctrine of the incarnation. It seems to me, however, that this honor belongs instead to the Apocalypse of John. For the God of Revelation is quintessentially incarnational: God become human—or, more to the point, as I have sought to show in this essay, God simply become *man*.

Notes

1. Just as the first word could only be pronounced as *Da*. "Does he really talk like that?" you may idly be asking yourself. To which I can smugly respond in the negative. Ten years at Trinity College, Dublin (founded in 1592 to educate the scions of the Irish colonial aristocracy, its porticos and quadrangles echoing with faux-English accents even to this day), certainly cured me of dat.

2. But let me numb you with it one more time: "My ... father ... was a butcher.... As a child, the inner geographical boundaries of my world extended from the massive granite bulk of the Redemptorist church squatting at one end of our street to the butcher shop guarding the other end. Redemption, expiation, sacrifice, slaughter.... There was no city abattoir in Limerick in those days; each butcher did his own slaughtering. I recall the hooks, the knives, the cleavers; the terror in the eyes of the victim; my own fear that I was afraid to show; the crude stun-gun slick with grease; the stunned victim collapsing to its knees; the slitting of the throat; the filling of the basins

with blood; the skinning and evisceration of the carcass; the wooden barrels overflowing with entrails; the crimson floor littered with hooves" (Moore 1996, 4).

3. John is obsessed with the number seven, as is well known (although also with the numbers three, four, twelve, and their multiples).

4. I was forcibly reminded of this forgotten drawer while perusing one of Jeff Staley's adventures in autobiographical criticism. Staley quotes a haiku that he composed when he was nineteen, adding: "This unpublished poem is entitled *The Artist's Studio* and can be found downstairs in the top drawer of my beige metal file cabinet. I've decided that the only way I will ever get any of my poems published is by putting them in scholarly articles" (Staley 1995a, 178 n. 13).

5. The earliest known title of the book, and the one it bears in modern editions of the Greek New Testament, is *Apokalypsis Iōannou*, "Revelation of John." For the historical vicissitudes of the title, see Aune 1997, 3–4.

6. Bauckham 1993b, 68–69.

7. See Butler 1990, 134–41.

8. Washington 1997, 329–30, paraphrasing Cooke 1993, 177.

9. Clines 1995, 218–19.

10. Bauckham 1993b, 69.

11. "So he carried me away in the spirit into a wilderness, and I saw a woman sitting on a scarlet beast ... and on her forehead was written a name, a mystery: 'Babylon the great, mother of whores and of earth's abominations'" (Rev 17:3-5). That the epithet "Babylon," used six times in Revelation (14:6; 16:19; 17:4; 18:2, 10, 21), is a cipher for imperial Rome is a hoary and venerable critical consensus (even Swete 1906, 180-81 and Charles 1920, 2.14 are relatively late expressions of it), which means that it is voluptuously ripe for definitive debunking (but not by me: most of the essays in the present volume would implode without it).

12. Washington 1997, 330.

13. Hoffner 1966, 327, quoted in Washington 1997, 330.

14. Washington 1997, 346. This theme is by no means limited to the Hebrew Bible, of course. Classicist Barbara Kellum (1996, 171) writes of "the timeworn analogies between the penetration of a woman's body and the breaching of enemy fortresses," referring us to *Iliad* 22.468-70 (cf. 16.100); *Odyssey* 13.388; Euripides, *Hecuba* 536-38; Euripides, *Trojan Women* 308-13; and Ovid, *Amores* 1.9.15-20.

15. Pippin 1992a, 57-68; 1999, 83. Catherine Keller, too, remarks of 17:16: "In God's name, a powerful, sexual, bejeweled woman is stripped, humiliated, and devoured by hairy and horny beasts. Vision becomes voyeurism: a pious snuff picture unfolds" (1996a, 76). Thus convicted, John stands squarely in the sordid biblical tradition appositely dubbed "pornoprophetics." See Setel 1985; Carroll 1995; Exum 1996, 101-28; and Brenner 1997, 153-74—all studies of the pornoprophets of the Hebrew Bible (Hosea, Ezekiel, Jeremiah, Nahum, et al.). For attempts to trace John of Patmos's links to them, see Selvidge 1996; Kim 1999, 61-62, 65-66, 69-74. Also required reading on this topic is Pippin's "Pornoapocalypse" (in Pippin 1999, 92-97).

16. Bauckham 1993b, 69-70; see also 1993a, 229-32.

17. The notorious 14:4: "It is these who have not defiled themselves with women [*meta gynaikōn ouk emolynthēsan*], for they are virgins; these follow the Lamb wher-

ever he goes" (the "these" referring to "the one hundred forty-four thousand who have been redeemed from the earth"). It was Ernst Lohmeyer, apparently, who first suggested (1926, 120) that 14:4 is an allusion to "holy war," which, according to Deut 23:9-10 (cf. Lev 15:16; 1 Sam 21:5; 2 Sam 11:11; 1QM 7.3–7), requires sexual abstinence of its participants, a suggestion that has been dutifully recycled by commentators ever since, whether to endorse it or to quibble with it.

18. The no less infamous 17:3-6 stigmatizes the enemy, who is "drunk with the blood of the saints and the blood of the witnesses to Jesus," as "Babylon the great, mother of whores [*hē mētēr tōn pornōn*]," who "hold[s] in her hand a golden cup full of abominations and the impurities of her fornication."

19. Bauckham 1993b, 77.

20. In common with a host of twentieth-century commentators, ranging from Charles (1920, 2:135) to Beale (1999, 960) and beyond.

21. Here I have modified the NRSV translation. Traditionally, Christian commentators on Revelation have displayed no disquiet even with such statements as 14:9–11, so unqualified has their tacit endorsement of Revelation's theology (and ideology) been. Typical is Beale. First he informs us that "the apocalyptic belief was that the wicked would be punished often by fire, in the presence of the righteous (1 En. 48:9; 62:12; 108:14–15; Wis 5:1–14; 4 Ezra 7:93; Targ. Isa. 33:17) forever (Isa. 66:22–24; 1 En. 27:2-3; cf. 1 En. 21)" (1999, 760 n. 443). Then comes the standard disclaimer: "Even this belief did not underscore gleeful revenge but drew attention to the truth formerly denied by the righteous" (760 n. 443). What Beale omits to mention, among other things, is that Rev 14:9–11 was a pillar of the ancient and medieval Christian doctrine (championed by Tertullian, Augustine, and Aquinas, among others) that part of the bliss of souls in heaven consists in contemplating the torments of the damned.

22. Schaberg 1998, 275: "The Gospel of Luke is an extremely dangerous text, perhaps the most dangerous text in the Bible."

23. Bauckham 1993b, 82–83.

24. "I saw one like the Son of Man, clothed with a long robe and with a golden sash across his chest. His head and his hair were white as white wool, white as snow; his eyes were like a flame of fire, his feet were like burnished bronze, refined as in a furnace, and his voice was like the sound of many waters. In his right hand he held seven stars, and from his mouth came a sharp, two-edged sword, and his face was like the sun shining with full force" (1:13–16). Most of the details of this portrait are copied from Dan 10:5–6 (except for those that are copied from Dan 7:9, 13), where they are used to describe an angelic being, probably Gabriel: "I looked up and saw a man clothed in linen, with a belt of gold from Uphaz around his waist. His body was like beryl, his face like lightning, his eyes like flaming torches, his arms and legs like the gleam of burnished bronze, and the sound of his words like the roar of a multitude." Adela Yarbro Collins notes how "most Christian readers downplay or ignore the angelic elements" in Rev 1:12–16 (1993b, 103); she argues that "the 'one like a son of man' in Revelation 1 is an angelic figure" (1993b, 102). See further Rowland 1980; Collins 1992, esp. 548–51; Barker 1992, 200–203; and esp. Carrell 1997, 129–74.

25. Clement of Alexandria condenses an entire epoch of gender ideology when he

declares: "To do [*to dran*] is the mark of the man; to suffer [*to paschein*] is the mark of the woman" (*The Instructor* 3.19.2).

26. That John's target audience has experienced persecution in the recent past is suggested by internal evidence (see esp. 1:9; 2:3, 9–10, 13; 3:8; 6:9). The puzzle for scholars is the extent of the persecution to which John's churches are subject as he writes: is the extensive persecution to which he alludes (see also 6:9; 7:14; 12:17; 13:7–10, 15–17; 17:6; 18:24; 19:2; 20:4) a present reality or (more likely) a prophetic alert? Ruminations on and around the problem include Collins 1984, 84–110; Thompson 1990, esp. 15–17, 171–97, 202–10; Aune 1997, esp. lxiv–lxix; and Beale 1999, 5–16.

27. Do you happen to know 4 Maccabees? It is a Jewish (proto)martyrological text from the early second century C.E. It tells the hortatory tale of how one Eleazar, an elderly Jew, and seven unnamed Jewish boys, together with their unnamed mother, inflict moral defeat on the Gentile Syrian tyrant Antiochus IV Epiphanes, who has subjected them to torture so as to compel them to renounce their religion. Actually, *4 Maccabees* contains some of the most sickening accounts of physical torture ever to bleed from a pen (easily eclipsing the *Acts of the Christian Martyrs*, Butler's *Lives of the Saints*, and even pp. 303–6 of Bret Easton Ellis's *American Psycho* [1991]). Here's a typical slice of the action: "Then at [Antiochus's] command the guards brought forward the eldest [of the seven boys], and having torn off his tunic, they bound his hands and arms with thongs on each side. When they had worn themselves out beating him with scourges, without accomplishing anything, they placed him upon the wheel. When the noble youth was stretched out around this, his limbs were dislocated, and every member disjointed.... They spread fire under him, and while fanning the flames they tightened the wheel further. The wheel was completely smeared with blood, and the heap of coals was being quenched by the drippings of gore, and pieces of flesh were falling off the axles of the machine" (9:11–14, 19–20)—and so on, *ad nauseam*, through six more chapters and six more boys, their mother all the while forced to look on, seeing "the flesh of her children melting in the fire, and their toes and fingers scattered on the ground, and the flesh of their heads right down to the jaws exposed like masks" (15:15, my translation).

28. As Janice Anderson and I attempted to show (1998). See also Young 1991.

29. That the observation might be a gloss does not reduce its interest, as Rajak remarks (1997, 52 n. 47).

30. The labeling of a male opponent as feminine was a stock polemical slur in Greco-Roman antiquity (see Kraemer 1994).

31. See Keller 1996a, 77: "we may behold the Whore of Babylon as a great 'queen' indeed: imperial patriarchy in drag." See further 143–46 below.

32. Ellis 1991, 326–28. The woman forced to watch the video herself becomes the main course in a further nightmarish repast. This occurs in the chapter titled "Tries to Cook and Eat Girl." The "tries" applies solely to the cooking, which is not the psycho's forte. He has no trouble eating large uncooked quantities of the woman, however, whom he has tortured to death (with the aid of a very large rat), and whose dismembered and eviscerated corpse has littered his exquisite Manhattan apartment for days or possibly weeks. The novel is narrated by the psycho himself, who is young, rich, and exceptionally handsome (he is regularly taken for a male model). By day he

works on Wall Street; by night he dines in chic restaurants with his peers. But day or night, whenever his busy schedule allows it, he kills (or, just conceivably, fantasizes that he kills) small animals (preferably beloved pets), children, homeless men, men in general, and women—especially women—with an insouciant savagery that makes Hannibal Lecter look positively humanitarian (and vegetarian). At its publication, *American Psycho* was the subject of much critical acclaim; it is a literary tour de force. But it outraged other readers who saw it as a virtual how-to manual for the serial rape, torture, and murder of women.

33. Although Ezek 23:25–30 shares the backcloth with Ezek 16:37–41; Hos 2:3; Jer 10:25; 41:22 LXX; Mic 3:3; and Nah 3:4–5, 15.

34. The cattle raid of the title is the invasion of the Irish province of Ulster by the armies of the province of Connaught, along with their allies from the remaining two provinces, Leinster and Munster, in pursuit of the *Donn Cúailnge*, a colossal brown bull that Medb, queen of Connaught, is determined to possess to match the colossal white bull owned by her husband Ailill. The *Táin* is the centerpiece of the second major cycle of ancient Irish myth and legend, the first cycle consisting of tales of the *Tuatha Dé Danann* ("people of the goddess Danu"), the mythical first inhabitants of Ireland, and the third cycle consisting of tales of Finn and the Fianna, the elite bodyguard of the high king of Ireland. In terms of content, all three cycles are pre-Christian (which, in the Irish context, means pre-fifth century C.E.), although the end of the third cycle heralds the arrival of the new religion, the last surviving member of the Fianna, Finn's son Oisín, whose life has been magically preserved for centuries after all his comrades have died, being converted by Saint Patrick before he too expires. Traditionally the *Táin* is said to be set in the time of Christ.

35. One of the three goddesses of war who feature in the *Táin*.

36. Later the narrator will be at pains to stress that, although hideous in combat, Cúchulainn cut a stunningly handsome figure ordinarily: "You would think he had three distinct heads of hair—brown at the base, blood-red in the middle, and a crown of golden yellow. This hair was settled strikingly into three coils on the cleft at the back of his head. Each long loose-flowing strand hung down in shining splendour over his shoulders, deep-gold and beautiful and fine as a thread of gold. A hundred neat red-gold curls shone darkly on his neck, and his head was covered with a hundred crimson threads matted with gems. He had four dimples in each cheek—yellow, green, crimson and blue—and seven bright pupils, eye-jewels, in each kingly eye. Each foot had seven toes and each hand seven fingers, the nails with the grip of a hawk's claw or a gryphon's clench" (Kinsella 1970, 156–58).

37. The boy-troop, together with Cúchulainn himself, still a boy of seventeen, was left to defend Ulster against Medb's invading army after a curse had left all the adult men of Ulster bedridden. Now the boy-troop has been massacred and Cúchulainn has set out to avenge them.

38. Dogs and horses, women and children…. A patriarchal gem, to be sure. Yet the *Táin* does not lack formidable women (by its own lights, at any rate), such as the memorable Queen Medb, distinguished by her greed, her prowess in battle ("Then Medb took up her weapons and hurried into battle. Three times she drove all before her until she was turned back by a wall of javelins"), and her "friendly thighs"; or

the woman-warrior Scáthach, "the Shadowy One," from whom the boy Cúchulainn received his training in arms, so that "he could beat any hero in Europe"; or Aife, "the hardest woman warrior in the world," although not quite as hard as the phallic super-warrior Cúchulainn ("Cúchulainn leaped at her and seized her by the two breasts. He took her on his back like a sack, and brought her back to his own army. He threw her heavily to the ground and held a naked sword over her"); or Cúchulainn's own woman Emer, who, although not a warrior herself, likes nothing better than to see Cúchulainn strut his stuff ("'That was a great deed,' Emer said, 'to kill one hundred armed angry men'"). The quotations are from Kinsella 1970, 247, 169, 28, 32, 32–33, and 37 respectively. Further on the ambiguous status of women in ancient Irish mythology, see Bitel 1997 and Findon 1997.

39. This translation of the *Táin* by the poet Thomas Kinsella is far superior to the one that I read, or had read to me, as a child—Lady Augusta Gregory's *Cuchulain of Muirthemne* (1902). Lady Gregory's translation frequently amounted to a sanitized paraphrase of the *Táin*, as I later discovered. "I left out a good deal I thought you would not care about for one reason or another," she explained to "the people of Kiltartan," the peasant tenants of her Galway estate to whom she dedicated her translation and on whose dialect she modeled it (1902, 5). As a boy, therefore, I was spared both the excitement and the bafflement that exchanges such as the following would have elicited in me: "Cúchulainn caught sight of the girl's breasts over the top of her dress. 'I see a sweet country,' he said, 'I could rest my weapon there.' Emer answered him by saying: 'No man will travel this country until he has killed a hundred men at every ford from Scenmenn ford on the river Ailbine, to Banchuing … where the frothy Brea makes Fedelm leap'" (Kinsella 1970, 27).

40. Bauckham's chapter is titled "The Book of Revelation as a Christian War Scroll."

41. The first group of (generally older) scholars includes Laughlin 1902; Scott 1928; Torrey 1958; Mussies 1971; and Thompson 1985. The latter group (which represents a spectrum of related views) includes Porter 1989; Callahan 1995; Beale 1997; and MacKenzie 1997.

42. The GPO (as it is colloquially known) is a three-storied early nineteenth-century edifice with a classical columned portico rising to its roof. It is still the most prominent building on O'Connell Street, Dublin's main thoroughfare.

43. The 1916 rebels were made up mainly of members of the Irish Volunteers under Pearse, who had cofounded the force in 1913, and members of the Irish Citizen Army under the socialist labor leader James Connolly.

44. Kiberd provides a fascinating reading of the uprising as street theater. Pearse was not the only rebel leader in dramatic garb: Éamon Ceannt wore a kilt and played bagpipes, university don Thomas MacDonagh wore a cloak and carried a swordstick, and poet Joseph Plunkett was bedecked with Celtic rings and bracelets. And although the GPO proved a disastrous choice militarily for the rebel stand, with its imposing Ionic pillars it was the perfect dramatic setting for this "poets' rebellion," as the uprising would derisively be called in the months and years ahead. Michael Collins, who himself fought in the GPO, noted that the entire revolt had "the air of a Greek tragedy." See further Kiberd 1995, 203–5, 223–24. He adds: "So it was fitting that the printing

press on which the Proclamation of the Republic was done should have been hidden in the Abbey Theatre" (1995, 204).

45. Earlier in the same speech/essay we read: "I hold that before we can do any work, any *men's* work, we must first realise ourselves as men. And we of this generation are not in any real sense men, for we suffer things that men do not suffer, and we seek to redress grievances by means which men do not employ. We have, for instance, allowed ourselves to be disarmed; and, now that we have the chance of re-arming, we are not seizing it" (1924a, 97). How gratified, then, Pearse would have been to know that he and his fellow rebels would be referred to ever after in postindependence Ireland as "the *men* of 1916."

46. *Fenian* was a generic term for the Irish rebels, one that also evoked Ireland's heroic age. The term derived from the *Fianna*, the warrior bands of Irish myth and legend. One of the two most powerful (and most conservative) political parties in contemporary Ireland still bears the comically incongruous name *Fianna Fáil*, "warriors/soldiers of destiny."

47. The Irish Free State was formed in 1922, although it remained a British Dominion and was partitioned from Northern Ireland, which remains part of the United Kingdom to this day. In 1937 the Irish Free State became Éire (the ancient name for the island, derived from that of the goddess Ériu), and in 1949 Éire in turn became the Republic of Ireland, only then officially seceding from the British Commonwealth.

48. Pearse would certainly have approved. The uprising had originally been scheduled for Easter Sunday—another fine theatrical gesture—but had to be postponed until Easter Monday due to logistical complications.

49. And one of its *seven* signatories, a touch John of Patmos would surely have appreciated. Their names were drummed into me at an early age, and I can still rattle them off at will (as effortlessly as the Seven Sacraments). The Easter Rising itself seems to have been a significant seven in the minds of the authors of the proclamation, which, as we saw earlier, states: "In every generation the Irish people have asserted their right to national freedom and sovereignty; six times during the past three hundred years they have asserted it in arms. Standing on that fundamental right and again asserting it in arms in the face of the world, we hereby proclaim the Irish Republic."

50. It is high time I inserted the conventional caveat, noting that whereas John does repeatedly employ the Greek term *martys* ("witness": 1:5; 2:13; 3:14; 17:6) and its cognates *martyria* ("witness," "testimony": 1:2, 9; 6:9; 11:7; 12:17; 19:10 [twice]; 20:4) and *martyreō* ("bear witness," "testify": 1:2; 22:16, 20), *martys* is not yet the *terminus technicus* it will become in later Christian literature (beginning with the *Martyrdom of Polycarp*, apparently). Yet we are well on the way to such usage—see especially Rev 2:13 ("you did not deny my faith even in the days of Antipas my *martys*, my faithful one, who was put to death among you"), 6:9 ("I beheld under the altar the souls of those who had been slain on account of the word of God and their *martyria*"), 17:6 ("And I beheld the woman [Rome], drunk with the blood of the saints and the blood of the *martyres* of Jesus"), and 20:4 ("I also beheld the souls of those who had been beheaded on account of their *martyria* to Jesus")—so that Revelation can aptly be regarded as a protomartyrological text, at least.

51. See 35–36 above.

52. Paradoxically, however, the enemies of God in Revelation embrace a course of action that is structurally parallel to Christian martyrdom. Certain Christians refused to curse Christ even under pain of death, as we know from Pliny the Younger's letter to Emperor Trajan (*Letters* 10.96–97), composed around 112 C.E., when Pliny was governor of Bithynia, the neighboring province to John's home province of Asia. But the unrepentant in Revelation dare to curse God under pain of death (16:9, 11, 21)—even eternal death by torture (14:9–11), unlike the temporary torment endured by the Christian martyrs—their gesture thereby exceeding that of the latter (at which point John's narrative rhetoric begins to elude his control and conspire behind his back).

53. Assuming that 4 Maccabees is a free expansion of 2 Macc 6:12–7:42.

4
Hypermasculinity and Divinity*

This is the oldest essay in the volume. The *JSNT* article from which it ultimately stems was my first published foray into Revelation. The problem with which the essay grapples is the same as that with which the previous two essays tussle: theological anthropomorphism. The focus, however, of the previous two essays was the *activity* of the one seated on the throne: his imperial activity and his martial activity, conducted through his angelic agents and his messianic regent. The focus of the present essay is the *passivity* of the one seated on the throne: his status as a static object of adoration for an innumerable audience of worshipers. What is it that we worship when we worship "God" (those of us who do)? This is the question that animates this essay, and Revelation is the ideal text in which to explore it in the Christian context. The long central section of Revelation (4:1–22:5) begins and ends with divine worship on an epic scale, and is punctuated with it throughout.

More even than most topics of theological debate, worship has the potential to be soporifically anodyne. Defamiliarizing intertexts were needed to provoke reconsideration (first of all in myself) of the archetypal scenes of worship staged in Revelation's throne room. Those intertexts were provided by two contemporary cultural spectacles: the bodybuilding posing exhibition and the "big reveal" of the makeover reality TV show. The essay's pop-cultural proclivities make it an

---

* This essay first appeared under the title "The Beatific Vision as a Posing Exhibition: Revelation's Hypermasculine Deity," *JSNT* 60 (1995): 27–55; it subsequently appeared in expanded form under its present title in Levine 2009, 180–204; it is reprinted in newly revised form with permission.

exercise in cultural studies, one of my first experiments in that critical genre (a genre constantly challenged by the ephemerality of its objects of analysis; Dorian Yates, for instance, the champion bodybuilder who plays a cameo role in the essay, has since been eclipsed by yet more alarming man-mountains of muscle). It was also my first foray into masculinity studies, and my first flirtation (still shy) with queer theory.

What is it that we worship when we worship "God"? If while prostrating ourselves abjectly before Revelation's great white throne (20:11) we peer through the rainbow that surrounds it and shield our eyes from the lightning that flashes from it (4:3, 5), we may find that what we worship is hegemonic hypermasculinity: in a word, "man."

In *The Vision of God*, an extraordinarily erudite tome first published in 1931, Kenneth E. Kirk, bishop of Oxford, sets out to demonstrate that the dictum "the end of life is the vision of God," which he takes to be a New Testament doctrine, has, through the ages, "been interpreted by Christian thought at its best as implying in practice that the highest prerogative of the Christian in this life as well as hereafter, is the activity of *worship*" (1931, ix).[1] And Kirk does succeed admirably in showing that elite Christian theologians, at least until the Reformation, tended overwhelmingly to view the vision of God as being indissolubly bound up with the worship of God.

But what is the New Testament basis for the belief that the end of life is the vision of God? Having disposed of various "Old Testament anticipations" of the dictum, along with sundry pagan anticipations of it, Kirk proceeds to survey the New Testament at some length, before turning to the daunting expanse of postbiblical Christian theology and following selected currents upstream to his own time. But although he devotes significant discussion to the Synoptic Gospels and the Fourth Gospel, the Letters of Paul and the Letter to the Hebrews, and even the Letter of James, he scarcely mentions the book of Revelation except in passing—a curious omission indeed, for in what other New Testament text are the vision and

---

1. The phrase "the end of life is the vision of God" is adapted from Irenaeus, *Against Heresies* 4.20.

worship of God so fully fused? Celestial life, according to Revelation, is the beholding of God, and the beholding of God irresistibly elicits the unending worship of God (4:8–11; 5:13-14; 7:9–12, 15; 8:3-4; 11:15–18; 14:1–3; 15:2-4; 19:1–18; 21:22–23; 22:3–5).[2]

Heaven Can Be Hell

> And the throne of God and of the Lamb shall be in it, and his slaves shall serve him. (Rev 22:3)

What does it mean to worship God? The essence of divine worship, for Kirk, is an overwhelming sense of one's own smallness, a profound sense of one's own insignificance, a painful sense of one's own imperfection, relative to the immensity, power, and perfection of the deity. Gazing at God, the worshiper "sees himself to be nothing.... Worship tells us much good of God, but little good of ourselves.... For that we may praise Him, but it leaves us nothing upon which to pride ourselves" (Kirk 1931, 448).

A more recent book by Richard Bauckham (1993b) returns repeatedly to the topic of worship in Revelation.[3] Bauckham's understanding of worship echoes that of Kirk (and innumerable other theologians, hence its interest for me). Commenting on the vision of the heavenly throne room in Revelation 4, for example,[4] Bauckham remarks:

> Especially prominent in the vision is the continuous worship by the four living creatures and the twenty-four elders. It is a scene of worship into which the reader who shares John's faith in God is almost inevitably drawn. We are thereby reminded that true knowledge of who God is is inseparable from worship of God. The song of the four living creatures and the hymn of the twenty-four elders express the two most primary forms of awareness of God: the awed perception of his numinous holi-

---

2. See also 5:8–12; 12:10–12; 14:7; 16:5–7; 19:10; 22:8–9. Unending worship is a frequent feature of apocalyptic depictions of heaven (see also, e.g., *1 Enoch* 39:12–14; 40:2; 71:7; *2 Enoch* 21:1; *Testament of Levi* 3:8).

3. This slim book is a companion to Bauckham's bulkier *The Climax of Prophecy: Studies on the Book of Revelation* (1993a). The former is distinguished from the latter by a strong apologetic thrust. It attempts to argue the relevance of Revelation for the current theological scene, one frequently inimical to it.

4. A scene that Jürgen Roloff has rightly described as "the theological center of the book" (1993, 68).

ness (4.8; cf. Isa. 6.3), and the consciousness of utter dependence on God for existence itself that is the nature of all created things (4.11). These most elemental forms of perception of God not only require expression in worship; they cannot be truly experienced except as worship. (1993b, 32–33)

There was a time when I myself would have endorsed such sentiments enthusiastically, indeed gambled heavily on their veracity. For what do the Pentecostal prayer meeting, with its ecstatic cacophony of tongues, or the Cistercian cloister, with its ethereal chorus of plainsong, purport to be if not antechambers to the celestial throne room, and I have lingered long in both waiting rooms. But I must confess that my reactions to such sentiments have long since been refashioned by a series of texts (what can one do in a waiting room but read?), subtle philosophical texts and not-so-subtle psychoanalytic texts, texts as ingenious and insidious as Derrida's "Différance" (1982a, 1–27) and as crude and rude as Freud's *The Future of an Illusion* (1927). I must be a simple fellow, for the latter in particular spoke powerfully to me, or, rather, roared in my ear. Hurriedly honing a blunted blade that had once belonged to Feuerbach, Freud argues that the worshiper in his or her relationship to the divine parent faithfully mirrors the child's relationship to its own human parents—its pervasive sense of its own smallness relative to the towering stature of its parents, its painful sense of its own powerlessness relative to the apparent omnipotence of its parents, its profound sense of its own dependence on its parents for its day-to-day survival, indeed for its very existence (1927, 17–30).[5] If God has so often been regarded as a Father in our culture, Freud slyly implies, it is because the father has so often been regarded as a god in our homes. Issuing from such a domestic shrine myself, this line of argument proved extremely seductive to me.

---

5. Freud's argument is anticipated in his *Leonardo da Vinci and a Memory of His Childhood* (1910, 123) and a 1910 letter to Jung (McGuire 1974, 183–84), and reechoed in his *New Introductory Lectures on Psycho-Analysis* (Freud 1933, 168). The "illusion" in *The Future of an Illusion* alludes to Feuerbach's famous claim that the relation of reason to religion "amounts only to the destruction of an *illusion*" (Feuerbach 1841, 408). In later years Freud recanted, suggesting that his characterization of Jewish and Christian religion as the infantile projection of omnipotence onto a divine Father had been ill-founded (1925, 72; the statement occurs in a 1935 postscript to the work). By then Freud was paving the way for his deeply personal and largely positive revaluation of Judaism in *Moses and Monotheism* (1939).

4. HYPERMASCULINITY AND DIVINITY

Bauckham may come from a happier home. He dubiously cites the tendency of some recent theologians, feminist theologians in particular, to castigate traditional images of the sovereignty of God as projections of patriarchal domination. For Bauckham, Revelation is entirely innocent of such charges.

> Revelation, by avoiding anthropomorphism, suggests the imcomparability of God's sovereignty. In effect, the image of sovereignty is being used to express an aspect of the relation between God and his creatures which is unique, rather than one which provides a model for relationships between humans. Of course, the image of the throne derives from the human world, but it is so used as to highlight the difference, more than the similarity, between divine sovereignty and human sovereignty. In other words, it is used to express transcendence. Much of the modern criticism of images of this kind seems unable to understand real transcendence. It supposes that the relation between God and the world must be in every respect comparable with relations between creatures and that all images of God must function as models for human behaviour. It is critical of images of transcendence, such as sovereignty, but *it takes transcendence to mean that God is some kind of superhuman being alongside other beings*. Real transcendence, of course, means that God transcends all creaturely existence. As the source, ground and goal of all creaturely existence, the infinite mystery on which all finite being depends, his relation to us is unique. (1993b, 44–45, emphasis added)

And yet my suspicions persist. What if at the core of all these subtle scholastic formulations there were nothing but a superhuman being after all—an embarrassingly muscular being, insatiably hungry for adulation, but subjected to a stringent diet throughout centuries of (unsuccessful) Christian apologetic aimed at stripping away its all-too-robust flesh? The God of Revelation might be just such a being, just such a creature, just such a projection.

To begin with, Revelation is not as free of anthropomorphism as Bauckham suggests. After all, the being seated on the throne *is* human in form: "Then I saw in the right hand [*epi tēn dexian*] of the one seated on the throne a scroll" (5:1; cf. 5:7; 6:16; 22:4).[6] "Revelation 5.1 is closely

---

6. The scholarly insistence that Revelation circumvents anthropomorphism has a long pedigree. R. H. Charles, for instance, twice avers that John "avoids anthropomorphic details" (1920a, 113; cf. 115). Ironically, however, Charles later elucidates *epi tēn dexian* in the following terms: "The book-roll lies on the open palm of the right hand, not in the hand" (1920a, 136).

modelled on Ezekiel 2.9–10," as Bauckham himself observes (1993a, 246; cf. 248–49). The latter passage begins: "I looked, and a hand [*yād*; LXX: *cheir*] was stretched out to me, and a written scroll was in it." What is more, "John's account of his vision of God [Rev 4] is considerably indebted to Ezekiel's vision of the divine throne (Ezek. 1)" (1993a, 246). What are we then to conclude? That the enthroned figure that John "saw" in his vision was one and the same as the enthroned figure that Ezekiel "saw" in his? That, at least, is what John would want us to conclude. And what Ezekiel saw, seated on the throne, was *dĕmût kĕmarĕ'ēh 'ādām* (literally, "a likeness like the appearance of a man/human being," 1:26).[7]

In fairness to Bauckham, however, it must be admitted that John does refrain from attempting to describe the divine physique, preferring to focus attention instead on the adulation and self-abasement of the celestial audience eternally privileged to behold it (4:8–11). But if Revelation leaves the details of that heavenly physique to our imaginations, how best to imagine it? We may take our cue from the audience's reactions to it. Among contemporary cultural and subcultural spectacles, it is the bodybuilding posing routine that provides the most illuminating analogue to the celestial scene Revelation sets before us.[8]

---

7. Which the Septuagint renders as *homoiōma hōs eidos anthrōpou*.

8. For practical reasons, my remarks on bodybuilders will be confined to the male of the species. The criteria governing women's competitive bodybuilding are ambiguous in the extreme. For male competitors, the invariable formula for success is muscle mass, symmetry, and "definition." The latter term is bodybuilding shorthand for an extreme physical condition produced by exceptionally developed musculature in combination with exceptionally low fat levels. In contest condition, the champion bodybuilder is an ambulatory three-dimensional anatomy chart. Each muscle, however minor, is clearly visible through the skin, which, stripped of almost all subcutaneous fat through stringent dieting, adheres to the muscles as closely as cling film. A male bodybuilding competitor cannot have too much muscle mass provided it is symmetrical and defined. For female competitors, however, there is an intangible fourth element—femininity—that sets strict limits on the amount of muscle that can be amassed. *Flex*, the premier hardcore "musclemag," regularly features centerfolds of female bodybuilders posing nude. The shots are prefaced by the following statement, which says it all: "Women bodybuilders are many things, among them symmetrical, strong, sensuous and stunning. When photographed in competition shape, repping and grimacing or squeezing out shots, they appear shredded, vascular and hard, and they can be perceived as threatening. Offseason they carry more body fat, presenting themselves in a much more naturally attractive condition. To exhibit this real, natural side of women bodybuilders, *Flex* has been presenting pictorials of female competitors

First consider the static quality of the figure who is the principal focus of worship in Revelation. From his first to his last appearance in the book, he sits immobile and almost aphasic on his throne (he speaks only in 1:8 and 21:5–8).[9] Alan M. Klein, in his incisive study of bodybuilding subculture, has remarked on the "static, statuesque nature of bodybuilders in competition" (1993, 257).[10] The typical posing routine is less a spectacle of motion than a succession of stills: the bodybuilder hits and holds a pose, wringing every last drop from it, massive muscles straining, face frozen in a grimace posing as a smile, before proceeding to the next pose. The God of Revelation is similarly engaging in a posing exhibition. He is the static, statuesque embodiment of absolute power, and his celestial audience cannot get enough of him. Indeed, seen in this (heavenly) light, he looks very much like an idol (a matinee idol?), despite the author's iconophobic attempts to prevent this very thing from happening.

God's silence further accentuates both his statuesque demeanor and his likeness to the bodybuilder. "Like the cartoon without a caption, the hypermuscular body … is supposed to communicate without an act; its presence is its text" (Klein 1993, 274).[11] Klein quotes an unnamed bodybuilder who confides, "I wanna be the biggest thing. I wanna walk on stage … without even posing, and people would just—(opens eyes in wonder).… I won't even have to pose, I'll be so awesome" (1993, 273). Presumably the one seated on the "great white throne" (20:11) is himself beyond posing, although he would have ample room to strut and flex should he choose to. In the ancient Mediterranean world, as Joan Massyngberde Ford reminds us, "the absolute ruler sat on an ornate throne. Archaeological discover-

---

in softer condition. We hope this approach dispels the myth of female-bodybuilder masculinity and proves what role models they truly are." For an incisive analysis of this glaring double standard, see Bolin 1992.

9. Unless the anonymous interjections in 11:12 ("Come up here!") and 16:7 ("It is done!") also be attributed to the deity.

10. It is not for nothing that the (Caucasian) bodybuilder's tan-in-a-bottle is known as "bronzer," for the bodybuilder is a bronze statue (see Fussell 1994, 45). Compare Rev 1:15, in which the eye-popping figure of "the one like a human being" (1:13), the second sanctioned object of worship in the book (see esp. 5:8–14), has feet "like burnished bronze" (*homoioi chalkolibanō*). The original model for the statue that Jesus has become, namely, the angel of Dan 10:4–6, is said to have "arms and legs like the gleam of burnished bronze [LXX: *chalkou stilbontos*]" (cf. Dan 2:32; Ezek 1:7).

11. "No inane comments, only total majestic silence" is Gerhard Krodel's admiring assessment of the taciturn figure on Revelation's throne (1989, 155).

ies show such thrones with a high back, a base decorated with pictures of conquered peoples and several steps leading up to it" (1975, 70).[12] As such, Revelation's representation of the deity anticipated the publicity poster for the 1984 sword-and-sandal epic *Conan the Barbarian*, which showed a still buff but brooding Arnold Schwarzenegger slumped on a massive throne, his equally massive bulk artfully draped over its contours. Arnold's early action movies, together with those of Sylvester Stallone, portrayed the male bodybuilder as the ultimate warrior. "The more exaggerated the musculature, the more it had to explain itself in mounds of dead bodies" (Simpson 1994, 24). Compare the mountains of dead bodies that litter the landscape of Revelation,[13] irrefutable proofs that the exaggerated majesty of the one seated on the throne is warranted: he can kill at will without lifting a finger—or moving a muscle.

Arnold was not the only champion bodybuilder with imperial inclinations. "It's Dorian's world; we're just visiting it," reads the caption to a telling article in *Flex* magazine (McGough 1995).[14] Dorian Yates, who dominated professional bodybuilding in the 1990s much as Arnold dominated it in the 1970s, also deployed the image of the throne to represent the (made-in-heaven?) marriage of absolute power and utter submission.[15] "The evening's entertainment highlight at the 1995 Night of Champions in New York was the guest appearance of a near 300-pound Dorian Yates," the article begins (1995, 197). Bodybuilder Bev Francis, herself no pencilneck, remarked on an earlier epiphany of Yates: "At the show, I had goose pimples as I announced Dorian, 'cos backstage I'd seen what he looked like—nobody has ever carried that much muscle. When he walked onstage there was such a collective intake of breath from the 1,000 or so crowd that all the oxygen left the auditorium" (quoted in McGough 1994, 114).[16]

---

12. The throne is ubiquitous in Revelation from chap. 4 onward (e.g., 5:6; 7:9; 8:3; 12:5; 14:3; 16:7; 19:5; 20:11; 21:3; 22:1, 3). G. B. Caird comments: "The final reality which will still be standing when heaven and earth have disappeared is the great white throne" (1966, 62).

13. Rev 6:4, 8; 8:9, 11; 9:15, 18; 11:13; 14:19–20; 15:2–10; 16:18–21; 19:11–21; 20:9, 15; 21:8; cf. 6:15–17; 8:4–6; 11:18; 14:9–11; 18:8, 19, 21; 19:2–3; 22:18.

14. The caption occurs in the issue's table of contents.

15. "My relationship to power and authority," Schwarzenegger once confessed, "is that I'm all for it.... Ninety-five percent of the people in the world need to be told what to do and how to behave" (quoted in Butler 1990, 34)—by the other 5 percent, presumably.

16. Yates's rock-solid 295 pounds was balanced, a little unsteadily, on a five-foot-ten-inch frame.

4. HYPERMASCULINITY AND DIVINITY 83

At the 1995 Night of Champions, the curtain drew back to reveal Dorian pensively perched on an ornate throne, and resplendent in an ermine-trimmed crown and robe. Other accessories included a pair of angelic attendants who abased themselves at Dorian's feet. "The girls then divested the three-time Mr. Olympia of his imperial accouterments, as a prelude to Dorian's posing before a raucous 3000 strong standing-room-only crowd" (McGough 1995, 197–98). Hardly the "myriads of myriads and thousands of thousands" that throng Revelation's throne room (5:11), but Dorian does not (yet) claim to be God.

For Klein, there is something profoundly unsettling, indeed fascist, about the spectacle of the bodybuilder on the posing dais. "Bodybuilding leads in various sociocultural directions," he writes, "but none is quite so disturbing or dramatic as its connection to fascist aesthetics and cultural politics. The fetishism for spectacle, worship of power, grandiose fantasies, … dominance and submission in social relations are all essential characteristics shared by bodybuilding and fascism" (1993, 254; cf. 253–67 passim; also Dutton 1995, 206–9). But these are also the essential characteristics of Revelation (if "dominance and submission in social relations" is extended to embrace divine-human relations), each so ubiquitous as to beggar documentation.

The avoidance of anachronism is not, perhaps, my strongest suit as an exegete. Indeed, I frequently employ anachronism deliberately as an exegetical tactic (taking my cue from the fact that anachronism is what biblical scholars fear most, that fear is but the obverse of fascination, and that the fascinating merits pursuit more than flight). And yet my description of Revelation as fascist is not intended altogether anachronistically. We should not presume on too narrow a definition of this singularly useful term. The *New Webster's Dictionary*, for example, having tracked the term to Mussolini's Italy, proceeds to define it more generally as "any political or social ideology … which relies on … the brutal use of force for getting and keeping power," a definition that enables the term to reach back and delineate a much earlier Italy, one at the center of a vast empire. I would argue, moreover, that the theology or ideology of Revelation is anything but a simple inversion, reversal, or renunciation of the political and social ideology of imperial Rome. Instead, it represents the apotheosis of this imperial ideology, its ascension to a transhistorical site.[17]

---

17. See 28–30 above on Revelation's imperial mimicry.

This conclusion is implicit in much of what critical commentators write on Revelation, although some work hard to circumvent it. M. Eugene Boring, for example (1989, 103), claims that the repeated accolade, "Worthy art thou [*axios ei*]" (4:11; 5:9; cf. 5:12), directed to God and the Lamb, "reflects the acclamation used to greet the [Roman] emperor during his triumphal entrance," while the title "our Lord and God [*ho kyrios kai ho theos hēmōn*]" (4:11; cf. 4:8; 11:17; 15:3; 16:7; 19:6) "is paralleled by Domitian's insistence that he be addressed by this title" (1989, 103; cf. 21).[18] David E. Aune, for his part, claims that the detail of the twenty-four elders casting their crowns before the divine throne (4:10) "has no parallels in Israelite-Jewish literature," but is comprehensible only in light of the custom of presenting crowns to a sovereign, which was "inherited by the Romans from the traditions of Hellenistic kingship" (1983, 12–13). Boring cites Tacitus's description of how the Parthian King Tiridates laid his diadem at the feet of Nero's seated effigy in order to offer suitably obsequious homage to the Roman emperor.[19]

What of the number twenty-four itself? Aune has an intriguing suggestion to make. "Roman magistrates were permitted to be accompanied by the number of lictors bearing fasces which corresponded to the degree of imperium which they had been granted," he begins. (As it happens, the fasces, a bundle of rods bound together with the blade of an ax projecting, later gave Italian *Fascismo*, which adopted it as an insignia, its name.)

> Consuls were permitted twelve lictors. Augustus apparently had twelve lictors from Actium ... though it is possible that he had twenty-four lictors until 27 BCE. At any rate the standard number of twelve lictors, indicative of the degree of imperium, was doubled by Domitian to twenty four. These lictors, of course, were not crowned, nor did they wear white robes. They did, however, constitute part of the official crowd of public servants which constantly surrounded the emperor. (Aune 1983, 13)[20]

---

18. These are tirelessly recycled arguments. For instance, the conjecture that the "Worthy art thou" acclamation is intended to evoke Roman imperial court ceremonial goes back at least to Erik Peterson (1926, 176–80). Like the majority of scholars, Boring believes that Revelation was written during the latter years of Domitian's reign. Did Domitian really insist on being addressed as "Lord and God"? Probably not, as we shall see, although the title does seem to have been applied to him nonetheless.

19. Boring 1989, 103, adducing Tacitus, *Annals* 15.28.

20. Aune's primary source here is Dio Cassius, *Roman History* 67.4.3.

Self-abasing celestial officials, the twenty-four elders prostrate themselves repeatedly before the divine throne (4:10; 5:8; 7:11; 11:16; 19:4). "The practice of the ritual of *proskynesis* before the early Roman emperors is incontrovertible," adds Aune, and he goes on to substantiate his claim (1983, 13–14).

Distantly related to Domitian's double allotment of lictors were Nero's Augustiani, an elite corps of presentable young men whose principal function was to lead the applause whenever the emperor deigned to make an appearance—an imperial cheerleading squad, if you will. Suetonius estimates the size of this squad at "more than five thousand."[21] Dominique Cuss has argued that the acclamations led by the Augustiani were designed to "underline the imperial claims to divinity" (1974, 77; cf. Aune 1983, 16). Paraphrasing Tacitus, she states: "Day and night, the applause and acclamations of these young men echoed around the palace, using such extravagant terms while describing the beauty and voice of the emperor, that they could have been applied to the gods" (1974, 78).[22] Or to God. Compare Revelation 4:8–11: "Day and night without ceasing [*anapausin ouk echousin hēmeras kai nyktos*] they sing"—and the words of the song follow, the singers in question being the "four living creatures" supported by the twenty-four elders. Together they drum up a chorus that swells until it encompasses every voice in the heavenly throne room (5:8–12), and then every voice in the universe (5:13), acclamation after acclamation washing over "the one seated on the throne," and his *Divi Filius*, "the Lamb."[23]

Such parallels could be multiplied; we have not yet exited Revelation 4–5, much less examined the extent to which the heavenly throne room

---

21. Suetonius, *Nero* 20.3. The Augustiani are also mentioned in *Nero* 25.1; Tacitus, *Annals* 14.15; Dio Cassius, *Roman History* 61.20.4–5; cf. 63.8.2–3; 63.15.2; 63.18.3. See further Bartsch 1994, 8–9.

22. The sentence from Tacitus begins: "Days and nights they thundered applause [*Ii dies ac noctes plausibus personare*], bestowed the epithets reserved for deity upon the imperial form and voice" (*Annals* 14.15, LCL trans.).

23. Strangely, Cuss does not appear to have Rev 4–5 in mind as she paraphrases Tacitus. Instead she is in the midst of a lengthy gloss on the "blasphemous names" on the head of the beast from the sea (13:1; cf. 17:3). Aune, however, does connect the Augustiani with the perpetual chorus of Rev 4–5 (1983, 16–18). The Augustiani's association with Nero might be said to bolster his argument, since Nero is a brooding presence in Revelation, as has long been recognized, and if not an Antichrist, then at least a Counterchrist, a dying and rising figure (see Rev 13:3, 12). The perpetual chorus of Rev 4–5 might then be said to form a corps of counter-Augustiani.

in Revelation as a whole (in which it function as the principal setting) mirrors the Roman imperial court.[24] What do the scholars who routinely argue these parallels make of them? Aune's conclusions are typical.

> John's depiction of the ceremonial in the heavenly throne room has been significantly influenced in its conceptualization by popular images of Roman imperial court ceremonial. For the most part, the individual constituents of that ceremonial used by John in his depiction of the heavenly ceremonial have been heightened, expanded and given even greater cosmic significance. The result is that the sovereignty of God and the Lamb have been elevated so far above all pretensions and claims of earthly rulers that the latter, upon comparison, become only pale, even diabolical imitations of the transcendent majesty of the King of kings and Lord of lords. (1983, 22; cf. 5)

Is this Aune's own view of the matter, or merely his rendition of John's view? Less guesswork is required in the case of Boring, who states: "The correlation of imagery from the imperial cult with that used to express faith in the sole sovereignty of God simply shows that all earthly claims to sovereignty are only pale imitations and parodies of the One who sits upon the one throne. Christians dare not give this homage to another" (1989, 103; cf. 185, 187, 192–93, 214–15, 211). Boring's summation does echo Aune's. The difference, however, is that Boring's popularly pitched but splendidly competent commentary on Revelation is everywhere punctuated by professions of faith, generally implicit but frequently explicit, so that Boring's theological viewpoint seems to blend completely with that of John. Boring stands staunchly by John's shoulder throughout, not only as his interpreter, faithfully translating even his most alien sentences and sentiments into contemporary theological idiom, but also as his disciple. Boring everywhere apologizes for John, and nowhere criticizes him.

The same is true of Bauckham—at any rate, in his *Theology of the Book of Revelation* (1993b). He refrains from attempting to connect the protocol of the divine throne room with imperial court ceremonial. Nevertheless, his view of the relationship between the two thrones—the two empires—matches that of Aune and especially Boring.

---

24. Nor is the imperial court the only place to look for such parallels. Ernest P. Janzen (1994) has argued, for instance, that Jesus' wardrobe in Revelation is modeled on that of the Roman emperor.

4. HYPERMASCULINITY AND DIVINITY   87

> The Roman Empire, like most political powers in the ancient world, represented and propagated its power in religious terms. Its state religion, featuring the worship both of the deified emperors and of the traditional gods of Rome, expressed political loyalty through religious worship. In this way it absolutized its power, claiming for itself the ultimate divine sovereignty over the world. And so in effect it contested on earth the divine sovereignty which John sees acknowledged in heaven in chapter 4. The coming of God's kingdom on earth must therefore be the replacement of Rome's pretended divine sovereignty by the true divine sovereignty of the One who sits on the heavenly throne. (1993b, 34; cf. 39, 43, 44–45, 59, 143, 159–60, 162–63)

Hundreds of other such claims could be cited. They are the bread and butter of church-oriented critical commentaries on Revelation. But what do these claims amount to, these accusations of pretention, these assertions that Roman imperial power is but a parody or pale imitation of divine power? What are Boring and Bauckham actually saying? Simply that God's imperial splendor far exceeds that of the Roman emperor, just as the emperor's splendor far exceeds that of any of his six hundred senators, and just as the senator's splendor far exceeds that of any provincial plebian, and so on down the patriarchal line to the most subdued splendor of the feeblest father of the humblest household? If so, the difference between Roman sovereignty and divine sovereignty would be quantitative rather than qualitative in Revelation.

Bauckham is not unaware of the problem. "It would subvert the whole purpose of John's prophecy," he admits, "if his depiction of the divine sovereignty appeared to be a projection into heaven of the absolute power claimed by human rulers on earth." But this danger "is averted by the kind of apophaticism in the imagery," he claims, "which purges it of anthropomorphism and suggests the incomparability of God's sovereignty" (1993b, 43). Apophaticism is "negative theology," he explains, and it "radically distinguishes God from all creaturely being by conceiving him in negative terms: he is *not* what creatures are" (1993b, 43 n. 8).

How does Revelation rate as negative theology? Encumbered by the exaggerated masculinity of its deity, it limps awkwardly indeed beside the consummately restrained and exquisitely delicate theological footwork of a Pseudo-Dionysius, a Meister Eckhart, or other Christian thinkers more commonly termed apophatic. Has John really succeeded in "purging" his text, as Bauckham claims? Has he really evacuated it, voided it of whatever modern theological sensibilities might deem unseemly or unsightly, specifically the glorification of absolute power? Has his apophatic purgative not been too mild for that?

For power of an alarmingly pure kind is what God's reign in Revelation boils down to. Here is Bauckham's (accurate) account of the life awaiting the blessed in the heavenly city (Rev 21:1–22:5):

> As for the image of God's rule in the eschatological kingdom, what is most notable is the fact that all the implication of distance between "the One who sits on the throne" and the world over which he rules has disappeared. His kingdom turns out to be quite unlike the beast's. It finds its fulfillment not in the subjugation of God's "servants" (22:3)[25] to his rule, but in their reigning with him (22:5). The point is not that they reign over anyone: the point is that God's rule over them is for them a participation in his rule. The image expresses the eschatological reconciliation of God's rule and human freedom, which is also expressed in the paradox that God's service is perfect freedom (cf. 1 Pet. 2:16). Because God's will is the moral truth of our own being as his creatures, we shall find our fulfillment only when, through our free obedience, his will becomes also the spontaneous desire of our hearts. Therefore in the perfection of God's kingdom theonomy (God's rule) and human autonomy (self-determination) will fully coincide. (1993b, 142–43; cf. 164)

A Foucauldian nightmare, this vision of heaven (but whose?) represents the absolute displacement of outward subjection, tangible coercion, by inner self-policing, which is now so deeply implanted in the believer as to be altogether indistinguishable from freedom. Revelation does present the individual with an option, of course—to be "tortured [*basanisthēnai*] with fire and sulfur" instead "in the presence of the holy angels and in the presence of the Lamb," "the smoke of [one's] torment" ascending "for ever and ever" (14:10–11; cf. 9:5; 20:15; 21:8).[26] We are deep within the

---

25. God's slaves (*douloi*), actually. Bauer remarks on this rendering of *doulos*: "'servant' for 'slave' is largely confined to Biblical transl. and early American times" (BAGD, s.v. *doulos*).

26. An option stated even more baldly in 4 Ezra 7:36–38: "Then the pit of torment shall appear, and opposite it shall be the place of rest; and the furnace of Hell shall be disclosed, and opposite it the Paradise of delight. Then the Most High will say to the nations that have been raised from the dead, 'Look now, and understand whom you have denied, whom you have not served, whose commandments you have despised! Look on this side and on that; here are delight and rest, and there are fire and torments!'" (*OTP* trans.; cf. Matt 25:31–46).

dystopian netherworld of Foucault's *Discipline and Punish* (1977): on one side, the absolute monarch publicly exacting frightful physical punishment on all who oppose his will; on the other side, a more "benign" realm in which the rack, the wheel, and the stake have been rendered obsolete—but only because the ruler's subjects are no longer capable of distinguishing his will from their own.[27]

For Foucault, the modern "disciplinary societies," with their insidious strategies of coercion and control,[28] have succeeded and surpassed the premodern "societies of the spectacle" with their rituals of dismemberment, disembowelment, and immolation enacted in the public square. Both these regimes coexist, however, in Revelation's "new heaven and new earth" (21:1; cf. 20:11)—not unexpectedly, since the two regimes represent the two faces of power, one scowling, the other smiling, and Revelation's climactic vision (21:1–22:5) is a vision (a projective fantasy?) of power absolutized. On the one hand we read, "Blessed are those who wash their robes, so that they will have the right to the tree of life and may enter the city by the gates" (22:14). Bauckham has told us what awaits them inside. On the other hand we read, "Outside [*exō*] are the dogs [*hoi kynes*] and sorcerers and fornicators and murderers and idolaters, and everyone who loves and practices falsehood" (22:15; cf. 21:27).[29] What will be their fate as outsiders? John has already told us that: "But as for the cowardly, the unbelieving, the polluted, the murderers, the fornicators, the sorcerers, the idolaters, and all liars, their place will be in the lake that burns with fire

---

27. See Krodel's telling remarks (1989, 156–57) on the "four living creatures" (Rev 4:6–8): "They are God's pets within the heavenly court.... Readers of this commentary should not be upset that I speak of God's pets. They should remember their own faithful dogs and cats whose joy it is to live in their presence, to please them and adore them. Worship ought to be just that.... God's pets in John's vision are the symbol of harmony and worship yet to come, when God shall dwell among his people (21:1–22:5)."

28. The supreme example of such a strategy would be television, although this is a "disciplinary technology" that Foucault himself never examined, preferring to focus instead on such phenomena as modern medicine (especially psychiatry), modern prisons, and modern sexuality.

29. Who are these "dogs"? "The term 'dog' is used in Scripture for various kinds of impure and malicious persons," contends Robert H. Mounce. He appeals in particular to Deut 23:17–18, in which "the term designates a male cult prostitute" (1998, 408). I reflect on Revelation's anathematized dogs below (239–40).

and sulfur [*en tē limnē tē kaiomenē pyri kai theiō*], which is the second death" (21:8; cf. 19:20b; 20:14–15; Gen 19:24).

The Beatific Vision

> On him, on him alone had I leisure avidly to gaze. (Statius, *Silvae* 4.2.40, on Domitian)

Who, then, is the God of Revelation? As I have been implying, he is revealed not through Jesus Christ so much as through the Roman emperor. For many of the emperors, the temptations of the flesh assumed a unique form: the temptation to become divinized flesh. To be or not to be a god? Tiberius sternly rejected the divine honors dangled enticingly before him. So did Augustus (less energetically, to be sure), Claudius, and Vespasian, although all three were deified after death, as was Titus. Caligula greedily seized the opportunity to become a god, wallowing in divinity. So did Julius Caesar, to a lesser extent, as well as Nero, and possibly Domitian.

Domitian's exploits and excesses are of perennial interest to scholars of Revelation, most of whom date the book to the latter years of his reign, which began inauspiciously in 81 CE and ended ignominiously in 96.[30] Did Domitian, ravenous for adulation, gluttonous for deification, really gorge on forbidden fruit, appropriating for himself the title "Lord and God," as his biographer Suetonius claims? In the *Lives of the Caesars* (13.2) we read: "With … arrogance he began as follows in issuing a circular letter in the name of his procurators, 'Our Lord and God bids that this be done [*Dominus et deus noster hoc fieri iubet*].' And so the custom arose of henceforth addressing him in no other way even in writing or in conversation."[31] This

---

30. Here, for once, critical scholarship on the authorship of a New Testament book is in step with church tradition—or, more precisely, largely dependent on it. Irenaeus claimed that John had his visions "toward the end of the reign of Domitian" (*Against Heresies* 3.50.3; cf. Eusebius, *Church History* 3.18.1; Victorinus, *Commentary on the Apocalypse* 10.11). For the classic presentation of the internal case for a Domitianic date for Revelation, see Collins 1984, 54–83. I realize, of course, that the hypothesis is not unassailable, but I would prefer to avoid staging yet another rehearsal of a convoluted (and irresolvable) debate, and so will proceed to construct a reading based on the majority opinion.

31. LCL trans., with slight modifications. Similar claims concerning Domitian occur in Pliny, *Panegyric* 2.33.4; 78.2; Dio Chrysostom, *Discourses* 45.1; Dio Cassius, *Roman History* 67.4.7; 67.13.4; Aurelius Victor, *On the Caesars* 11.2.

would have been a bold self-designation indeed, even by the standards of the imperial court, for a deified emperor, or *divus*, did not a *deus* make. "The best an emperor could expect after death was to be declared a *divus*, never a *deus*," explains a more recent biographer of Domitian, and "a living one had to make do with even less" (Jones 1992, 108). And if Domitian could so overcome his natural modesty, "why should he hesitate to proclaim it publicly (and epigraphically)?" (1992, 108). But the title has yet to turn up in any inscription, coin, or official document. It does seem to have been applied to Domitian nevertheless; his contemporary, Martial, does so, for one, and clearly implies that others did so as well.[32] Dio Cassius, too, tells of a certain Juventius Celsus, who, accused of conspiring against Domitian, saved his skin by prostrating himself before the emperor and addressing him "as 'Lord and God,' names by which he was already being called by others" (Dio Cassius, *Roman History* 67.13.4). Domitian "obviously knew that he was not a God," claims Jones, but "whilst he did not ask or demand to be addressed as one, he did not actively discourage the few flatterers who did" (1992, 109).

Of course, the *adulatio* lavished on Domitian by his most fervent flatterers was by no means limited to the hyperbole *Dominus et Deus*. Witness, for example, the laudatory immoderation of Martial and his fellow court poet Statius, as reported by Kenneth Scott:

> Statius calls him *sacratissimus imperator, sacrosanctus, sacer*, and *verendus*. His home is described as *divina*. Indeed all that pertained to the monarch is named sacred.... The emperor's person is sacred, his side, breast, ear [Martial], and feet [Statius], and the rebellion of Saturninus against him is sacrilegious [Martial]. His name is *sacer* [Martial], as are his secrets [Statius]. His banquet is "sacred" or "most sacred"; the golden wreath which he bestows as prize in the Alban contest is "sacred"; the day on which he feasts the people is *sacer*; his fish are "sacred" [Martial]; his treasures are *sanctae*, and the nectar which he drinks is *verendum* [Statius]. (1975, 100)

The devoted duo also insists that Domitian is a *deus praesens*—a Jupiter *praesens*, what is more—proximate, tangible, and accessible, as distinct

---

32. Martial, *Epigrams*, esp. 5.8; 7.34; 8.2; 9.66; 10.72; cf. 5.5; 7.2, 5; 8.82; 9.28. Does *Epigrams* 5.8 lend support to Suetonius's claim that Domitian began an official letter with the self-designation "Lord and God"? The passage reads: "As Phasis in the theater the other day ... was praising the edict of our Lord and God [*edictum domini deique nostri*], whereby the benches are more strictly assigned" (LCL trans.).

from the distant Olympian (Scott 1975, 107–8, 137–38). But it was not only in poetry that Domitian was hailed as Jupiter. The compliment also occurs in epigraphic finds, notably an Attic inscription that bestows the title Zeus Eleutherios on the emperor (Scott 1975, 139). Numerous coins, too, struck by Domitian, depict him enthroned as "father of the gods" (Jones 1980, 1033).

What of Roman Asia, the region to which Revelation is addressed (1:4; cf. 1:11; 2:1–3:22)? The province had two temples dedicated to Domitian, one at Ephesus, the other at Laodicea. A massive marble statue of the emperor erected in the Ephesian temple became a focal point of the imperial cult in Asia.[33] "Some impression of the scale [of the statue] is given by the fact that the lower part of an arm is the height of a man," observes S. R. F. Price. "The height of the whole, to the top of the spear which the standing figure was probably holding, was some seven to eight metres"—on the same scale, that is to say, as cult statues of the gods (1984, 187; cf. 255). Statues of Apollo, Artemis, and Leto, for example, each of them seven to eight meters tall, stood in Apollo's temple at Claros, a scant few miles from Ephesus; while Josephus tells of "a colossal statue [*kolossos*] of the emperor [Augustus], no smaller than Zeus at Olympia, which served as its model," which dominated the temple of Roma and Augustus at Caesarea (*Jewish War* 1.21.7 §414, my trans.; cf. *Antiquities* 15.9.6 §339). Tacitus, for his part, tells how the Senate, following a minor military victory, resolved to present Nero "with a statue of the same size as that of Mars the Avenger, and in the same temple" (*Annals* 13.8, LCL trans.). Of Domitian's Ephesian colossus Price remarks, "This is the most extreme form of the modelling of the emperor on the gods, no doubt with awesome impact on the population" (1984, 188).[34]

A colossal hard body—Greco-Roman culture and contemporary body-building subculture converge strikingly in their respective conceptions of the godlike physique. A recent advertisement in the muscle magazines for GIANT MEGA MASS 4000, a weight-gaining product, features a "hard and massive" Dorian Yates flexing alongside the caption, "The Giant That

---

33. Although the cult's official center was Pergamum, with its temple dedicated to Roma and Augustus.

34. Price, a classicist, suggests that "the establishment of the provincial cult of Domitian at Ephesus, with its colossal cult statue," looms behind Rev 13:11–15, so that the beast from the land would be "the priesthood of the imperial cult, particularly … in the province of Asia" (1984, 197).

Won This Year's Mr. Olympia" (Mr. Olympia being the most prestigious title in professional bodybuilding). Further into the four-page advertisement another elite bodybuilder, Gary Strydom, is labeled "A Rock-Solid 260 lbs.," the giant container of GIANT MEGA MASS 4000 that he holds triumphantly aloft suggesting that he owes his statuesque condition to the miraculous product.[35] And what of the Mr. Olympia title itself? "In 1965 I created the Mr. Olympia contest," explains Joe Weider, princeps of the bodybuilding empire. "The name seemed appropriate. The time had come to enter the hallowed ground of the ancient Greek gods with incarnate image. We live among them."[36] And in the course of a somewhat surreal exchange between former Mr. Olympia Frank Zane and Michael Murphy, founder of Esalen Institute in Big Sur, California, and "a leader in the human potential movement," the latter enthuses, "One of the things I've admired about your attitude, Frank, is your experimentation with somatic mutability, which has enabled you to change your body at will. The further reaches of training point to glimpses of divination of the body, a new kind of flesh" (Zane 1994, 247).

Of course, the flesh in question is penile flesh; for what does it mean to say that size and hardness are the sine qua non of the bodybuilding physique except that the bodybuilder is an outsized penis in the state of permanent erection?[37] "He hones his hard body (to be soft is anathema)," as one commentator remarks (Fussell 1994, 46). His body is "'pumped up,' 'rock hard' and 'tight,'" observes another (Simpson 1994, 33). And the current rage for "vascularity" in competitive bodybuilding, calling for minimal subcutaneous fat, means that "the road map of veins is clearly visible, standing out from the flesh in a fashion alarmingly reminiscent of an erect penis" (1994, 33). On stage, the bodybuilder is "turgid," "constantly at attention, ready to explode," his entire body engorged like an enormous organ (Fussell 1994, 46–47; cf. Dutton 1995, 43; Miles 1991,

---

35. Reflecting on his own statuesque physique, the young Arnold remarked: "You don't really see a muscle as part of you.... You look at it as a thing.... You form it. Just like sculpture" (quoted in Gaines and Butler 1974, 52; cf. 106, 108; also Dutton 1995, 312–15).

36. Quoted in Klein 1993, 258. Klein also quotes Weider as saying, "The modern bodybuilder has followed in the footsteps of the Greek Olympian gods. Obsessed with heroic proportions as they were, how far would the Greeks have taken physical development had they our knowledge of weight training?" (1993, 259; cf. Dutton 1995, 21).

37. See Klein 1993, 247: "The fear of size loss ... is the converse of the bodybuilder's search for size and hardness."

111; Simpson 1994, 22). Arnold tells of the "pump" that he gets when the blood is flooding his muscles: "They become really tight with blood. Like the skin is going to explode any minute."[38] Indeed, former Mr. Universe Steve Michalik entertained a fantasy that he would someday literally explode on stage, showering his worshipers with the viscous contents of his phallic physique.[39]

Domitian's cult too was preposterously phallic; how else should we interpret the massive marble-hard image of his power erected at Ephesus, to cite one of the more Priapean pillars of his cult? Unfortunately for Domitian, his splendid cult did not survive his assassination but swiftly wilted, the Senate according him a *damnatio memoriae* and having his statues—glorious statues, if his biographers are to be believed—destroyed or rededicated. It appears that the Ephesian colossus was allowed to stand until late antiquity, however, although the temple housing it was rededicated to Vespasian and the statue itself passed off as a representation of the latter, the Flavian paterfamilias strapping the monstrous monument onto his own withered loins. Indeed, were it not for the author of Revelation, Domitian's divinity might have died with him.

For in and through Revelation, the emperor ascends into heaven and becomes a god, and the god he becomes is none other than the Father of Jesus Christ. John's attempt to counter the magnificent imperial cult with the image of a yet more magnificent heavenly cult (the latter modeled in no small part on the former, as we have seen) has resulted in a fascinating (con)fusion of figures, the Roman emperor coalescing into the Christian God. And so Domitian is assumed into heaven, together with all his court. He is decked out for battle, for Revelation is a military epic.[40] But his body armor or cuirass, sculpted to simulate a heavily muscled male torso, in the Roman manner, has vanished.[41] In its place is divinized flesh, the unimaginably muscular and exquisitely sculpted physique of the God of Israel,[42]

---

38. Quoted in Butler 1990, 124. Hence Arnold's famous dictum, "A good pump is better than coming" (in Gaines and Butler 1974, 48).

39. Recounted in Fussell 1994, 50.

40. See 42–65 above.

41. "It is highly likely that the statue of Domitian at Ephesus, of which numerous fragments survive, was cuirassed" (Price 1984, 182). Cuirassed statues of the Roman emperors were extremely common. Nude statues of the emperors were also common—fat free, needless to say, and with excellent muscle tone, symmetry, and definition—evoking the traditional representations of the gods.

42. On which see Moore 1996, 86–102.

4. HYPERMASCULINITY AND DIVINITY

whose appearance is of precious stones: "And the one seated there looks like jasper [*iaspis*] and carnelian [*sardion*]" (Rev 4:3).

Why precious stones? Because precious stones are hard, and (phallic?) hardness is the sine qua non of a godlike physique, as we have seen. But why these precious stones in particular? G. R. Beasley-Murray explains:

> The appearance of God was as jasper and carnelian. The former could vary in appearance from a dull yellow or red or green, or even translucent like glass (as apparently in 21.11). In view of the later passage we may take the last to be in view. Carnelian, or sardius (originating from Sardis), was red. The divine appearance, therefore, was as it were, transparent "white" and red. (1978, 113)

The divine physique is characterized both by transparence and redness, meaning that the raw musculature of the deity is entirely visible through the skin. "Thin as Bible paper," the latter is "so translucent one can visibly see raw tissue and striated muscle swimming in a bowl of veins beneath" (Fussell 1994, 49). The Heavenly Bodybuilder is thus the mythological—and myological—figure, the ambulatory anatomy chart, that all earthly bodybuilders aspire to be. "When you hit a most-muscular pose," Arnold advises the bodybuilding neonate, "[you] should look like an anatomy chart—every area developed, defined, separated, and striated" (Schwarzenegger and Dobbins 1985, 291).[43]

"He was tall of stature," Suetonius writes of Domitian, "handsome and graceful too [*praeterea pulcher ac decens*], especially when a young man," although "in later life he had … a protruding belly, and spindling legs … thin from a long illness" (*Domitian* 18.1, LCL trans.). Yet the deified Domitian could hardly have been astonished at the physical metamorphosis that came with his apotheosis. After all, in the temple to Hercules that he had erected on the Appian Way, the cult statue of the massively muscled god had been fashioned with the facial features of the emperor himself (Scott 1975, 143). And prescient Martial, noting how Hercules had ascended to the "starry heaven" by virtue of the punishment he had inflicted "on the wrestler of the Libyan palaestra, and by the throwing of ponderous Eryx in the Sicilian dust," had exclaimed how insignificant such feats appear

---

43. In the "most-muscular" pose, the bodybuilder leans forward, brings his or her clenched fists together in front of his or her crotch, bares his or her teeth, and flexes every muscle group simultaneously.

when measured against Domitian's own prodigious prowess: "How many weights heavier than the Nemean monster fall! How many Maenalian boars does your spear lay low!" (*Epigrams* 5.65; cf. *Spectacles* 6B, 15, 27).[44] For these and other heroic deeds—"three times he shattered the treacherous horns of Sarmatian Hister; he three times bathed his sweating steed in Gethic snow"—Domitian shows himself to be a *maior Alcides* (a "greater Hercules"), far outstripping the *minor Alcides*,[45] and deigning to supplement the latter's brutish physique with his own effulgent features.

A mighty warrior needs a mighty weapon. "Around the throne is a rainbow [*iris*]" (Rev 4:3; cf. Ezek 1:28), we read, and this turns out to be Domitian's weapon. As numerous commentators have noted, this rainbow evokes that of Genesis 9:13: "I have set my bow [*qešet*] in the clouds, and it shall be a sign of the covenant between me and the earth." A common interpretation of the latter passage, in turn, suggests that the bow set in the clouds after the flood represents the war bow (also *qešet*) with which Yahweh wages his battles (e.g., Deut 32:23, 42; 2 Sam 22:15; Pss 7:12; 18:15 [14]; 77:18; 144:6; Lam 2:4; 3:12; Hab 3:9-11; Zech 9:14), so that his placing it in the clouds would signify the cessation of his warlike hostilities against humanity. Thus it is that the deified Domitian suddenly finds himself in possession of Yahweh's awesome weapon. Now the bow also happened to be the mortal Domitian's weapon of choice, according to Suetonius:

> He was ... particularly devoted to archery [*sagittarum ... praecipuo studio tenebatur*]. There are many who have more than once seen him slay a hundred wild beasts of different kinds on his Alban estate, and purposely kill some of them with two successive shots in such a way that the arrows gave the effect of horns. Sometimes he would have a slave stand at a distance and hold out the palm of his right hand for a mark, with the fingers spread; then he directed his arrows with such accuracy that they passed harmlessly between the fingers. (*Domitian* 19, LCL trans.)

Now divinized, Domitian's aim is still more deadly—and so are his intentions, as it happens. The rainbow in Revelation 4:3, far from representing the cessation of the deity's warlike hostilities against humanity, signifies

---

44. Here and in what follows I am quoting Scott's translation of the *Epigrams* (1975, 142, 145), which often reads better than Bailey's LCL translation.

45. Martial, *Epigrams* 9.101; cf. 9.64. Statius, too, casts Domitian in the role of a *maior Alcides* (*Silvae* 4.2.50).

instead their resumption, as the ensuing annihilation of the earth and its inhabitants testifies.

All in all, then, the God of Revelation is a hypermasculine God. But masculinity in excess tends to teeter over inexorably into its opposite. Towering on stage, engorged muscles ready to explode through his taut skin, the male bodybuilder seems a veritable caricature of the ultravirile male. In all probability, however, as ex-bodybuilder Sam Fussell discloses, "he's pumped so full of steroids that he's literally impotent" (1994, 52). "But not only is he less of a man at his moment of majesty," continues Fussell, "he's actually more of a woman. Faced with a flood of surplus testosterone, the body reacts by temporarily shrinking the testicles (with a resultant sperm count drop) and releasing an estrogen counterbalance," which, over time, can engender a pair of pubescent breasts (the condition known as gynecomastia, or "bitch tits" in gym vernacular) (1994, 52; cf. Fussell 1991, 110, 120). As Mark Simpson sagely observes, hardcore bodybuilding frequently epitomizes the inherent instability, the ambiguity, the flux of masculinity "right at the moment it is meant to solidify it in a display of exaggerated biological masculine attributes" (1994, 30).

Somewhat to his surprise, therefore, the emperor finds that he has also acquired a pair of female breasts in the course of his apocalyptic apotheosis. Well might he now accept the offer effusively extended to him by extravagant Martial in the seventh book of the *Epigrams*: "Accept the rough breastplate [*crudum thoraca*] of the warrior Minerva, O you whom even the wrathful locks of Medusa fear" (7.1). Domitian's ample Minervan bosom enhances rather than hinders the eternal posing routine that his life has now, blessedly, become. The beholding of his wholly hypermasculine, altogether Priapean, and hence queerly feminine form elicits utter adulation, indeed outright adoration, in the vast audience (the myriad of Martials, the slew of Statiuses) that eternally throngs the heavenly temple.[46] For just as the temple is, or might as well be, a temple to the divine physique, a Go(l)d's Gym, the worship that resounds within it is, or might as well be, hero worship. And that is precisely what the God of Revelation craves. Indeed, this vast audience of idolizers—nameless, faceless, countless (see 5:11–13; 7:9–10; 19:1–8)—is actually nothing more than an infinite row of mirrors lining the interior walls of the heavenly city, which

---

46. The throne room in Revelation doubles as a temple (see esp. 7:15; 11:19; 14:17; 15:5–6, 8; 16:7). In the ancient world, palace and temple were often closely connected.

turns out to be a perfect cube some 12,000 stadia (approximately 1,500 miles) high, broad, and long (21:16; cf. 1 Kgs 6:20).[47] And the sole purpose of this vast mirrored enclosure is eternally to reflect the divine perfection back to the divinity himself. The emperor has become his own love object.

Afterword: Revelation's Big Reveal

When I saw her, I was greatly amazed. (Rev 17:6)

As I explained earlier, it was the conjoining of Christian worship and the adoring gaze that initially caught my eye as I brushed off Bishop Kirk's *The Vision of God* (1931), the dusty tome that set this essay in motion, while rummaging aimlessly in the university library one day. But Kirk's study sparks at least one further question that I have not yet addressed. Those classic theologians whom Kirk eulogizes, and who wrote so rhapsodically on the beatific vision, were all, to a man, men. Apart from the beatific vision, it is in the sphere of male sexuality that the gaze is intertwined most intimately with bliss (see Mulvey 1999). The question then arises: what might the relationship be between the beatific vision and the voyeuristic vision?

The male voyeuristic gaze can be said to shuttle between two poles and two goals: it can abject its object or exalt it. In the case of the beatific vision, the object seen elicits—or, better, exacts—worship. Exaltation, not abjection, is thus the dominant register (in contrast, say, to Revelation 17:1–6, in which Rome, misogynistically decked out in drag, becomes the object of an abjecting gaze: "Come and I will show you the comeuppance of the great whore").[48] In male voyeurism, the worshipful mode finds its stereotypical expression in the following scenario: a boy or man, himself unseen, feasts his fevered eyes on the spectacle of an unsuspecting woman in a state of undress—a woman who, moreover, meets his own personal criteria for archetypal, awe-inspiring femininity—and he climaxes in an ecstasy of worship. This scenario, however, is not the canonical one; for the object of the gaze in the book of Revelation, and in the theological tradition of the beatific vision more generally, is male. An awe-inspiring male, then, and an altogether queerer vision.

---

47. I revisit this remarkable cube below (236–38).
48. Further on that forced drag performance, see 143–46 below.

But have I, perhaps, been overhasty in suggesting that the vision of God in Revelation, and in the beatific vision at large, is essentially voyeuristic? For if voyeurism classically entails an *unseen* observer, then the deity in Revelation is never the object of a voyeuristic gaze on the part of those who abject themselves unceasingly before him. It is only the reader of Revelation who, strictly speaking, is placed in the position of a voyeur—and irrespective of her or his extratextual gender, or sexual proclivities. John, the "seer" of Revelation (as he is so aptly named), holds the heavenly door open for us ("I looked, and there in heaven a door stood open!," 4:1; cf. 3:8) and implicitly invites us to peek over his shoulder at the secret spectacle unfolding within. For those already inside the throne room, however, "the myriads of myriads and thousands of thousands" assembled around the throne (5:11), the looking is open and unabashed. Not voyeurism, then, technically speaking. Yet occupying a stance, nonetheless, that is intimately proximate to that of the voyeur. For the act of looking, watching, staring, gazing, gaping, gawking, gawping … eternally transmuted into an act of enraptured worship, defines the audience of the heavenly throne room no less than the stereotypical voyeur, crouched behind a bush and transfixed by his own beatific vision. The worshipful gaze in Revelation, it is tempting to conclude, is thus a gendered gaze. More precisely, it is a masculine gaze, to the degree that voyeurism is commonly coded as masculine. (It cannot unequivocally or absolutely be coded as masculine, of course; otherwise *Playgirl* magazine would exist entirely outside the realm of the explicable and enjoy the status of a numinous object and sacred text.)

In pondering the visual agency of the heavenly audience of Revelation, I am playing a variation on the previous "ending" of this essay, which focused the heavenly throne room and heavenly city through the eyes of the enigmatic figure at their center, "the one seated on the throne." I suggested that the innumerable throng of worshipers constitutes a vast mirrored enclosure the sole purpose of which is to reflect the divine perfection back to the deity, who thereby becomes his own eternal love object. But perhaps it is less a case of a single reflecting surface than of two mirrored surfaces facing and reflecting each other. For is not the deity also a mirror? (A mirror whose frame is anthropomorphism and before which theological "reflection"—also aptly named—is endlessly enacted. Dismantle this frame and what remains of "the biblical God"? The image disappears. Rather than resist the anthropomorphic lure, therefore, Christian theology has traditionally swallowed it hook, line, and sinker: "man" becomes

"God" in the very moment that "God" becomes "man.") This divine mirror reflects me, the Christian worshiper, back to myself. But the me I see in this celestial mirror cannot merely be the mundane me. This me has undergone *metanoia* ("Repent then"—Rev 2:16; see also 2:5, 21–22; 3:3, 19; 9:20–21; 16:9, 11), and, in consequence, metamorphosis (see 7:13–14: "Who are these clothed in white robes?")—or it better have if I am ever to set foot in the heavenly throne room and join the ranks of those who have been perfected (see 3:2). To translate all of this into a more contemporary idiom—and my sole interest in what remains of this afterword will be in (further) relating Revelation to contemporary popular culture—this supramundane me, whom I see reflected in the celestial mirror, has undergone a *makeover*. Yet the marvels of the makeover in Revelation pale in relation to reality, as we are about to discover.

In contemporary U.S. culture, "reality" is less the ontological ground of being than the name of a TV genre. The makeover, originally a feature of the Oprahesque talk show, has come to assume the proportions of a veritable theophany within the reality genre. I have in mind particularly the reality series *The Swan*, together with its less lyrically titled twin *Extreme Makeovers* (both shows doubtless long canceled by the time you read this, but numerous other variations on the makeover formula of social death and certain resurrection will have surged in to take their place). "I consider that these present sufferings are not worth comparing with the glory that is to be revealed," the apostle Paul attests (Rom 8:18), and that verse might well be embroidered and framed above the beds of the self-immolating martyrs undergoing weekly metamorphosis in *The Swan* and *Extreme Makeovers*. So mummified in bandages are they that their brutalized faces are barely visible within the cocoon, following the torturous series of flesh-slicing, bone-breaking, and fat-sucking "procedures" to which they have been subjected before the prime-time audience in the operating theater become amphitheater: the liposuction, the tummy tuck, the breast augmentation; the eye lift, the brow lift, the facelift; the rhinoplasty, the otoplasty; the jaw surgery, the chin surgery, the cheekbone surgery, the oral surgery; and so on. Afterward they are in more pain than Christ on his cross. And to what end?

Here we veer from Paul of Tarsus back to John of Patmos. For the vindicating moment of glory for these prime-time martyrs to the cult of youth and beauty is what the makeover reality show terms *the Reveal*. On the night of her "Big Reveal," the revealee, whose transfigured face we have not yet been permitted to behold, is filmed from behind in soft focus and slow motion as she glides into the sacred space in which the Reveal will be consummated—an ample space lined with spectators whose expectant faces instantly light up with unfeigned awe as the revealee makes her triumphant entrance. The indispensable instrument of the Big Reveal, however, is the outsized mirror on the other side of the room, veiled for now by a curtain. At a signal from the trembling revealee—who, like us, has been denied the vision of her face all the while she has been in the tomb—the veil is ceremoniously raised. The revealee's verbal response to the image in the mirror never varies; it is as predictable as any verbal response in any liturgical celebration. And it is intoned over and again as though for absolute emphasis: "Omigod! Omigodomigodomigod! Oh. My. God." And as she turns to face the room, ecstatic tears coursing down her resculpted cheeks, the audience now takes up the awed refrain: "Omigod!"

It is tempting to view this ritualistic scene as the quintessential moment of theophany in contemporary U.S. culture: theophany as self-revelation, the beatific vision as the beautific vision. For it is in this moment of utter physical transformation that the metaphysical is most unequivocally revealed (at least in TV-land—but is not that, as much as anything, now the American homeland?), eliciting instinctive, spontaneous acknowledgment of that revelation in the ritual's participants. The apparent distance between "My Lord and my God!" (John 20:28; cf. Rev 4:11) and "Omigod!" all but vanishes in such moments. God may, among many other things, be absolute goodness, but beauty is the privileged symbol of goodness, or virtue if you prefer, in contemporary U.S. culture—not that contemporary U.S. culture is original or unique in that regard, except in one way. The virtues most deafeningly and most incessantly trumpeted in the U.S. popular media, and, as such, the cardinal virtues of U.S. popular culture, are not four, as in Greco-Roman antiquity, but merely two: dieting and exercise.

Revelation's Big Reveal—or Biggest Reveal, rather, for it is merely one of many—occurs, appropriately enough, in the climactic chapter of the book and is announced with an admirable economy of words: "they shall see his face [*kai opsontai to prosōpon autou*]" (22:4; cf. Exod 33:18–23; Gen 32:30; Ps 42:2). The singling out of the deity's *face* for this climactic act of seeing in this scopophilic tale told by a seer implies, perhaps, that it is the *beauty* of the Godhead that is in view (see Ps 27:4; 96:6; Isa 28:5; 33:17), or possibly his glory (see Exod 33:18, 22)—but can glory be distinguished meaningfully from beauty in this instance? This glorious beauty has its locus in the divine visage. This beauty, then, is no ananthropomorphic abstraction, no anemic Platonic idea. Rather, it is an embodied beauty, a gendered beauty, a *male* beauty, the eternal beholding of which is the quintessence of bliss within Revelation's world, or rather its heaven. The bea(u)tific vision in Revelation would thus seem—yet again, to echo the previous "conclusion" of this essay—a narcissistic male fantasy enacted in a claustrophobic mirrored enclosure. Of the essence here is the hyperidealized male image—absolute power residing in a body all but undescribed (minimalistically possessed of a hand, a face, buttocks on which to be seated, vocal cords with which to utter a few terse words), but nevertheless coded as male, and beautiful enough, apparently, to merit an eternal, worshipful gaze. Everything else in Revelation's vision of heaven—the celestial city, the celestial throne, the celestial audience—is there merely to provide assorted pedestals for this hyperidealized male image, and also to provide the illusion of transcendence. But there is no room for real transcendence (even assuming that such a thing is ever possible) in the claustrophobically constricted space between the quasi-voyeuristic, implicitly male (see 14:4) gaze and the reflected male image. Yet there is at least one element of truth in this celestial hall of reciprocally reflecting mirrors. The thunderous clamor of the heavenly chorus, its eternally reiterated ejaculation, does perfectly encapsulate the essence of this narcissistic transaction: "Omigod! Omigodomigodomigod! Oh. My. God."

5
THE EMPRESS AND THE BROTHEL SLAVE*

Co-authored with Jennifer A. Glancy

The spotlight now shifts from the bodybuilder to the sex worker. If Revelation's hypermasculine heavenly monarch was the main focus of the first three essays (with a joint focus on the messianic superwarrior in the second essay), Revelation's femme fatale, "the great whore," is the principal preoccupation of the next three essays (also with a joint focus on Jesus in the second essay; John's fondness for neat numerical structures is infectious, it seems).

Like every essay in this collection, this one began life as a conference paper, specifically, a joint paper with Jennifer Glancy titled "How Typical a Roman Prostitute Is Revelation's 'Great Whore'? (The) John and the Working Girl," which we presented at the 2009 SBL annual meeting. Jennifer and I shared a fascination with the woman Babylon in Revelation. For me, she is the most vivid female character in the New Testament writings. If her transmutation into a city is regarded as an integral part of her tale, then she receives roughly as much narrative space as Luke's Mary the mother of Jesus and considerably more than John's Mary Magdalene, Babylon's only real competitors as (moderately) fleshed-out female characters in the New Testament canon. Babylon (or "Babs," as Jennifer and I took to calling her) is, of course, a symbolic character, a hyperbolic character, a fantastic character. Yet she is also an

---

* This essay first appeared under the title "How Typical a Roman Prostitute Is Revelation's 'Great Whore'?" *JBL* 130 (2011): 543–62. It is reprinted in lightly revised form with the permission of the co-author and the publisher.

abject character plucked from the Roman underclass. That, at any rate, was what we discovered when we plugged *pornē*, the epithet John uses for Babylon, into the engrossing body of work on ancient Greek and Roman prostitution that has ballooned in the field of classics in recent years. Babs, it turned out, was a brothel slave. Yet also, somehow, an empress. Holding these two antithetical elements of her identity in tension and uncoiling the gender and political ideologies tightly intertwined within them is the project of this essay.

John of Revelation famously introduces the woman Babylon as a *pornē* ("Come, I will show you the judgment of the great *pornē*," 17:1; cf. 17:5, 15–16; 19:2). But would early readers or hearers of Revelation have tended to see Babylon, based on John's description of her, as a typical Roman *pornē*? What were the typical, or stereotypical, traits of a *pornē* by the latter half of the first century CE? And how well does Babylon fit the profile? These are the principal questions that animate this essay. In pursuing them, we argue that John's representation of Babylon as a prostitute mimics a pattern of gender-based derision characteristic of coeval Roman writings, a pattern contingent on features of ancient prostitution that we elucidate. Imagining Rome as a prostitute who declares herself empress, John relies on the same logic that informs Roman authors who characterize imperial figures as pimps and whores. In Revelation, however, it is not the empress who is characterized as a prostitute but the empire itself.

Ancient Mediterranean sex workers came in two principal types: the *pornē* and the *hetaira*. The first term might be translated "brothel worker," "brothel slave," or, more colloquially, "streetwalker," depending on the context.[1] The second term is best translated "courtesan."[2] Although John terms Babylon a *pornē*, scholars have tended to treat her as a *hetaira*.[3] By the beginning of the common era, however, the *hetaira* was largely a literary construct—"not a historical entity, but a cultural sign," as classicist Laura

---

1. More on which below.
2. See, e.g., Kapparis 2011, 223: "*hetaira* (female companion), euphemistically describing a high-class courtesan."
3. Typical examples include Aune 1998b, 935; Rossing 1999a, 77–82; Koester 2001, 154–58; Osborne 2002, 611; DeSilva 2009, 108.

McClure puts it (2003, 5).⁴ Even more typical of the scholarly approach to the *pornē* Babylon is recourse to the topos of the harlot in the Hebrew Bible and the Septuagint. A parade example is G. K. Beale's fourteen-page treatment (1999, 847–61) of Rev 17:1–16, which is dense with textual references to Old Testament harlots and their attributes.⁵ Our point is not that Jewish Scripture is irrelevant to the depiction of the woman Babylon in Revelation, or even that the courtesan topos is irrelevant to it. Our point is rather that the scholarly approach to Babylon has been excessively "bookish." Reading the myriad commentaries, monographs, and articles on Revelation, one might be forgiven for supposing the primary, indeed only, knowledge (cultural, if not carnal) that John's original audiences had of prostitutes was derived either from elite Greek or Latin literary texts or from Jewish Scripture.⁶ By and large, the social realities of prostitution in the Roman world have not been adduced by such scholars to reconstruct the immediate connotations of the word *pornē* for such audiences—connotations far from the scriptorium or the symposium, as we shall see, and much closer to the *porneion* or brothel.⁷

This neglect is, however, understandable. It is only in recent years that the study of prostitution in the ancient Mediterranean world has

---

4. Cf. McClure 2003, 169: "The Greek hetaera had become a kind of historical relic, a literary figure associated with … bygone literary genres and vanished monuments." See further Kurke 1999, 178–87, a section titled "Inventing the *Hetaira*."

5. Apart from a passing reference to Seneca, Beale appeals only to the Hebrew Bible/LXX and ancient extrabiblical Jewish sources in these pages.

6. We agree with Elisabeth Schüssler Fiorenza that "the rhetorical-symbolic discourse of Revelation clearly understands [Babylon] as an imperial city and not an actual woman" (1998, 219). But we suspect that she imputes an overly cerebral experience of the *pornē* metaphor to the original audiences who coolly process it as a "figure of speech" principally drawn from "the prophetic language of the Hebrew Bible" (1998, 220). Drawing on conceptual metaphor theory, Lynn R. Huber has insisted with regard both to the imagery of the bride (Huber 2007) and the 144,000 male virgins (Huber 2008) that adequate engagement with Revelation's metaphoric language necessitates giving full weight to the associations evoked by the "source domain"— which, in this case, would be the familiar figure of the brothel worker or streetwalker.

7. Caroline Vander Stichele (2009, 117–20) does have recourse to real sex workers to interpret the figure of Babylon—those of the famed red-light district of Amsterdam. Avaren Ipsen (2009, 166–204) relates Revelation 17 to the violence experienced by contemporary sex workers; Marion Carson (2011) relates it to contemporary sex trafficking; and Jean K. Kim (1999) relates it to the Korean "comfort women" forced into sexual slavery by the Japanese Imperial Army.

fully taken off, finally becoming "[more than] a footnote to scholarship on ancient sexuality and gender" (McClure 2006, 3). The present essay is thus an attempt to recontextualize and reread the fraught figure of Babylon, *hē pornē hē megalē*, in light of the burgeoning body of work on ancient Greek and Roman prostitution. Recognition of the difference between a *pornē* and a courtesan is essential for appreciating the bite of ancient Roman invective. We shall find the woman Babylon's closest analogue in an altogether unlikely place, the *Annals* of Tacitus, which will help us comprehend the political valence of John's reliance on the trope of prostitution.

1. Between the Symposium and the Brothel: Locating Babylon's *Porneia*

Why Babylon Is Not a True Courtesan

By the late fourth century BCE, the *hetaira* had become a readily recognizable type in Athenian comic drama in particular, "one that traveled well to non-Athenian theaters scattered throughout the ancient Mediterranean world," and survived into the Roman era to become a fixture in the comedies of Plautus and Terence and the literary symposia of the Second Sophistic authors (McClure 2006, 15).[8] Classically, such women are celebrated figures, even celebrated wits, who consort with illustrious men and are "distinguished by a famous name" (McClure 2003, 12; cf. McClure 2006, 7). Classically, too, as McClure observes, the *hetaira* is "maintained by one man in exchange for his exclusive sexual access to her.... Alternately seductive and persuasive, providing her services in exchange for gifts, the *hetaira* perpetually left open the possibility that she might refuse her favors" (2006, 7; cf. McClure 2003, 11). More specifically, the *hetaira*, "by definition an unmarriageable woman not under the control of a father, husband or pimp," made her way in the world "by sexually attracting propertied men. Because of her social and economic independence, her affections had to be won by a prospective lover and could not be permanently controlled" (McClure 2006, 11; cf. Davidson 1997, 109–36 passim; Corner 2011, 74–75).

---

8. Further on the role of courtesans in the Second Sophistic symposium, see Anderson 1993, 183–85.

Already in the classical period, there is frequent slippage between the terms *hetaira* and *pornē*, as Leslie Kurke and others have noted (Kurke 1999, 178),⁹ a slippage with which we later find certain of Athenaeus's archaic Athenians wrestling—when they are not gleefully exploiting it: "And so in this instance, you happen to be in love not with a *pornē*, as you say, but a *hetaira*. But is she really so simple?" (Athenaeus, *Deipnosophistae* 572b).¹⁰ Edward Cohen styles the terms *pornē* and *hetaira* a "complementary antithesis" (2006, 96-97), an elegant formulation that, however, risks eliding the contempt in which the *pornē* was held.

As James Davidson observes, the term *pornē* connoted the depersonification, reification, and commodification of women, "their bodies, their time and their services." The cultural discourse in which the term is enmeshed "is primarily a discourse of contempt" (1994, 142, quoted in Kurke 1999, 180). If the *hetaira* belonged to the symposium, the *pornē* "belonged to the streets: she was the *hetaira*'s nameless, faceless brothel counterpart, and participated in a type of commodity exchange that continually depersonified and reified" (McClure 2006, 7).¹¹ Athenaeus vividly evokes the brutal reality of the *pornē*'s starkly commodified existence: "The women stand naked that you be not deceived. Look at everything.... The door is open. One obol. Hop in. There is no coyness, no idle talk, nor does she snatch herself away. But straight away, as you wish, in whatever way you wish. You come out. Tell her to go to hell. She is a stranger to you" (*Deipnosophistae* 569e-f). It was not only the *type* of sexual relationship, however, that distinguished the *hetaira* from the *pornē*; it was the *number* of relationships in addition. As McClure phrases it, the *hetaira* "expected relative permanence in her liaisons with men and professed fidelity," whereas the *pornē* was distinguished by the anonymity and sheer number of her sexual partners (2003, 14-15; cf. Davidson 1997, 125, 132-33).

---

9. See also Dover 1989, 21; McClure 2003, 18; Glazebrook and Henry 2011a, 4-8; Corner 2011, 75-78.

10. The translation of Athenaeus's *Deipnosophistae* used here and throughout is McClure's (2003), unless noted otherwise. The *Deipnosophistae* (late second or early third century CE) is an extensive compendium of Greek literary sources unified around the theme of the symposium. There is no standard English translation of the title *Deipnosophistae*. Renderings range from *Dinner Table Philosophers* to *Banquet of the Learned* to *The Gastronomers*.

11. See further Davidson 1997, 118-19. The *hetaira*, in contrast, participates in an economy of gift exchange, according to Davidson.

And this is the primary reason why the woman Babylon in Rev 17 is not a *hetaira* in any simple, straightforward sense. John provides no details of her sexual liaisons. He does, however, emphasize that her sexual partners are many. The very first mention of Babylon in Revelation, in which the city is already personified as female, declaims her spectacular promiscuity, even though the epithet *pornē* is not yet applied to her: "Fallen, fallen is Babylon the great, she who made all the nations drink from the wine that induces lust for her prostitution/fornication [*hē ek tou oinou tou thymou tēs porneias autēs pepotiken panta ta ethnē*]" (14:8).[12] When next she is displayed she is labeled *hē pornē hē megalē* "with whom the kings of the earth have committed fornication [*eporneusan*], and with whose wine that induces prostitution/fornication [*ek tou oinou tēs porneias autēs*] the inhabitants of the earth have become drunk" (17:2). Even when the woman subsequently transforms into a city, the image of her prodigious promiscuity persists: "because all the nations have drunk her wine that induces prostitution/fornication [*porneias*], and the kings of the earth have committed fornication [*eporneusan*] with her" (18:3); "And the kings of the earth who committed fornication and lived sensuously with her [*hoi met' autēs porneusantes kai strēniasantes*] will weep and wail over her" (18:9). The note of spectacular, earth-encompassing promiscuity is also sounded in the final mention of Babylon in Revelation: "he judged the great *pornē* who corrupted the earth with her prostitution [*hētis ephtheiren tēn gēn en tē porneia autēs*]" (19:2). In short, and in a fashion that is thoroughly circular, John constructs the sexual activity of Babylon in such a way that the epithet *pornē* will be seen to stick to her. She is a *pornē*, in John's discourse of contempt, because she has had many sexual partners, and she has had many sexual partners because she is a *pornē*.

Into the Streets

A significant feature of Roman prostitution was its high visibility. "What is striking about the topography of Roman prostitution," notes Thomas

---

12. All translations of Revelation are ours. We are, however, following BAGD 365 and Beale 1999, 755 in reading *tou thymou* and *tēs porneias* as genitives of cause, purpose, or result. Rossing translates *porneia* consistently as "prostitution," the first definition of the term given by LSJ, as she notes (1999a, 65 n. 11), which seems like a sound principle to us, maintaining as it does the etymological link between *pornē* and *porneia*.

McGinn, "is the complete absence of any evidence for ... moral zoning" (2006, 162).[13] Apparent instead is a wide pattern both of public solicitation and nonbrothel prostitution in places of public entertainment ranging from circuses, theaters, amphitheaters, and other public buildings such as temples and baths to markets, fairs, festivals, and public spectacles of every kind, and to military encampments and even circuit courts (McGinn 2006, 162; 1998b, 22–28). Other common nonbrothel venues for prostitution included inns, taverns, lodging houses, and eating and drinking establishments of every kind, many of them containing dedicated *cellae meretriciae*, one-room venues for commercial sex (McGinn 2006, 163; 1998b, 15–20). Brothels themselves, meanwhile, tended to be freely distributed throughout the Roman city (McGinn 2006, 163; 1998b, 20). Concomitant with the unrestricted distribution of prostitution in the Roman city was the sheer profusion of prostitutes within it.[14]

McGinn's analysis of the material evidence for prostitution across the Roman world turns up far more in the way of recurrent sameness than of regional difference, suggesting a common culture of Roman prostitution, at least in urban areas (1998b, 220–39).[15] There is no reason to suppose that sex workers were significantly thinner on the ground in the cities of Asia where the intended audiences of Revelation lived, or that they were less distributed and hence less visible. John's contemporary Dio Chrysostom, himself a native of the neighboring province of Bithynia, laments the profusion of prostitutes "in dirty booths which are flaunted before the eyes in every part of the city, at the doors of the houses of magistrates and in market-places, near government buildings and temples, in the midst of all that is holiest" (7.133–134, LCL trans.). Inscriptions from the Roman era designating or mentioning brothels are rare, yet one such inscription has apparently been found in Roman Ephesus.[16] Given the profusion and distribution of sex workers in Roman cities, it seems reasonable

---

13. An alternative version of McGinn's essay appears as ch. 3 in his *The Economy of Prostitution in the Roman World* (1998b).

14. As suggested by the material evidence. See McGinn 1998b, 167–219 passim; also McClure 2006, 18; Clarke 2003, 63–64.

15. Rebecca Flemming notes (1999, 43) that while prostitutes and brothels were common throughout the empire, significant archaeological evidence for brothels is limited to Italy. We can safely assert that Ephesus, say, had its share of brothels, but we do not know whether they were constructed along lines similar to those of Pompeiian brothels.

16. Carved on an architrave, this fragmentary inscription "mentions a latrine, in

to us to suppose that the term *pornē* would have conjured up first and foremost in the minds of the urban Christians addressed in Revelation a certain category of flesh-and-blood person that one encountered with considerable frequency in the streets, a fixture of the urban landscape, as opposed to a figure of high literature, or a literary or philosophical topos, or a scriptural type. In other words, the term *pornē* would have evoked a brothel worker in the first instance, not a *hetaira*, or a courtesan, or an "OT harlot." But what further and more specific associations might have accompanied that identification?

A second significant feature of Roman prostitution is that it was an activity almost universally associated with slaves or other persons on the bottommost rungs of the social ladder. The glamour attaching to certain courtesans in certain of the literary sources should not blind us to the fact that the typical location of prostitutes on the Roman socioeconomic scale was exceedingly low (Flemming 1999, passim; McGinn 1998c, 71). In common with many other classicists, Edward Cohen contends that, throughout antiquity, the term *pornē* was, for all intents and purposes, a virtual synonym of *doulē*, "[female] slave" (2006, 103–8).[17] Roman prostitutes catered to clients who themselves were typically of low status (Flemming 1999, 45). Roman males with sufficient economic means to own slaves "had little reason to frequent brothels," as John Clarke remarks. "They purchased slaves to fulfill their sexual desires" (Clarke 2003, 63). With regard to the numerous graffiti that have been found at Pompeii advertising prostitutes, Clarke observes: "Analysis of the names of the prostitutes reveals that they were all slaves (both male and female). And their owners were usually ex-slaves" (2003, 64). Such appears to have been the pattern throughout the empire (McGinn 1998c, 60; cf. Flemming 1999, 43). The servile associations of prostitution are crucial to Roman barbs indicting the imperial family for involvement in prostitution, as we shall see. They are also crucial, we will argue, for appreciating the full force of

---

connection with *paidiskēia* (scil. 'brothel facilities')" (McGinn 1998b, 209; cf. 225). See further Jobst 1976–77. The inscription probably dates from the late first century CE.

17. See McGinn 2011, 264: "The origins of Greek prostitution are located in a context of extreme exploitation, one of sexual slavery and rape"; Kurke 1999, 178: "The *pornē* … derives her name from the verb *pernēmi*, 'to sell (especially slaves)'"; McClure 2003, 15: "The term *pornē* originally denoted a brothel slave"; Glazebrook 2011, 35: "*Pornai* are technically slave prostitutes"; McGinn 1998b, 59: "When the ancient evidence registers the status of a [Roman] prostitute, more often than not she is a slave."

John's ironic portrayal of Babylon as a *pornē* who services the kings of the earth.

What can be known of the living conditions of such prostitutes? "Slave prostitutes seem to have been fairly tightly controlled," writes McGinn, "and the sources often suggest an environment of coercion. It is clear that these prostitutes were expected to live and eat in the brothel and were perhaps permitted to leave only rarely" (1998b, 37; cf. 236–37). The prostitute ordinarily worked in a booth or small room (*cella*) within the brothel (1998b, 39; cf. Clarke 2003, 63). Above or next to the *cella*'s entrance, a small sign (*titulus*) advertising her price was sometimes placed (McGinn 1998b, 39). The prostitute inside the *cella* was either nude or scantily clad (1998b, 40). Kelly Olson remarks: "Nudity was the marker of the lowest whore, a woman who was said to be ready for every kind of lust. The whores in a squalid brothel would also be naked, and Juvenal describes this sort of harlot as 'the whore that stands naked in a reeking archway'" (Olson 2006, 195). The literary sources describe the brothels as filthy places, although whether the reference is to moral or material filth is sometimes hard to determine (McGinn 1998b, 40).[18] Most indicative of the low status of the clients to which such brothels catered were the prices paid for the use of their inmates. Of the sixty-six known examples of such pricing, thirty-eight name prices that are 2.5 *asses* or lower, with 2 *asses*, at twenty-five examples, being far and away the most typical price (McGinn 1998b, 42).[19] To set these sums in context, 2 *asses* was approximately the price of a loaf of bread, or one-sixth of the daily wage of a male laborer, or slightly more than half of the daily discretionary income of a legionary soldier (1998b, 47, 54).

These then would have been some of the principal associations, or cultural connotations, conjured up by the term *pornē* in Rev 17:1–16 and 19:2. To put it mildly, the term would not, in and of itself, have automatically evoked a courtesan in the classical mold, a *hetaira*. The term would instead have summoned up a denizen of a far more squalid, far more

---

18. Certain of the sources seem to allow for little ambiguity; Horace, for instance, refers to the "evil-smelling cell" of the brothel (*Satire* 1.2.30). To view the prostitute's world as degraded and squalid, however, is to adopt the perspective of our elite sources. Flemming (1999, 45–46) makes the point that we do not know how prostitutes and their customary clients viewed that world.

19. See further McGinn 1998b, 267–68, 278–81; Flemming 1999, 48.

sordid reality—a grimy, street-soiled, social reality. But it is not the term alone that would have conjured up that domain.

At least one highly prominent feature of Babylon's description comports with her being labeled a *pornē*, a "brothel worker," or, better, a "brothel slave." "On her forehead,"[20] John tells us, "a mysterious name" has been inscribed: "Babylon the great, mother of *pornai* and earth's obscenities" (*kai epi to metōpon autēs onoma gegrammenon, mystērion, Babylōn hē megalē, hē mētēr tōn pornōn kai tōn bdelygmatōn tēs gēs*," 17:5).[21] Classicist C. P. Jones suggested more than two decades ago that "the author [of Revelation] perhaps imagines the Woman not only as a whore, but as a whore of the most degraded kind, a tattooed slave" (1987, 151). The suggestion is ventured in the course of a celebrated article that, among other things, argues convincingly that human branding was virtually unknown in the Roman world but tattooing was far more common than hitherto suspected. In Roman culture, a tattoo was ordinarily "a sign of degradation" (1987, 143). Slaves were a class of persons with whom tattoos were especially associated.

The forehead is one of the body parts most frequently mentioned in connection with tattoos in ancient Greek and Latin sources (1987, 142–43). Tattooed inscription could at times be extensive, as in the description of the freedman "who had, not a face, but a narrative on his face [*suggraphēn epi tou prosōpou*], the mark of his master's harshness" (Diogenes Laertius, *Lives* 4.46). So established is the practice of punitive facial inscription by the late first century CE that Martial can mine it for a vivid metaphor with little fear of being misunderstood: "whatever the heat of my anger brands [*inusserit*] on you will remain forever and be read throughout the world" (*Epigrams* 6.26)—words that might aptly be addressed to Babylon in Revelation.

Given the profusion of evidence presented by Jones for tattooing in the ancient Mediterranean world, and the close associations with slavery and

---

20. Ezek 9:4–6 is regularly adduced as the major source for this motif. John M. Court (1979, 41), however, sees the reference to the "forehead of a whore" in Jer 3:3 as lying behind it.

21. LSJ lists as among the primary meanings of *bdelygma* "filth, nastiness." Cf. BAGD: "gener. someth. that causes revulsion or extreme disgust, a 'loathsome, detestable thing.'" *TDNT* adds: "a shameless attitude" (Foerster 1964, 598). All of this we have attempted to evoke with our translation of *bdelygmata* as "obscenities." Further on *bdelygma*, see 166–67 below.

general degradation entailed in the practice, his suggestion that the *pornē* of Revelation 17 is to be regarded "as a whore of the most degraded kind, a tattooed slave" (1987, 151) seems compelling, at least to us.[22] McGinn (1998b, 37 n. 159), furthermore, and apparently independently of Jones, takes for granted the fact that facial tattooing (and even branding) would have been part and parcel of the brutal reality of slave prostitution in the Roman world, so that a brothel slave with a punitive facial tattoo would not have been anomalous.[23]

A Paradoxical *Pornē*

It might, however, be objected that other facets of Babylon's description seem nonetheless to justify the traditional preoccupation of commentators with the courtesan topos in identifying her. First and most obvious is her dress: "The woman was arrayed in purple and scarlet, and bedecked with gold and jewels and pearls" (17:4). Babylon lacks the garment that, rightly or wrongly, is often seen as the distinguishing dress, not of the *hetaira*, but of the common Roman prostitute. Certain Roman authors (Horace, Martial, Juvenal, Cicero, Tibullus, Acro, and Titinius) state or, more generally, imply that prostitutes, together with condemned adulteresses, wear the toga.[24] And there is nothing to suggest that Babylon is *togata*.

Kelly Olson, however, an expert on Roman costume and sumptuary norms, argues that the toga was not the standard dress of such women but

---

22. More compelling, certainly, than the "headband" hypothesis. R. H. Charles's assertion in his standard-setting commentary that "Roman harlots wore a label with their names on their brow" (1920b, 65; cf. Hauck and Schulz 1964, 594: "Like the city harlots of the day, [Babylon] bears its name on a golden head-band"), has been periodically recycled by commentators ever since in explicating Rev 17:5 (for a recent example, see Witherington 2003, 219), despite doubts that such a practice ever existed having been voiced even within Revelation scholarship itself (e.g., Ford 1975, 279).

23. Alone among recent commentators, so far as we can discover, David E. Aune (1998b, 936) cites Jones's proposal that Babylon's forehead inscription signifies that she is to be regarded as a tattooed slave. But he does so only in passing, possibly because his research has not adequately prepared him to hear the term *pornē* as already implying slavery, so that the facial tattoo merely makes explicit what is already implicit in the epithet.

24. In the case of prostitutes, probably to suggest that their bodies, like those of men (although on different grounds), are located in the public domain; so, e.g., Duncan 2006, 270; and D'Ambra 2007, 4.

only one of many types of dress that they could adopt. Roman prostitutes, indeed, could "appear in everything from expensive clothing down to little (or no) clothing at all" (Olson 2006, 194; cf. Olson 2008, 47–51). A character in Plautus's *Epidicus* enthuses over a certain *meretrix* (here, "streetwalker") who happens to be dressed rather similarly to the *pornē* Babylon: "But the way she was dressed, bejeweled, bedecked, sir—so charmingly, so tastefully, so stylishly!" (191–92, quoted in Olson 2006, 194). This *meretrix*, however, is plying her trade at the city gate,[25] a locale more suggestive of the common prostitute than the classic courtesan. Lavish female dress was not only compatible with slave status; such attire was among the tricks of the sex trade. In the *Ephesian Tale* of Xenophon of Ephesus, for instance, the heroine Anthia, sold into slavery, is compelled to exhibit herself in front of the brothel decked out in beautiful clothes and weighed down with jewelry (5.7). In short, Babylon's lavish dress does not cancel out the servile connotations of the term *pornē* or the (tattooed?) inscription on her forehead.

What of Babylon's cup? "The woman ... was holding in her hand a golden cup [*potērion*]" (17:4; cf. 18:6). Somewhat surprisingly, the scholars who have recourse to the courtesan topos to explicate the details of Babylon's portrayal do not tend to highlight the cup as contributing to it.[26] What could be more emblematic of the symposium, the drinking party, and hence, by extension, of the courtesan? The courtesan topos, however, called for the *hetaira* to be a model of decorum at table, at least as far as table manners, including moderation, were concerned (James 2006, 241; McClure 2003, 119). Babylon does not typify a courtesan in her relationship to the cup, however (even aside from the cup's singularly unappetizing contents), for that relationship is marked by drunkenness: "And I saw the woman drunk [*methuousan*]" (17:6).

Of course, it is not as *hetaira* but as *pornē* that Babylon is labeled in the text. What might the cup connote in relation to the latter term? *Pornai* in the Roman world were hardly strangers to situations in which alcohol flowed freely and drunkenness reigned—and not only because taverns and other drinking establishments were places where *pornai* plied their trade. *Hetairai* may have been regular fixtures at dinner parties in literary

---

25. "When I come to the gate I—yes, sir—I see *her* waiting there, and four flute girls [*tibicinae*] along with her" (*Epidicus* 217). The flute girl is regularly a type of prostitute in both Greek and Latin comedy (McClure 2003, 21–22).

26. Rossing (1999a, 77–80 passim) comes closest, as far as we know.

and philosophical works, but *pornai/scorta* contributed to the edgy, sexually charged atmosphere of many actual Roman dinner parties (McGinn 1998b, 27). Vulgar Babylon, emblazoned with her degrading tattoo permanently proclaiming her availability to service all and sundry for a price, and drunkenly clutching her wine cup, would be a better fit with such a rowdy dinner party than with the more refined dinner parties staged in the elite literary sources.

And yet discrepancies remain. In the final analysis, the figure of Babylon is not simply reducible to the lowest-status Roman prostitute, the tattooed brothel slave. Even apart from the fact that her clientele includes "the kings of the earth," there is the line—the one line—attributed to her, which must also be factored into any elucidation of her portrait: "I sit as a queen/empress [*kathēmai basilissa*], and I am not a widow, and I shall never see mourning" (18:7; cf. Isa 47:8). Not only does she sit, indeed, but she is also enthroned. Barbara Rossing notes, "'The one seated' or 'enthroned' (*hē kathēmene*, Rev 17:1, 3, 9, 15) is the most frequent label for Babylon in Revelation 17–18, culminating in Babylon's audacious boast that 'I sit as a queen'" (1999a, 66).[27] Revelation 17–18 presents us, then, with the paradox of an enthroned *pornē*, and it is precisely the combination of lowly and exalted elements in that paradoxical portrait that any adequate construal of it must ultimately hold in tension. This brings us to the final section of our study.

2. Imperial Whore, Savage Whore: Babylon and Messalina

What the paradoxical figure of Babylon would have evoked for first-century audiences, we would argue, is not the social type of the brothel slave, pure and simple, not yet the literary topos of the courtesan, pure and simple—although in contrast to most previous scholarship we would see her as closer to the former than the latter. Rather, we have come to understand the fraught figure of Babylon as something more akin to the Roman empress Messalina as refracted through the prurient popular imagination and its literary distillations and elaborations. Juvenal, for instance, labels

---

27. The verb *kathēmi* is a theologically and politically pregnant one in Revelation, God preeminently and paradigmatically being "the one seated [*ho kathēmenos*] on the throne" (Rev 4:2, 9–10; 5:1, 7, 13; 6:16; 7:10, 15; 19:4; 20:11; 21:5). Rossing thus suggests that "the portrait of the enthroned prostitute in Rev 17:1–4 is structured as a deliberate contrast to God's great throne room scene in Revelation 4" (1999a, 67).

Messalina *meretrix augusta*, "whore-empress," as we shall see, an epithet no less applicable to the complex amalgam of exalted and abject elements that is the *basilissa/pornē* of Revelation 17–18. "To adapt a conceit familiar from anthropology," Roman authors as different as Juvenal, Tacitus, and John of Revelation "found prostitutes convenient to think with,"[28] and to think about empire in particular. To appreciate the impact of associating an empress—or emperor—with prostitution requires us once again to recognize that in the Roman world prostitutes were primarily associated not with the rarefied world of the *hetaira* but with the commodified sale of sex.

Although representations of Messalina offer the most vivid and direct analogues to John's Babylon, such representations acquire resonance from their location in a wider pattern of Roman discourse associating the emperors and their families with prostitution. The younger Seneca, for instance, claims that Augustus's daughter Julia was so promiscuous that the emperor—famously the promulgator of family-values legislation—had her categorized as a prostitute (*On Benefits* 6.32.1). Seneca salaciously claims that Julia met men for anonymous sex in the very Forum where Augustus had proposed his strict legislation on adultery. Sexual invective is standard fare in Roman political discourse.[29] Suetonius's *Lives of the Caesars* reads at times like a catalog of depravity. Concerning Caligula, Suetonius writes, "The story so far has been of Caligula the emperor, the rest must be of Caligula the monster" (*Caligula* 22).[30] Tales of Caligula's pleasure in sexual humiliation run parallel to tales of his pleasure in cruelty. Suetonius's claim that Caligula profited from prostitution, however, occurs not in the context of Caligula's manifold sexual excesses but in the context of his manifold schemes to swell his coffers, schemes ranging from outright robbery to a tax on prostitution (*Caligula* 38–42). According to Suetonius, however, Caligula did not stop there. Thinking to turn a tidy profit, Caligula established a brothel (*lupanar*) in the imperial palace on the Palatine (*Caligula* 41).[31] Respectable women and young men offered

---

28. The quoted phrases are from McGinn 1998c, 348 (referring not to Roman authors, however, but to Roman legislators).

29. For a broad introduction to the topic, see Jennifer Wright Knust's essential *Abandoned to Lust* (2001, esp. 15–50). See also Rauh 2011.

30. Catherine Edwards's translation (Suetonius 2000) here and in what follows.

31. For an analysis of this narrative and other ancient references to it, see McGinn 1998a.

their services, supposedly, or were coerced to do so, in the brothel's well-appointed cubicles.³²

Through this lurid description, Suetonius effectively characterizes the emperor as a pimp. We are not concerned here with the question of whether or not the report reflects historical events.³³ For purposes of our argument, what is significant is Suetonius's reliance on the degrading associations of the sex trade in his litany of Caligula's acts in an orgy of greed—greed that tellingly assumes an erotic sheen: "Finally, seized with a passion for handling money, he [Caligula] would often walk with bare feet on the huge heaps of gold pieces he had piled up in the most public places and sometimes he would even roll about in them with his whole body" (*Caligula* 42). Suetonius's characterization of Caligula as pimp of pimps emerges seamlessly from his critique of the monster-emperor who has reduced the respectable citizens of Rome to the status of slaves and even of brothel slaves, to be used and abused at his pleasure. John's characterization of Babylon as *pornē* both participates in and disrupts this pattern of discourse, as we shall see.

Even more memorable than the pimp-emperor is the whore-empress. In an elegant and incisive essay, classicist Sandra Joshel traces the process whereby "the name of a particular woman, Valeria Messalina, becomes the proper name for uncontrolled female sexuality" in ancient Roman literature, above all that of Tacitus and Juvenal (Joshel 1997, 222).³⁴ Uncovering the "historical" Messalina is not Joshel's aim, however, even if that were possible; rather, it is "the politics of a particular intersection of gender and empire" (1997, 222) that intrigues her (and us).

This "whore-empress" (*meretrix augusta*), declaims Juvenal, speaking transparently of Messalina, would leave her husband Claudius snoring unsuspectingly in their bed while she stole forth from the palace by night.

---

32. McGinn writes: "Dio elaborates on the particular about the *matronae ingenuique*, describing them as 'the wives of the leading citizens and the children of upper-class men' and adding a spicy detail: some were willing, some not" (1998a, 95).

33. McGinn is more inclined than we are to accept the tale of Caligula's brothel as historical fact.

34. As the subtitle of Joshel's essay ("Tacitus's Messalina") suggests, it is Tacitus's rather than Juvenal's Messalina that is its principal subject, although Juvenal's also plays an important role in it. Specifically, Joshel's primary focus is Tacitus, *Annals* 11.1–4, 12, 26–38, written more than half a century after Messalina's death.

Like that, with a blonde wig hiding her black hair, she went inside a brothel [*lupanar*] reeking of ancient blankets to an empty cubicle—her very own. Then she stood there, naked and for sale, with her nipples gilded, under the trade name of "She-Wolf" [*tunc nuda papillis / prostitit auratis titulum mentita Lyciscae*], putting on display the belly you came from, noble-born Britannicus. She welcomed her customers seductively as they came in and asked for their money. Later, when the pimp [*leno*] was already dismissing his girls, she left reluctantly, waiting till the last possible moment to shut her cubicle, still burning with her clitoris inflamed and stiff [*adhuc ardens rigidae tentigine vulvae*]. She went away, exhausted by the men but not yet satisfied, and a disgusting creature, she took back to the emperor's couch the stench of the brothel. (*Satire* 6.116–135)[35]

Note the details of a rank-and-file Roman sex worker's circumstances evoked by Juvenal: the cubicle, the sign announcing the brothel worker's name (or at least her nom de guerre), her displayed nudity, and the presence of her pimp. Joshel argues that although Juvenal's Messalina appears as the climax to a catalog "of women's flaws," the poet nonetheless "projects elements of an imperial discourse onto her" (1997, 248). In other words, he stages a pornographic display of a politicized female body. The insatiable lust for power and territory endemic to empire is figured in the transgressive body of a "whore-empress" who turns tricks not in order to survive but only to feed her voracious lust.

The political critique implicit in Tacitus's representation of Messalina is still more sharply honed. The signal feature of his Messalina is that she is a figure of excess (Joshel 1997, 230). Bored by adultery, which she finds all too easy and unchallenging, Messalina is, in Tacitus's terms, "flowing out into untried lusts [*ad incognitas libidines profluebat*]" (*Annals* 11.26.1). Ultimately, says Tacitus, she desires "the magnitude of the *infamia* that is the last source of pleasure for the licentious" (11.26.6). Although empress, she burns to embrace the social status of the *scortum* or *prostibula*, and to wallow shamelessly in it.[36] As such, her desire is also *chaotic*, in accordance with "a commonplace of Roman moral rhetoric that associates

---

35. LCL translation, modified.

36. This is a role that Messalina also plays in Pliny, *Natural History* 10.172; Juvenal, *Satire* 6.115–132; and Dio Cassius, *Roman History* 61.31.1, as Joshel notes (1997, 231). On Roman prostitutes as *infames*, "without reputation," see Edwards 1997; Duncan 2006.

uncontrolled female sexuality with chaos" (Joshel 1997, 231). Such chaotic excess connotes a collapse of social hierarchy. In this case, the topmost reaches of the social order have toppled into the gutter (cf. Joshel 1997, 231).

Messalina's desire is also a *violent* desire, on Tacitus's construction. He has an anonymous character, a textual mouthpiece, label her a "savage whore [*saevienti impudicae*]" (*Annals* 13.43.5). Tacitus's representation of Messalina is of a hyperpromiscuous woman whose sexual voracity veers into outright savagery: another index of the excessiveness of her desire. Her former lover, Mnester, bears the scars of floggings apparently administered by the empress herself (11.36.1–2). Men and even women "lose their lives to her lust for bodies and things" (Joshel 1997, 232).[37] "Many murders [are] perpetrated on Messalina's orders [*multasque mortis iussu Messalinae patratas*]" (*Annals* 11.26.4).

In terms of genre, the *Annals* of Tacitus is, of course, cultural worlds apart from the Apocalypse of John—and that is not all that divides them. However critical Tacitus may be of individual Roman emperors and even of the principate itself, he is devoted to Rome and its glory, while John, to put it mildly, is not. Where the two works intersect, however, is in their respective treatments of sex and gender as they relate to empire. Notwithstanding the opaque apocalyptic genre in which it is enmeshed, Revelation's treatment of these themes is, paradoxically, more transparent than that of the *Annals*. What is embedded in Tacitus's historical narrative, requiring a skillful critic such as Sandra Joshel to prize it out, is displayed on the surface of Revelation's apocalyptic narrative: in contrast to the *Annals*, Revelation's tale of the wicked woman is *explicitly* said to be an allegory of the evil empire: "And the woman that you saw is the great city which has *imperium* [*basileian*] over the kings of the earth" (17:18).[38] In the implicitly allegorical figure of Tacitus's Messalina as in the explicitly allegorical figure of John's Babylon, "empire is mapped on the uncontained, overflowing body of a woman. Its lack of limits echoes the extension of the geographic boundaries of empire as well as the long-held Roman vision of empire as boundless" (Joshel 1997, 247). The allegory entails gender transformation in accordance with an implacable cultural script that reflexively codes excess as feminine. In both the *Annals* and the

---

37. Further on Messalina's savagery, see *Annals* 11.28.2; 11.32.6; cf. Dio Cassius, *Roman History* 14.3; 15.5, 18; 22.4–5; 60.8.5.

38. Further on *imperium*, see 15 and 84 above.

Apocalypse, absolute imperial power—which is also hyperbolic *masculine power*—is represented as a promiscuous, voracious, and violent *feminine desire*. The monstrous spectacle of a sexualized woman utterly out of control serves as a trope for imperial autocracy—absolute power exercised to excess, entirely without restraint.

The violent nature of Babylon's desire invites further reflection. Entire nations are drunk with lust (*thymos*) for Babylon as a result of her boundless promiscuity (Rev 14:8; 17:2; 18:3, 9; 19:2), as we noted earlier. She herself, however, is drunk on violence. The golden cup she clutches in her hand is "full of abominations/obscenities and the impurities/filth of her prostitution [*gemon bdelygmatōn kai ta akatharta tēs porneias autēs*]" (17:4). That loathsome decoction is almost immediately decoded as deadly violence: "And I saw the woman drunk on the blood of the saints and the blood of Jesus' witnesses [*methuousan ek tou aimatos tōn hagiōn kai ek tou haimatos tōn martyrōn Iēsou*]" (17:6; cf. 19:2). This savage feast of blood is subsequently and hyperbolically extended—in a paroxysm of excess, as it were—to "all who have been slaughtered on earth [*pantōn tōn esphagmenōn epi tēs gēs*]" (18:24), corresponding to the earlier hyperbolic identification of those who have become drunk on the wine of her prostitution simply as "those dwelling on earth [*hoi katoikountes tēn gēn*]" (17:2). Her lust is a lust for violence that intoxicates and infects the entire inhabited world.

Although a figure of lust, then, Babylon is not (necessarily) an active subject of sexual desire. Along with her desire for blood, she also craves acclaim and wealth: her sins are characterized as self-magnification and luxurious living (18:5–7). She is voracious, but is she sexually voracious? Babylon is a thoroughly ambiguous figure. She effectively manipulates the sexual desires of men to feed her bottomless appetites for blood and luxury goods.[39] Juvenal creates an empress so sexually hungry—and perverse—that she turns tricks in a common brothel. Yet this does not imply that Roman sex workers were widely believed to be in the sex trade to satisfy their erotic longings.[40] To read Babylon as seeking sexual gratifica-

---

39. McGinn observes that "in many cultures prostitution can serve as a metaphor for a voracious, almost limitless mode of consumption that merges the sexual and the material" (1998b, 53–54).

40. Roman writers who accorded even scant attention to the question of why women entered prostitution understood the rationale as economic. For discussion and exceptions, see Flemming 1999, 41–43.

tion through promiscuity is to read her through later, largely Christianized lenses. John rather presents her as a degraded woman who achieves international infamy and a sumptuous style of living by preying on the desires of powerful men. Although she is marked as a slave—perhaps the slave of the beast?—there are nonetheless hints of the pathetic in her profession. Does she whore to sate her own lust (for sex or for blood), or to amass finery, or for fear of a savagely violent pimp (if that indeed is what the beast is)[41]—a well-founded fear, as it turns out?

Returning to Tacitus's Messalina, we note that Tacitus's strategies for encaging and containing the rampaging female monster he has created are also those of John of Revelation. "Tacitus oscillates between his need for a powerful female voice and his reluctance to describe or quote female speech," notes Joshel (1997, 234). On the one hand, Tacitus's narrative "cannot work without the power of Messalina's voice." On the other hand, his narrative seeks to minimize the power of her voice by reducing it to its effects: seduction, murder, and general disorder (1997, 234). As a result, Messalina's speech "is banished to the interior spaces" of Tacitus's narrative world (1997, 234).

Babylon, too, is all but mute. John allows her only one line, as we noted earlier. Ironically it is a line that declares her sovereignty—"I sit as a queen/empress [*basilissa*]" (18:7)—even as her otherwise voiceless role declares the stark limits of that sovereignty, at least within the world of the narrative. And yet John's characterization of her presupposes not just a voice but also an incomparably powerful voice, the most authoritative voice in John's world: "And the woman whom you saw is the great city that has *imperium* over the kings of the earth" (17:18). Like Messalina's voice in the *Annals*, however, Babylon's voice is banished to the interior spaces of Revelation's narrative world, so that we only hear its echoes. To put it another way, John is careful to ensure that the *basilissa* is subsumed in the *pornē*. Ancient Roman sex workers "cannot tell us their story," notes McGinn (1998c, 9), speaking of the real women and girls

---

41. The dragon, the sea beast, and the land beast exist in a hierarchical relationship, as is commonly observed: the sea beast serves the dragon (Rev 13:2b, 4), while the land beast serves the sea beast (13:12–17). Where does the woman Babylon fit in the hierarchy? Does her forehead inscription (17:5), evocative of slave tattoos, as we have seen, together with the fact that she is eventually the victim of deadly violence on the part of the (sea) beast and its underlings (17:16–17) suggest that she is subservient to the beast? Structurally, is their relationship one of slave-prostitute and pimp-owner?

behind the male-generated literary and legal sources. Even when they are fictional characters within these sources, they may be systematically silenced, even—or especially—when simultaneously endowed with terrifying power. Such is the case with the *pornē* Babylon.

Silencing, however, is but the beginning. Not only does Tacitus set forth the problem—the category error of an empress who is a brothel worker—but he also provides the solution. He structures a narrative that not only silences this female embodiment of anarchic disorder but annihilates her altogether. Like John of Revelation, Tacitus is adept at the creation of monsters. He parades for "the gaze of his male readers the all-encompassing wickedness of the castrating woman." The fear evoked by this monstrous image, however, "is relieved by a narrative that hunts her down. Step by step, Tacitus drains her of sexuality and life until she becomes a corpse" (Joshel 1997, 247). Driven to suicide but too diminished even to complete the act, Messalina is run through by a tribune's sword. When her death is announced at dinner, Claudius does not bother to ask its cause but keeps on eating. This final depersonalizing indifference is but the logical conclusion of a narrative designed to evoke, exorcise, and erase the empire-unraveling chaos of an empress turned whore.

John's strategy is structurally parallel to that of Tacitus, but it also has its own distinctive features. The story of Babylon's demise is the story of a great many sex workers in every age, including our own. She is the victim of deadly violence on the part of her clients and, we have suggested, her pimp, the beast: "And the ten horns that you saw, they and the beast will loathe the whore, and they will ravage her and strip her naked, and they will devour her flesh and burn her with fire [*houtoi misēsousin tēn pornēn, kai erēmōmenēn poiēsousin autēn kai gymnēn, kai tas sarkas autēs phagontai, kai autēn katakausousin en pyri*]" (17:16).[42] The ten horns have earlier been identified as "ten kings" (17:12) and thus may plausibly be num-

---

42. Tina Pippin was the first to accord this troubling verse the attention it deserves, beginning with her *Death and Desire* (1992a, 57–68 passim). See further Vander Stichele 2000, which extends Pippin's trajectory. In common with the present essay, these readings contrast with that of Schüssler Fiorenza, who resists the attribution to Revelation of a pernicious gender ideology (see, e.g., 2007, 130–47 passim). Rossing's position is related: "While not disagreeing that Revelation's violence poses ethical problems, I argue that the violence [in Rev 17:16] is directed primarily against a city's landscape, not against a woman" (1999a, 90 n. 89; cf. 87–97 passim). For the limitations of this approach, see n. 6 above.

bered among "the kings of the earth" who have previously availed of the sexual services of the "great *pornē*" (17:2–3; 18:3, 9). They now have their revenge on their seductress, and even her owner has no further use for her, apparently, other than providing this grim object lesson in how to deal with a whore.

3. Conclusions: Empress as Brothel Slave

John's Babylon is no ordinary *pornē*. As prodigious in her allure as in her appetites, she could not be confused with the impoverished, servile women and girls who typically serviced male clients in brothels throughout the Roman Empire. We have nonetheless argued that it is important to come to terms with John's identification of her not as *hetaira* but as *pornē*. Unlike a *hetaira* loyal to a single man, Babylon is spectacularly promiscuous, servicing all the kings of the earth. With her name emblazoned on her forehead, she is implicitly marked as a slave, thereby sharing the status of so many flesh-and-blood brothel workers of the empire. Yet at the same time, Babylon calls herself a queen/empress. Paradoxically, the *pornē* Babylon's very claim to imperial status underscores the importance of locating her in the context of the servility and squalor of the Roman sex trade.

Sexual invective was standard in Roman political discourse. A recurrent feature of that invective was the association of members of the imperial family with prostitution, a pattern of invective contingent on identifying the most powerful figures of the empire with debased and dishonorable sexual practices. Babylon is thus a "whore-empress," both like and unlike Messalina, who was labeled *meretrix augusta*. Juvenal relies on quotidian details of a Roman sex worker's existence—the stench of the brothel, the commodifying display of its denuded, human wares—to emphasize the empress's moral turpitude. Similarly, the impact of John's representation of Babylon is contingent on the audience's recognition of the degradations to which enslaved brothel workers were subjected, including tattooed foreheads and perpetual vulnerability to violence. John's representation of a whore seated as empress is designed to indict the empire itself, and this representation gains resonance from its location in the wider pattern of sexual invective characteristic of Roman political discourse. Understanding Babylon as *basilissa*, we argue, requires that we give full weight to John's designation of her as *pornē*.

6
Raping Rome*

What was Babylon before she was made a *pornē*, a brothel slave? The answer is that she was a goddess. This essay takes as its point of departure the now common suggestion that Revelation's "great whore" is a parodic representation of *Dea Roma/Thea Rhōmē*, the goddess who personified the city of Rome and, by extension, the Roman state. Relatively little has been written on Roma by classicists, even those interested in ancient Roman gender ideologies. That surprises me. Roma is an intriguing gender anomaly: an armed and armored warrior who represents Roman military might, but who also happens to be a woman. Revelation strips Roma of her armor, reclothes her as a prostitute, and sexually humiliates and annihilates her. Indeed, Rome is Roma/Babylon in Revelation only that it may be symbolically raped. In the process, however, Roma's already complex gender identity becomes yet more convoluted. Roman hypermasculine militarism is paradoxically symbolized by a female body in the masculine dress of a warrior, and that Amazonian warrior is now sexually degraded by being turned into a brothel slave—a case of triple transvestism that calls out for queer analysis, but which kind? Increasingly nowadays, queer theory is in a "post-Butlerian" phase (e.g., Nigianni and Storr 2009; Ruffolo 2009; Penney 2013); yet it is "early" Judith Butler, the Butler whose immensely influential

---

* First published under the title "Metonymies of Empire: Sexual Humiliation and Gender Masquerade in the Book of Revelation," in *Postcolonial Interventions: Essays in Honor of R. S. Sugirtharajah* (ed. Tat-siong Benny Liew; The Bible in the Modern World 23; Sheffield: Sheffield Phoenix, 2009), and reprinted in revised form with permission.

theory of gender performativity was catalyzed by the phenomenon of drag, to whom Roma/Babylon seems to be calling, and so the youthful Butler of *Gender Trouble* (1990), still unaware that she is the doyenne-to-be of queer theory, wades into Revelation 17 to untangle its knotty gender contradictions and ponder their (bloody) stakes. Yet the queerest twist of all in Revelation concerns its leading man, who, so often, is less or more than manly. Not least, the body of the risen Christ in Revelation 1:12–16 is an intersexed body, as we discover, and the (more mature) Butler of *Undoing Gender* (2004) helps us to reflect on it and relate it to the triply cross-dressed body of Roma/Babylon.

Imperial Rome is represented in Revelation as a woman (14:8; 17:1–18; 18:3–9, 16; 19:2), once again impelling the question, why? Is it because Babylon, the prototypical evil empire in Jewish tradition and the code name for Rome in Revelation (17:6, 9, 18), was already represented as female in that tradition (e.g., Isa 47:1–15; Jer 50:9–15; Zech 2:7)? Or is it because Rome was already represented as female in the cult of the goddess Roma, one with deep roots in Roman Asia (more on which below)? Assumedly there is no need to choose exclusively between these two alternatives. If I fixate on the latter alternative in the present essay, it is only because it represents the road less traveled. But all roads lead equally to Rome in and around Revelation.

Here, first, is what the main thoroughfare has looked like. As scholars have long surmised, Revelation renames Rome as "Babylon,"[1] confers the name of another city on it, because that city, also an empire, epitomizes in Jewish Scripture and tradition human empire at its most destructive,[2] but also at its most seductive. And that predatory and alluring empire (whose

---

1. As does 1 Pet 5:13; *4 Ezra* 3:1–2, 28–31; *2 Baruch* 10:1–3; 11:1; 67:7; and *Sibylline Oracles* 5.143, 159. Dissenters from the critical-scholarly consensus identifying Revelation's Babylon with Rome have been few. But even within the consensus there are unsuspected and unsettling spaces that have yet to be explored, as I attempt to show in this essay.

2. Leaving aside the question of whether Revelation is pre- or post-70 CE. If the latter, Rome is most of all Babylon because it has destroyed the rebuilt Jerusalem temple.

imagined antithesis is the empire of God) comes already sexed and gendered in the tradition that John of Revelation has inherited and internalized (even apart from the prophetic passages listed above), Babylon being a feminine noun in both Hebrew (*bābel*) and Greek (*babylōn*).

But is Babylon also the object of sexual shaming in the Jewish Scriptures? In the five oracles against Babylon found in the prophetic literature (Isa 13:1–22; 14:22–23; 21:1–10; Jer 25:12–14; 50:1–51:64), only Jer 50:12a, "your mother shall be thoroughly shamed/ashamed, she who bore you disgraced" (*bôšâ 'immĕkem mĕ'ōd ḥāprâ yôladtĕkem*), presents itself as a candidate for such interpretation, it seems to me—and then only if we take the referent of "your mother" to be Babylon and construe her shaming as sexual.[3]

More significant is the fact that the epithet "whore" (Heb. *zānâ*; Gr. *pornē*) is never leveled at Babylon in the Jewish Scriptures,[4] or, so far as I am aware, in any other extant Jewish source prior to Revelation. This may be coupled with a second observation. The image, recurrent in Jeremiah, of Yahweh's enemies being compelled to imbibe from the cup of his wrath (Jer 13:13–14; 25:15–29; 48:26; 51:39, 57; see also Isa 51:17–23; Lam 4:21), is an arresting one for John. He recycles it in Revelation 14:8, 17:2, and 18:3—but he also sexualizes it. In John's sweaty hands, the cup metaphor consistently becomes an allegory of Babylon/Rome's *porneia*: "Fallen, fallen is Babylon the Great who caused all nations to drink of the wine of her lustful passion" (*tou thymou tēs porneias autēs*, 14:8; the images of wine, drunkenness, and *porneia* recur in 17:2 and 18:3, and are implicit in condensed form in 19:2).[5] Revelation's pornoprophecy, far from simply being siphoned, already fully fermented, from the pornoprophecy of the Hebrew prophets, is John's own distinctive concoction.[6]

---

3. The translation of Jer 50:12a is mine. For discussion of the verse, see Stulman 2005, 373; Allen 2008, 513. Neither scholar, however, suggests that the shaming is sexual, nor have I been able to discover any that do. Such discussion tends to focus instead on Jer 2–3, where the emphasis is on the sexual shaming of Israel (e.g., Brenner 1995, which discusses Jer 2 and 3:1–3, together with 5:7–8).

4. It is Israel that is the primary recipient of the epithet (Jer 3:6–10; Ezek 16:15–22; 23:1–49; Hos 4:12–13; 5:3), while it is secondarily applied to Jerusalem (Isa 1:21), Tyre (Isa 23:15–17), and Nineveh (Nah 3:4).

5. Translations of Revelation throughout this essay are my own.

6. Feminist scholarship on Revelation has frequently focused on its relationship to this lewd and lurid strand of Hebrew prophecy; see especially Selvidge 1996; Vander Stichele 2009, 109–14.

The Babylon of Hebrew prophecy, then, in and of itself, provides little in the way of answers to the question of why Rome is represented in Revelation not just as a woman but also as a prostitute. For that we need to set foot on the less-traveled path to which I earlier alluded. The Rome-as-Babylon equation lacks a crucial middle term. Implicit in the Rome-as-Babylon motif, as we shall see, is a Rome-as-Roma-as-Babylon motif.[7] In other words, John's counter-imperial representation of Rome as a certain kind of woman is parasitic on, and parodic of, pro-imperial representations of Rome as an altogether different kind of woman. Parody of the Roman imperial order in Revelation reaches its scurrilous climax in the depiction of the goddess Roma, austere and noble personification of the *urbs aeterna*, as a tawdry whore who has had too much to drink.[8] And so to *Dea Roma* we now turn.

The Hardest Woman Warrior in the Roman World

That the Roman imperial cult is a major polemical target of Revelation has long been a commonplace of critical scholarship on the book. That polemic is customarily explained with reference to the importance of the province of Asia for the origins and evolution of the cult. Augustus and his successor Tiberius only permitted two temples to be established in their honor during their respective reigns. Both were Asian temples, Augustus's being erected in Pergamon, Tiberius's in Smyrna, two of the seven cities to whose churches Revelation is addressed (1:11; 2:8, 12).

That, however, is only part of the story of how the province became tutor to the metropolis on the art of constructing religious cults that would be consummately expressive of Roman hegemony.[9] The provincial temple that the Assembly of Asia received permission to build at Pergamon in 29 BCE was to be a shrine dedicated not only to Augustus but

---

7. This is not the only possible path, needless to say. We could also detour through the prophetic texts adduced in n. 4 above and arrive at Rome-as-prostitute that way, but that road is already too crowded. Ian Boxall lists as first of the "small number of issues of common interest" to modern commentators on Rev 17 "the Old Testament background to the vision" (2001, 53).

8. As I remark in passing on 22 above. In effect, the present essay is an elaboration of that statement.

9. See further 19–21 above.

also to Roma.¹⁰ This is not to imply, however, that the institutionalization of the cult of Roma in Asia coincided with the institutionalization of the imperial cult. A temple to *Thea Rhōmē* had stood at Smyrna since 195 BCE, the first known temple to Roma in the East, and hence anywhere.¹¹ The cultic veneration of Roman supremacy thus had primary expression in the province of Asia in the cult of *Thea Rhōmē* long before it found supplementary expression in the imperial cult.¹² And while the imperial cult does enable us to make sense of much of Revelation's anti-Roman invective, especially that found in chapter 13, *Thea Rhōmē* enables us to understand why it is that Rome is represented both as a woman and as a prostitute in a further major portion of that invective, namely, that concentrated in chapter 17.¹³

The cult of Roma reached deep into the civic life of Roman Asia, as evidenced by its sheer longevity. The Romaia, an elaborate series of religious rituals, athletic contests, and other public spectacles designed to honor the goddess, "were still celebrated at Smyrna four centuries after

---

10. See Friesen 2001, 25–32; Burrell 2004, 17–22.

11. Cf. Tacitus, *Annals* 4.56: Smyrna "had also been the first…to erect a temple in honor of Rome [*seque primos templum urbis Romae statuisse*], during the consulship of Marcus Porcius Cato, when Rome's power was already great, but not yet at its apex…" (my translation).

12. As in other provinces of Asia Minor, cults of Roma flourished not only in coastal cities and on the islands (including Chios, about twelve miles off-shore from Smyrna) but in a number of inland cites as well (see Price 1984, 43).

13. Ronald Mellor, in what has become the standard monograph on the cult of Roma, surmises in passing that "there are even allusions to the cult … in the attack on the Roman Empire in the Apocalypse of St. John" (1975, 127). Among New Testament scholars, David E. Aune (1998b, 919–28) has ventured farthest down this speculative path, as far as I know. Of particular interest to Aune is a coin minted in Asia in 71 CE that depicts Roma seated on Rome's seven hills with a personified river Tiber at her feet; he links these details with Rev 17:9 ("the seven hills on which the woman is seated") and 17:1 ("the great whore who is seated on many waters"; cf. 17:15). Aune's erudite discussion, however, lacks any element of gender analysis. (Symptomatically, his bibliography on Rev 17–18, which lists more than fifty works in four languages [1998b, 905–6], does not include a single feminist study of Revelation.) Aune's positing of a Roma-Babylon connection in Revelation was also anticipated by Robert Beauvery (1983) and Adela Yarbro Collins (1984, 121), among others. Among more recent studies, see especially that of James Knight (2005). Elisabeth Schüssler Fiorenza, for her part, identifies the woman clothed with the sun (Rev 12), rather than Babylon, with Roma (e.g., 1991, 80; 2007, 139).

the establishment of the cult" (Mellor 1975, 51).[14] Romaia were also staged at Sardis and Pergamon. A cult of Roma is likewise attested for Thyatira and especially for Ephesus (to confine ourselves to the Asian cities mentioned in Revelation), issuing in an Ephesian temple dedicated to Roma and Divus Julius in 29 BCE and one dedicated to Roma and Augustus sometime before 6 BCE (Mellor 1975, 57–58, 138, 167; Price 1984, 43; Friesen 2001, 25–27).

In essence, the cult of Roma represented a solution to a problem. The problem for the Hellenistic cities in which the cult originated was that of coming to terms with an irresistible power that emanated from a place far distant from the city, yet extended deep within it, and was a permanent source both of potential devastation and essential benefaction. In all of this, that power was structurally similar to that of the traditional gods, and the cults of the gods thus became the logical model for the inhabitants of the cities for managing and placating that power, but also for representing it to themselves, in terms that were not imposed from outside but adapted from their own traditions.[15]

At base, the cult of Roma was an atropaic celebration of strength. Moreover, it was a celebration of the strength of a city whose very name, in Greek at least, *meant* strength: *Rhōmē*. The more of the Greek world this city named Strength took possession of, the more the inhabitants of that world must have marveled at the aptness of the name (see Erskine 1995, 370), as is indeed suggested by various homonymic flourishes in the Greek literature of the period. A first-century BCE geographical poem, for instance, notes that Rome "has a name equal to its power."[16] Aelius Aristides, in his (in)famous encomium of Rome from the mid-second century CE, takes full opportunity of the city's felicitous name, likening Rome to

---

14. Correspondingly, the temple of Roma appeared on coins minted in Smyrna "well into the third century of the Empire" (Mellor 1975, 51).

15. I am adapting S. R. F. Price's influential interpretation of the symbolic function of the Hellenistic ruler cults and the Roman imperial cult in the east (1984, 29–30, 43–44, 52). Price himself deals with the Roma cult only in passing. Unlike Mellor, however, he treats it as an expression of authentic religiosity (Price 1984, 24, 43)—not surprising in a book that amounts to a thoroughgoing deconstruction of the binary opposition between religion and politics endemic to earlier scholarship. For an extended critique of Mellor's *thea rhōmē*, read through the lens of Price's *Rituals and Power*, see Knight 2005, 107–16.

16. Pseudo-Scymnus, *The Circumnavigation of the Earth*, lines 231–32, cited in Erskine 1995, 372.

"a man who surpasses everybody else in size and strength [*rhōmē*]," and adding: "Thus, the name of the city is significant and what you see around you is nothing but *rhōmē*."[17] Plutarch, for his part, writing a little earlier, implies that the city's name has been a subject for etymological speculation: "Some say that the Pelasgians, after they had wandered over most of the earth and overpowered most men, settled there, and on account of their strength [*rhōmē*] in arms they named the city in this way."[18] So widely remarked was the *rhōmē*/"strength" homonym, apparently, that it even spilled over into Latin literature, notwithstanding the fact that *Roma*, the city's Latin name, carried no such double meaning. "Rome, your name is fated to rule the earth," declaims the Latin poet Tibullus; and Ovid, Horace, and Livy likewise play compulsively on the name.[19]

Strikingly, however—and highly relevant for the reading of Babylon in Revelation toward which we are inching—the deity created to symbolize the irresistible strength of Rome was not a god but a goddess, one whose very name was "Strength." The two earliest extant visual representations of *Thea Rhōmē* display that strength differently. A headless statue of her from the island of Delos, probably stemming from the late second century BCE, is unarmed and pacific; while she is armed and warlike on the earliest Greek coin that depicts her, a didrachma from Locri Epizephyrii in Magna Graecia from around 204 BCE.[20]

As Roma matures, however, and representations of her begin to proliferate, her warlike aspect predominates. Modeled on Athena, goddess of war and wisdom, but also on the Amazons, Roma typically appears in military dress (sometimes with bared breast, Amazon-style: see fig. 1), wearing a crested Roman helmet, and occasionally holding a spear[21] but more often clutching a parazonium, the ceremonial short sword or

---

17. Aelius Aristides, *Roman Oration*, cited in Erskine 1995, 374.
18. Plutarch, *Life of Romulus* 1, cited in Erskine 1995, 372. Conceivably, Rev 18:10 also plays on *Rhōmē*, although not homonymically: "Woe, woe, you great city, Babylon, you strong city [*hē polis hē ischyra*]!"
19. Tibullus, *Elegies* 2.5.51, cited in Erskine 1995, 378. See also Ovid, *Loves* 2.9a.17–18; Horace, *Epodes* 16.2; Livy, *History of Rome*, preface 4; cf. 30.44.8.
20. See Mellor 1975, 145–47, for further details of these representations.
21. As she does, for example, in the fragmentary relief of her found in Caesarea Maritima, which appears to date from the late first century BCE or the early first century CE. Her other hand holds a shield, and her head, which is damaged, was probably helmeted. Caesarea also contained a colossal cult statue of the goddess in the temple to Roma and Augustus that Herod the Great had built there (Josephus, *Jewish War*

Fig. 1. Cancelleria Reliefs, Frieze A, late first century CE. Left to right: The goddess Athena/Minerva, the emperor Domitian (resculpted as the emperor Nerva), the goddess Roma, and the *Genius Senatus* (representing the Roman senate). Photograph courtesy of the VRoma Project (www.vroma.org).

outsized dagger that was symbolic of military rank. The parazonium also tends to feature in representations of the Roman emperors and of Mars and Virtus, the latter being the divine personification of military valor and manly virtue, and a figure who fuses with Roma: they are frequently all but indistinguishable in their representations (more on which below). Typically, too, Roma holds a miniature Nike (victory personified) in her right hand, again on the model of Athena (see fig. 2).

All in all, however, Roma is a more unrelentingly martial figure than Athena, who, after all, was also the goddess of household arts and crafts such as spinning and weaving, and other unwarlike activities. As befits the tutelary deity of a city whose military conquests dwarf those of Athens, Roma has resolutely set such pacific pursuits aside. Again and again on coins minted under Nero, Galba, Vespasian, and Titus, Roma is seated,

---

1.21.7; *Antiquities* 15.9.6), although that statue, modeled on Hera rather than Athena, seems to have represented her as unarmed. For more, see Gersht 2001, 73-76.

warrior-style, on her own cuirass, or alternatively, on a mound of shields (see Vermeule 1959, 31–33, 41, 66, 87). An invincible warrior, she is triumphantly enthroned on the weapons of the armies she has vanquished. A late first-century-CE marble frieze from Cumae, the first Greek colony on the Italian mainland, pushes further into triumphal hyperbole: the goddess is not seated but standing, because the mound of war booty behind her is too vast to form a throne (fig. 3).

Roma's military prowess and irresistible might are also the principal subjects of the hymn by Melinno of Lesbos honoring her, a panegyric of five Sapphic stanzas.[22] "Hail, Roma, daughter of Ares," it begins, "warlike mistress with a girdle of gold [*chruseomitra daïphrōn anassa*]"—the belt worn by Amazons. The poem proceeds to pile up masculinizing appellations, a feature all the queerer for that fact that, not only is it addressed to a female figure, but it is, so far as we know, also the work of a female poet. "Fate has given you the royal glory of everlasting rule so that you may

Fig. 2. Gold aureus coin, 65–66 CE. Obverse: Bust of the emperor Nero. Reverse (shown): The goddess Roma seated on her cuirass with her right foot on a helmet, holding her parazonium in her left hand and holding in her outstretched right hand Nike, who extends a wreath toward her. Photograph courtesy of the VRoma Project (www.vroma.org).

---

22. The hymn is analyzed by Mellor (1975, 121–24) and Erskine (1995, 368–69). The older analysis by Bowra (1957) is also worth consulting. The translation of the hymn that I cite is that of Erskine.

Fig. 3. Late first-century CE marble frieze from Cumae depicting the goddess Roma, standing, with mound of war booty. Photograph courtesy of The VRoma Project (www.vroma.org).

govern with lordly might [*ophra koiraneōn echoisa kartos agemoneuēs*]," Melinno declaims. "With a sure hand you steer the cities of men [*su d'aspheleōs kybernas astea laōn*]." In the final stanza, the warrior queen abruptly becomes a mother, but even the maternal imagery is torqued to extol her innate masculinity: "Certainly, out of all people you alone bring forth the strongest men [*kratistous andras*], great warriors as they are, just as if producing the crop of Demeter from the land."[23]

Roma and Demeter/Ceres also constitute a couple—a butch/femme couple, to be precise—on the Ara Pacis Augustae, the magnificent altar consecrated in Rome in 9 BCE to celebrate the *Pax Augusta* putatively established through Augustus's victories in Hispania and Gaul.[24] Nowhere is Roma more suggestively gendered, indeed, than on that altar, where the panel devoted to her forms a pair with that devoted to Ceres, goddess of grain and fertility and paragon of motherhood, chastity, and domestic feminine virtue in general. Although the Roma panel survives only in frag-

---

23. Tellingly, the hymn survives only because the fifth-century CE compiler John Stobaeus saw fit to include it in the section of his *Anthology* titled *Peri Andreias*, "On (Manly) Courage."

24. For an empire-critical analysis of the altar, see Kraybill 2010, 57–59.

mentary form, it is clear that the goddess is enthroned once again on a pile of weapons, denoting her martial might, while Mother Ceres is enthroned on the nurturing earth that is her element. Both goddesses are productive of the peace that the Ara Pacis celebrates. But whereas Roma represents the peace procured through force of arms and military conquest, Ceres represents the peace produced through agricultural abundance and fecund harvests.[25] As represented on the Ara Pacis, Ceres emblematizes mythic femininity, while Roma emblematizes mythic masculinity.

The full hyperbolic dimensions of Roma's paradoxical masculinity, however, as represented in the stereotypical images of her that survive, may best be gauged by employing the gender scripts of Roman literary sources to illuminate them, most especially those scripts for the successful enactment and embodiment of Roman masculinity. Craig A. Williams's monograph *Roman Homosexuality* (1999), the real topic of which is encapsulated in its subtitle, *Ideologies of Masculinity in Classical Antiquity*,[26] remains the most ambitious distillation to date from Roman literary sources of the rules of masculine performance.[27] Much of the chapter titled "Effeminacy and Masculinity" may be read as incisive commentary on the images of Roma we have been considering—incisive and inadvertent, as Roma is not so much as mentioned in Williams's book.

The language of masculinity in the elite male Roman authors whom Williams is mainly mining "often invokes such notions as *imperium* ['dominion'] and *fortitudio* ['strength']" (2010, 139). Dominion and con-

---

25. My interpretation of the Ceres/Roma contrast on the Ara Pacis is indebted to Barbette Stanley Spaeth (1996, 144). For Ceres as emblem of Roman feminine virtue, see Spaeth 1996, 103–24 passim.

26. A subtitle that mysteriously disappeared from the second edition of the book, even though "the five chapters of the first edition" remain "unchanged in substance" in that second edition (Williams 2010, xv). The second edition does contain a rewritten introduction, an afterword and other additional concluding material, and a foreword by Martha Nussbaum.

27. There is relatively little that is original in Williams's excellent book. In effect it is an encyclopedic compendium of twenty years of classical scholarship on ancient Greek and Roman sex and gender—masculine gender especially but not exclusively—work pioneered by such scholars as Kenneth Dover, Paul Veyne, and Michel Foucault and extended and refined by such scholars as John Winkler, David Halperin, Amy Richlin, Eve Cantarella, Maud Gleason, Judith Hallett, and Marilyn Skinner. Significant related work that has appeared since Williams's study includes Halperin 2002; Skinner 2005; Sissa 2008; and Ormand 2009.

trol, both over others and over oneself, are identified by Williams (building on Foucault) as "the central imperative" of Roman masculinity (2010, 139; cf. Foucault 1985, 65–77 passim). What then are we to make of the goddess whose quintessential capability is *imperium* (as Melinno of Lesbos so effusively acknowledged) and whose very name is "Strength" if not the very embodiment of the central imperative of Roman masculinity—the imperative, the command, *to* command? *Imperium* permeated the entire social fabric of ancient Roman culture. It was the dominion that a free Roman man exercised over social inferiors, that military commanders exercised over lesser ranks, that magistrates exercised over those bound by Roman law, and that the Roman emperor exercised—in principle, anyway—over the entire inhabited world (cf. Williams 2010, 146). In the temple erected at Pergamon in the province of Asia, however, the supreme symbol of Roman *imperium* was not one but double: the male Augustus and the female Roma—just one index of the intense internal contradictions attending the hegemonic Roman ideology of masculinity, as we shall see.

Another vital term in the language of Roman masculinity was *virtus*. "Etymologically nothing more than 'manliness,'" as Williams notes, this term connotes valor, even virtue, in certain contexts, "but it is always implicitly gendered. *Virtus* is the ideal of masculine behavior that all men ought to embody, that some women have the good fortune of attaining, and that men derided as effeminate conspicuously fail to achieve" (2010, 139). Roma, however, is a woman who has not so much had the good fortune of attaining *virtus*, but is, as much as anything else, the very personification of *virtus*—so much so, indeed, that, as we noted earlier, the official personification of the virtue, the deity Virtus, is notoriously difficult to distinguish from Roma in many of their extant visual representations. Both figures, for example, favor an Amazonian pose, and merge into each other to such a degree that Cornelius Vermeule, author of the standard work on Roma iconography, is reduced to referring to a "Roma-Virtus type" (1959, 71).[28] Myles McDonnell, in his study *Roman Manliness*, explains how this (con)fusion came about: "So close was the identification of *virtus* with Rome that when *virtus* was honored with a state cult, the image chosen for

---

28. See also Vermeule 1959, 29 ("The standing figure of the Amazon Roma in short skirt and slipped tunic, so like the figure of Virtus…"), 41, 65–67, 83, 87–88, 96–97.

the cult statue was the same as that of the goddess Roma herself: an armed amazon" (2006, 2).

While *virtus* tends to gravitate to male bodies, it also floats free of them, and of anatomical sex as such. Seneca, for instance, states that women who attain to extraordinary displays of *virtus* ascend to the level of "great men [*magni viri*]" (Seneca, *Dialogues* 12.16.2, cited in Williams 2010, 145). Roma, although strictly speaking the epitome of inaction—she simply stands or, more often, sits or even lounges—is, arguably, the most fully realized example of that relatively rare but highly symptomatic Roman gender type, the woman who "overcomes" her femininity to act as befits a man. Valerius Maximus holds up for our amazement and admiration Porcia, daughter of Cato the Younger, who commits suicide by swallowing live coals, thereby imitating "the manly death of her father [*virilem patris exitum*] with a woman's spirit," and Lucretia who likewise commits virtuous suicide, thereby revealing that she has a "man's soul [*virilis animus*]" imprisoned in a woman's body (Valerius Maximus, *Memorable Doings and Sayings* 4.6.5 and 6.1.1, cited in Williams 2010, 145–46). Then there is the unnamed Jewish mother in 4 Maccabees, she of the seven martyred sons, and herself a paragon of masculinity according to the author ("More noble than men in perseverance and more manly than men in endurance [*andrōn pros hypomonēn andreiotera*]!" 15:30), and who likewise takes her own life.[29] It seems that the manly woman of the Greco-Roman male literary imagination is irresistibly impelled to unleash deadly force on her own female body, it being the primary inertial mass impeding her ascent to the masculine. Roma is different; the implied violence with which so many of her representations seethe is outward-directed.[30] She is a manly woman who is sanctioned to kill others besides herself.

Seneca styles *virtus* as antithetical to "women's vices" (*mulebria vitia*), namely, *impudicitia* ("unchastity," "sexual profligacy," "shamelessness"),[31] "weakness for jewels and riches," and "excessive pride in appearance" (Seneca, *Dialogues* 12.16.2, cited in Williams 2010, 366 n. 25). Tellingly, these are also the very vices in which Revelation's female personification of imperial Rome wallows (17:1–2, 4–5; 18:3, 7, 16). Babylon epitomizes fem-

---

29. In 4 Macc 16:14 she is "found to be stronger than a man [*dynatōtera ... andros*]," and her manliness is further extolled in 15:23 (cf. 2 Macc 7:21). For an analysis of this motif, see Moore and Anderson 1998, especially 265–72.

30. Although the situation is also more complex than this, as we shall see.

31. For *impudicitia* as "shamelessness," see Ginsburg 2006, 128.

inine vice even as Roma epitomizes masculine virtue. Each would seem to be the antitype of the other, a relationship to which we shall later return.

Roma's acute queerness emerges into even sharper relief when she is contrasted with the female personifications of other nations found in Roman imperial iconography. The temple complex known as the Sebasteion, for instance, located at Aphrodisias in Asia Minor, contains a relief representing Britannia as a half-naked, unarmed female, arm outflung, held down and menaced by an armed Roman warrior representing the emperor Claudius (fig. 4); while certain of the famous Judea Capta coins minted to celebrate the Roman suppression of the Judean revolt represent Judea as an abject female captive, seated, unarmed, and in mourning, above whom towers the threatening figure of an armed Roman soldier, possibly representing the emperor Vespasian, whose weaponry includes the parazonium. Contrast the representations of Roma we have been considering, in which the female body has migrated into the military armor and thereby been transformed from passive, abject object to active, virile agent.

Also notable is the *Gemma Augustea*, the exquisite low-relief sardonyx cameo fashioned for Augustus, which depicts two scenes, one in the upper tier of the work, the other in the lower tier (fig. 5). In the lower tableau, a company of Roman soldiers is triumphing over a cluster of cowering, defeated barbarians, including a captive woman who is being ominously dragged into the scene by her hair. The other prominent woman on the cameo presents a rather different spectacle. *Dea Roma* is the central figure in the upper-tier tableau, in her customary helmet, one hand gripping a spear, the other caressing the hilt of her sword, her feet resting on the armor of her conquered foes. She is seated serenely at Augustus's right hand, and her company also includes other members of the Roman male superelite, Tiberius alighting from a chariot, with Germanicus, most likely, standing next to it (Galinsky 1998, 120–21), although only Roma is conspicuously armed. In short, the contrast between the woman who customarily represents Rome in imperial iconography and the women who customarily represent barbarian nations in it could not be starker or more striking.[32]

---

32. For incisive discussion of the Claudius-Britannia relief, the Judea Capta coinage, the *Gemma Augustea*, and other related images, see Lopez 2008, 26-55. Lopez does not factor Roma, however, into her analyses of Roman imperial iconography.

Fig. 4. Mid-first-century CE relief from the Sebasteion at Aphrodisias depicting the emperor Claudius subduing Britannia. Photograph © Steve Kershaw, licensed under Creative Commons Attribution NonCommercial ShareAlike terms.

Fig. 5. The *Gemma Augustea*, an early first-century CE sardonyx cameo gem depicting the goddess Roma enthroned with the emperor Augustus in the upper tableau and Roman soldiers erecting a *trophaeum/tropaion* (victory trophy) in the lower tableau. Photograph courtesy of Kunsthistorisches Museum, Vienna.

Virtus is a grammatically feminine noun—grammatically feminine but, as Williams argues, ideologically masculine: "Grammatical gender yields, of course, to the overarching imperative of masculine ideology" (1999, 134).[33] *Roma/Rhōmē*, too, is grammatically feminine, yet masculine

---

33. McDonnell, however, cautions that *virtus* is still more elusive of univocal definition: "Yet *virtus* is a notoriously difficult word to translate. As in most cultures, in ancient Rome the term for manliness had a number of different denotations. Yet it is striking that a word whose etymological connection to the Latin word for man is so apparent, can be attributed not only to women, but to deities, animals, abstract ideas, and inanimate objects. As a purely linguistic phenomenon this is noteworthy, but

on the level of ideology, or, alternatively, rhetoric. What is the relationship between grammar and rhetoric? One of tension, interruption, disruption, and subversion is the answer that has echoed forth most forcefully from poststructuralist literary criticism,[34] and it would seem fully borne out by the contradictory figure of Roma. The word *Roma/Rhōmē* itself, grammatically feminine but rhetorically, ideologically, and conceptually masculine, might be said to be iconic of the very process that the image of Roma enjoins: the sublation of femininity in masculinity. But that process is always necessarily incomplete, according to the hegemonic Roman gender scripts. Roman masculinity is always tenuous, fragile, fluid, always threatened, always incompletely achieved, ever under siege, ever liable to lose its footing on the greased gender gradient sloping precipitously down to femininity and hence irrevocable shame, irredeemable disgrace (see Gleason 1995, 59, 81; Skinner 2005, 12, 248, 254; Williams 2010, 155–56). Roma, then, is a figure in perpetual deconstruction. She holds the terms "femininity" and "masculinity" in constant, warring tension (no wonder she is so heavily armed), without ever reconciling them, without ever merging them into a harmonious synthesis. Each term perpetually threatens the other; each is the always unrealized negation of the other.

A prominent theme in ancient Roman literature "is that true Roman men, who possess *virtus* by birthright," are destined to "exercise their dominion not only over women … but also over foreigners, themselves implicitly likened to women," less than fully masculine, relative to the putative hypermasculinity of free Roman males (Williams 2010, 148). Greece, in particular, epitomized the "feminine" softness that contrasted stereotypically with the "masculine" hardness of Rome (2010, 148–49). Yet Greece was not the softest of the soft. Rather, argues Williams, it was the cities of Asia Minor that "seem to have represented to Romans the ultimate in decadence and luxury and consequently softness and effeminacy" (2010, 149). Sallust, for instance, laments how Sulla's army on entering Asia immediately began to lose its hardness, the soldiers' "fierce minds"

---

since *virtus* was regarded by the Romans as a preeminent social and political value, its wide and sometimes odd semantic range has implications that go beyond philological significance" (2006, 3–4).

34. For the classic statement, see de Man 1979, esp. 3–19. "To the extent that a text is grammatical," writes de Man, "it is a logical code or machine" (1979, 268). "Rhetoric," however, "radically suspends logic and opens up vertiginous possibilities of referential aberration" (1979, 10).

being "softened [*molliverant*] … in an environment of leisure [*voluptaria*]"; while Valerius Maximus declaims of the Spartan general Pausanias, "As soon as he adopted Asian ways, he was not ashamed to soften [*mollire*] his own fortitude with the effeminate [*effeminate*] Asian style."[35] Such stereotypes also find expression in Virgil and Cicero: in the *Aeneid*, the Trojans, themselves Asians, are labeled *semiviri*, "half-men," soft and effeminate, by Italian warriors and other enemies; while Cicero attributes to Cato the Younger the claim that the Romans in their war with the Asian King Mithridates had fought "with women" (*cum mulierculis*).[36] These are only secondarily gender stereotypes, however; first and foremost it is ethnic stereotyping that is being enacted in these texts. But that dynamic is already implicit in the concept of "*Roman* masculinity": ethnicity, no less than gender, needs its constitutive others in order to be performed and constructed.[37] How might the figure of Roma be situated in relation to these stereotypes?

Let us reframe this complex question. According to the hegemonic Roman gender script, masculinity was the quality of being in control of, exercising dominion over, others and also oneself, while femininity was the quality of ceding control of oneself to others. The notion of a man submitting to the domination of a woman is, therefore, according to the script, a "conceptual anomaly," a "self-contradictory impossibility" (Williams, 1999, 137). What, then, are we to make of *Rhōmē/Roma*, a female whose name is "Strength" and who is the very emblem of masculine *imperium*? What does it mean that *this* is the image of imperial Rome that the provinces choose to reflect back to the metropolis? What is actually being said here?

Like any complex image, Roma admits of multiple interpretations. A female body overlaid with the trappings of Roman military discipline,[38]

---

35. Sallust, *Conspiracy of Catiline* 11.5–6; and Valerius Maximus, *Memorable Doings and Sayings*, 2.6.1, cited in Williams 2010, 149.

36. Virgil, *Aeneid* 4.215–217, 9.598–620, 12.97–100; and Cicero, *On Behalf of Lucius Murena* 31, cited in Williams 2010, 149–50.

37. To the extent that the constitutive other in this instance is Asian, these elite Roman authors can be said to be engaged in a proto-Orientalizing discourse. Edward Said's primary focus in *Orientalism*, his classic postcolonial study, is the modern discursive construction of "the Orient" by the West. Said identifies the cultural stereotype, however, as a key Orientalizing strategy (e.g., 1978, 26, 152, 285, 321). Further on the ancient Roman stereotyping of Asians specifically, see Isaac 2006, 61–64, 70–72; cf. 39–41.

38. Williams writes of "that ultimate bulwark of Roman masculinity, military dis-

Roma may be read as a triumphant celebration of a masculinity that constructs itself through the unceasing suppression of femininity. As such, Roma may be read as a visual allegory of hegemonic Roman gender ideology. Roma's iconography may be interpreted to say—indeed, to trumpet incessantly—that masculinity is the defeat of femininity. Roma's armor and weapons would then be designed not only to ward off the external threat, the threat from without, but also to ward off the internal threat, the threat from within. Roma would represent primal femininity perpetually in the process of mutating into masculinity. In the hegemonic Roman gender script, femininity is the given, the a priori, the default state, while masculinity is what must be achieved and maintained. It is the hard-won product of (self-)conquest. Thus interpreted, Rome's tutelary deity does more than guard the city of Rome and the provincial cities that belong to Rome; no less jealously, she guards the sex-gender ideology of Rome. Roman masculinity is brittle, beleaguered, besieged; it demands constant vigilance. Is this why Roma is regularly depicted in military garb? Is she—or, in essence, *he*—armed first and foremost against himself?

More subtle readings are, of course, possible. Roma, for centuries a resident of Asia before taking up her abode in the metropolis,[39] might also be read as a slavish representation of the "proper" relations between Asian softness and Roman hardness: Roma is the overcoming of that unmanly softness through military discipline. Roma would then represent Asian internalization of, and acquiescence to, the Roman ethnic stereotype of Asian males. But interpretations less affirming of Roman superiority might also be ventured. Roma—female but not feminine, masculine but female—might instead be read as saying that the Roman ideology of masculinity is a self-contradictory and self-subverting impossibility. Or it might even be read as a satiric assertion that Roman masculinity is always threatening to shrivel back into the femininity on which it is erected, and which is always showing through its armor, Roma thus dismantling the

---

cipline" (2010, 142). And again: "Military discipline, pertinacity, endurance, and bravery in the face of death are all coded as masculine, and their absence as effeminate" (2010, 152; a substantiating quotation from Livy [*History of Rome* 5.6.4–5] follows). Skinner remarks: "Rome was a warrior society in which military prowess remained vital to the notion of masculinity even after the age of the citizen-soldier was long past" (2005, 208).

39. She did not acquire her first temple in the city of Rome until 137 CE (more on which below).

hard/soft, Roman/Asian dichotomy on which the denigrating stereotype depends. And, of course, it might also be read as saying all these things at once, in a cacophonous, self-contradictory babble. If Roma is always already positioned on the slippery slope that leads to Babel and, indeed, Babylon, then all that John of Revelation needs to do is give her a brisk shove. Which, as we are about to see, is precisely what he does.

The Self-Deconstruction of Roman Gender

Roma receives rough treatment in Revelation. When we first encounter her, she has been stripped of her armor and decked out as a prostitute: "The woman was clothed in purple and scarlet and bedecked with gold, jewels, and pearls, with a golden cup in her hand full of obscenities [*bdelygmatōn*] and the filth of her fornication [*ta akatharta tēs porneias autēs*]; and inscribed on her forehead was a mysterious name: 'Babylon the Great, Mother of Whores and Earth's Obscenities [*hē mētēr tōn pornōn kai tōn bdelygmatōn tēs gēs*]'" (17:4–5).[40] To restate the argument of the previous section in slightly different terms, the goddess Roma is Roman imperial patriarchy paradoxically embodied as a woman in the trappings of an invincible warrior. In other words, Roma is hegemonic Roman manhood encased in female flesh that is clad in hypermasculine garb. To simplify, Roma is a man dressed as a woman dressed as a man. Babylon in the passage just quoted, then—Roma stripped of her military attire and reclothed as a prostitute—would be a man dressed as a woman dressed as a man dressed as a woman. In other words, Babylon would be Rome in triple drag.[41] It is high time we called in Judith Butler.

---

40. See DeSilva 2009, 108: "John tells a very different story about the mission and the destiny of *Roma Aeterna* from the one heard in Virgil's *Aeneid* or Plutarch's *On the Fortune of Rome*.... John presents the goddess *Roma* in very different dress. No longer wearing the modest toga of the goddess worshiped in the temple of Rome and the *Augusti*, nor even the ascetic military garb in which she appears on the reverse of the famous *Dea Roma* coin, she is dressed in her evening wear, sporting the lingerie of a self-employed courtesan."

41. Here I am complicating Catherine Keller's suggestive formulation, "we may behold the Whore of Babylon as a great 'queen' indeed: imperial patriarchy in drag" (1996, 77). More single-mindedly queer reading of Revelation has included Pippin and Clark 2006 and Huber 2011. See in addition Clark 1999; Pippin 1999, 117–25; Huber 2008; and Runions 2008a. For the intersection of queer studies and postcolo-

6. RAPING ROME

In what remains of this essay, Butler's *Gender Trouble* (1990), the inadvertent charter document of queer theory,[42] will play a yet more inadvertent role: it will be read as commentary on the figure of Babylon in Revelation. *Gender Trouble* famously argues that gender identity is purely performative, the product of a socially scripted set of stylized actions, which combine and conspire, through sheer repetition, to generate retroactively the illusion that gender is natural and innate, an inner essence, merely "expressed" by the speech, gestures, and other behaviors that in fact produce or en-gender it (Butler 1990, esp. 134–41).[43] Famously, too, the book singles out drag as illustrative and exemplary of gender performativity: "The performance of drag plays upon the distinction between the anatomy of the performer and the gender that is being performed" (Butler 1990, 137). Drag thereby decouples gender from anatomical sex, and thus decoupled, gender becomes a vertiginous performance. Butler cites Esther Newton's *Mother Camp*, a text that partially anticipates, and partially catalyzes, her argument.

> At its most complex, [drag] is a double inversion that says, "appearance is an illusion." Drag says "my 'outside' appearance is feminine, but my essence 'inside' [the body] is masculine." At the same time it symbolizes the opposite inversion; "my appearance 'outside' [my body, my gender] is masculine but my essence 'inside' [myself] is feminine." (Newton 1971, 103, quoted in Butler 1990, 137)[44]

---

nial studies, see Campbell 2000; Hawley 2001a; 2001b; Romanow 2006; and Aydemir 2011; and for their intersection in biblical studies, see Punt 2008 and Moore 2013.

42. In an interview Butler confessed: "I remember sitting next to someone at a dinner party, and he said that he was working on queer theory. And I said: What's queer theory? He looked at me like I was crazy" (Butler, Osborne, and Segal 1994, 32).

43. The theory undergoes incremental elaboration and refinement in Butler's subsequent work, partly in response to criticism; see especially her *Bodies That Matter* (1993); her preface to the "tenth anniversary edition" of *Gender Trouble* (1999); and her *Undoing Gender* (2004).

44. The bracketed insertions in the quotation are Butler's own. Butler would later nuance the role of drag in her gender theory as follows: "*Drag* was a way of exemplifying how reality-effects can be plausibly produced through reiterated performances, but it was never meant to be the primary example or norm for gender subversion. Of course, it makes sense that it was taken up that way, but it has never had that particular place for me. It was meant to elucidate a structure that is at work in everyday performances of gender, and so make this reiterative production of reality-effects legible as a repeated practice in so-called ordinary social life" (2006, 282).

All of which brings us back to Roma and Babylon. In effect, Roma is announcing: "My outward appearance may be female but my inner essence is masculine. This I represent to you by adopting the attire not just of a man but of that most manly of men, a Roman warrior." Revelation, however, replies to Roma and hence to Rome: "Your adopted appearance may be masculine but your inner essence is feminine, indeed slavish. This I represent to you by stripping you of your masculine disguise and dressing you more fittingly, not just as a woman but as the epitome of fallen femininity. For inside you are nothing but a *pornē*, a whore, a brothel slave."

If the goddess Roma, then, is the implicit embodiment of one particular answer to the question of how Rome obtained an empire—through sheer military superiority—the prostitute Babylon is the implicit embodiment of an altogether different answer. Paramount for Revelation—or so it would seem—is not Rome's military might but the seductiveness of her culture. Revelation's stance on Christian participation in the civic life of Roman Asia and accommodation to its cultural norms, not least those associated with the imperial cult, is sternly antiassimilationist ("Come out of her, my people," 18:4).[45] This is why Rome—or, more precisely, Roma—is represented as a whore in Revelation: what better embodiment, for the sex-queasy seer, of repellent seductiveness?

John himself is not immune to Rome's seductions, as it turns out. Paradoxically, however, it is not her civic culture that seduces him but rather her military might. Essentially, Revelation's messianic empire ("The world empire [*hē basileia tou kosmou*] has become the empire of our Lord and his Messiah," 11:15) will be established by the same means through which the Roman Empire was established: war and conquest, entailing, as always, mass slaughter, but now on a hyperbolic scale.[46] All of which is to say that John is secretly in love with Roma, inaccessibly resplendent in her irresistible armored might, triumphantly enthroned on the mountain of war booty confiscated from the countless warriors she has slain. He loves her and hates her with equal passion. Which is why he deals so savagely with her.

As we saw, Roma is Rome in double drag: phallic masculinity masquerading as female flesh masquerading as hegemonic masculinity. And Babylon is Rome in triple drag: phallic masculinity figured as female and

---

45. See 32–34 above.
46. See 55–56 (left-hand column) and 82 above.

clothed as virtuous and victorious warrior, then reclothed as degenerate and defeated brothel slave. In Revelation, Roma is stripped of her hypermasculine attire and decked out as a whore. No longer the emblem of self-autonomy, she no longer belongs only to herself. No longer is her body a consummate instrument of control; now it is at the disposal of others. But this is only the first phase of Roma's two-stage ritual shaming in Revelation.

Roma is violently stripped of her manly garb and dressed as a prostitute—but only in order to be violently stripped once again, sexually shamed, and physically annihilated: "they will loathe the whore, and they will ravage her and strip her naked, and they will devour her flesh and burn her with fire" (*houtoi misēsousin tēn pornēn, kai erēmōnenēn poiēsousin autēn kai gymnēn, kai tas sarkas autēs phagontai kai autēn katakausousin en pyri*, 17:16).[47] "Gender is a performance with clearly punitive consequences," as Butler notes. "We regularly punish those who fail to do their gender right" (1990, 140). For what, then, is Babylon being punished? For being a failed man, or a failed woman, or a fallen woman, or a man fallen into femininity? The more fundamental question might be, why are sex and gender contiguous with empire, or intimately associated with it, in Revelation in the first place? In other words, why are sex and gender metonymies of empire in this text? Roma is only part of the answer, as we are about to see.

Of what is Roman sex, as represented in the overwhelming majority of the extant literary sources, quintessentially expressive? *Of social hierarchy* is the answer resoundingly offered by classicists in recent decades.[48] But what is the ultimate manifestation of social hierarchy in the Roman world? *The empire* would seem to be the answer (which is, of course, to say the Roman world itself; a tautologous answer, then, but accurate nonetheless). The Roman Empire would then be that to which Roman sex, at least

---

47. For a précis of prior scholarship on this verse, see 122 n. 42 above.

48. Williams, for instance, in the course of surveying recent scholarship on Roman sex, claims that "the relationships among phallic penetration, social relations, and whatever we mean by 'power' overwhelmingly tends to be constructed in the Latin textual tradition [as in the Greek textual tradition] in such a way as to align the distribution of penetrative roles with various social hierarchies"—although Williams is emphatic that textual exceptions to that rule also occur (2010, 261). Skinner contends: "Although the 'penetration model' does not cover the full spectrum of sexual acts performed in … Roman texts, a hierarchy of dominance and submission is assumed even in circumstances where participants deviate from the norm" (2013, 256).

as construed by elite Roman authors, ultimately gestures. It is probably no accident, therefore, that Revelation's denunciation of imperial Rome is simultaneously a denunciation of Roman sex and sensuality. But this double denunciation misses both of its marks, and for the same reason each time. While overtly resisting the Roman Empire, John covertly replicates it, constructing an empire of God on its model, as I have previously argued.[49] While overtly resisting the Roman sex/gender system, John also covertly replicates it, as I shall now argue.

In *Gender Trouble*, Butler sought to decouple gender from anatomical sex—to deprive gender, both masculine and feminine, of its "natural," "self-evident" basis in anatomical sexual difference—in the conviction that such decoupling would call into question the construction of sex and gender in hierarchical binary terms, of "man" over "woman."[50] That such decoupling does not, however, lead inevitably to such interrogation[51] is suggested by the book of Revelation, a particularly inflamed symptom, it seems to me, of the simultaneous rigidity and fluidity that appears to have characterized Roman sex/gender ideologies more generally.

First, the rigidity. Sexual violence in Revelation appears to be an affirmation of gender hierarchy. Aggression is turned inward as well as outward: outward toward Roman civic culture, epitomized by the imperial cult, but inward toward rival Christian communities willing to accommodate to that culture. "Jezebel" is the name for Christian assimilationism in Revelation (cf. Duff 2001, 51–60), just as "Babylon" is the name for what must not, under any circumstances, be assimilated, what must be rejected by being abjected. What this means is that inward-directed aggression in Revelation finds symbolic expression in imagined sexual violence directed against the female—that is to say, against Jezebel ("she doesn't wish to repent of her whoring [*tēs porneias autēs*]. Look, I am throwing her onto a bed [*ballō autēn eis klinēn*]," 2:21–22)[52]—just as outward-directed aggres-

---

49. See 28–31 and 35–37 above.

50. "I asked, what configuration of power constructs ... that binary relation between 'men' and 'women,' and the internal stability of those terms?" (Butler 1990, viii).

51. A position from which Butler herself subsequently retreats; for example: "there is no guarantee that exposing the naturalized status of heterosexuality will lead to its subversion" (1993, 231).

52. The notion that the expression "throw on a bed" is a Hebraism meaning "cause to become ill" has been recycled by scholars since R. H. Charles's commentary (1920a, 71–72) at least (cf., e.g., Swete 1906, 44), but the tide has begun to turn (e.g., Duff 2001, 160 n. 17, 165 n. 53; Frilingos 2004, 109; Conway 2008, 162; Marshall

sion also finds symbolic expression in imagined sexual violence directed against the female—that is to say, against Babylon ("they will loathe the whore, and they will ravage her and strip her naked," 17:16). The female is everywhere the object of sexual violence in Revelation, then—except where she assumes the patriarchally preapproved forms: virgin bride (19:6–8; 21:2, 9; 22:17) or self-sacrificing mother (12:1–6, 13–17).[53]

Now the rigidity with the fluidity. Danger in Revelation, as we have seen, is consistently figured as feminine—a "defiling otherness," as Butler might say (1990, 133). The consolidation of the masculine through the exclusion of the feminine is a preoccupation of Revelation; consider the company of the redeemed: "These are they who have not defiled themselves with women [*meta gynaikōn ouk emolynthēsan*], for they are virgins" (14:4). The derisive transmutation of masculinity into femininity is another preoccupation; look no further than imperial Rome as "Babylon the Great, Mother of Whores" (17:5). Gender in Revelation is at once utterly fluid and mobile and absolutely rigid and inflexible. It is fluid because masculinity and femininity are always in process, forever threatening to leak into each other. But it is rigid because masculinity is always flowing down from above the gender bar, transforming into femininity as it does so, while femininity is always flowing up from beneath the bar, becoming masculinity in the process. In other words, what is not explicitly called into question in Revelation is the hierarchical gender binary itself that sets masculinity over femininity, man over woman. But does the hierarchical opposition actually remain intact—can it possibly remain intact—in the midst of all this movement?

Paradoxically, it is in the representation of its protagonist, its hero, that Revelation troubles the gender binary most thoroughly. For what an unlikely leading man he turns out to be! In his first appearance in Revelation he comes replete with a pair of female breasts. "Girt about the paps with a golden girdle" is how the King James translators rendered *periezōsmenon pros tois mastois zōnēn chrysan* (1:13). Yet it is not the bra-like girdle that conceals this anatomical anomaly from contemporary audiences so much

---

2009, 21–22, 30; Thimmes 2009, 74). Punitive sexual violence seems implied by the fact that the bed in this instance is flanked by fornication on one side ("she doesn't wish to repent of her whoring/fornication") and adultery on the other ("and those who commit adultery with her"). Jezebel's sexual shaming would then mirror that of Babylon in 17:16.

53. As numerous feminist commentators on the book have noted.

as the normalizing tendencies of the translators. The KJV better captures the sense of the verse than most modern English translations, which nervously cover up the breasts, turning them into a manly chest.[54]

Paradoxically, John terms this androgynous being "one like a son of man" (*homoion huion anthrōpou*, 1:13). In accordance with standard ancient Greek usage, *anthrōpos* here makes man the generic, standard-human stand-in for humans of both sexes. For understandable and entirely commendable reasons, modern inclusive translations of *huios anthrōpou*, here and throughout the New Testament, airbrush this usage, rendering the expression as "son of humanity," or, more inclusively still, "the Human One."[55] Yet the contiguity of the *anthrōpos* and the *mastoi* in the Greek of Revelation 1:13 fissures naturalizing or normalizing notions of the human, whether ancient or contemporary. Butler's *Undoing Gender* (2004) will help us undo the gendered knots that hold this mammiferous son of man together. The question, what counts as a person? is culturally bound up with the question, what counts as coherent gender? as Butler (2004, 58) notes. Channeling Foucault she asks, "what happens when I begin to become that for which there is no place within the given regime of truth?" (2004, 58). She continues:

---

54. Even the New King James Version renders *mastoi* as "chest," as do the New Revised Standard Version, the New International Version, the Good News Bible, and other translations too numerous to name. Exceptions (all older) include the Darby Bible, the American Standard Version (both translate the phrase as "girt about at the breasts with a golden girdle," following the Revised Version), and the Bible in Basic English ("with a band of gold round his breasts"). The Revised Standard Version opts for the singular (and safer) "breast." The KJV inherited "paps" from Tyndale (1534) and the Bishop's Bible (1595). "Paps" had also been used in Douay-Rheims (1582), while Wycliffe (1395) had opted for "teats." Much earlier, the Vulgate had rendered *mastoi* as *mammillae* (*praecinctum ad mamillas zonam auream*). Early commentators on Revelation, such as Caesarius of Arles (*Exposition on the Apocalypse* 1.13, homily 1) and Andrew of Caesarea (*Commentary on the Apocalypse* 1.13), also understood the Christ of Rev 1:12–16 to be possessed of female breasts. Jesse Rainbow (2007) argues that Jesus owes these breasts to those of the male lover in the Septuagintal version of the Song of Songs on whom John has partially modeled him. "Your breasts [*mastoi sou*] are better than wine," the female lover tells the male lover of the Song (1:2 LXX).

55. "The Human One" is the Common English Bible translation of *ho huios tou anthrōpou* throughout the New Testament. The Inclusive Bible renders *homoion huion anthrōpou* in Rev 1:13 as "a figure of human appearance."

This relationship, between intelligibility and the human ... carries a certain theoretical urgency, precisely at those points where the human is encountered at the limits of intelligibility itself. I would like to suggest that this interrogation has something important to do with justice. Justice is not only or exclusively a matter of how persons are treated.... It also concerns consequential decisions about what a person is, and what social norms must be honored and expressed for "personhood" to become allocated, how we do or do not recognize animate others as persons depending on whether or not we recognize a certain norm manifested in and by the body of that other. (2004, 58)

What might all or any of this have to do with Revelation's inaugural vision? At the limits of intelligibility itself, John encounters an animate other whom he does and does not deem human: he sees one *like* a son of man, akin to a human being. Is it entirely incidental that this figure that flickers vertiginously at the edge of intelligibility is manlike yet woman-breasted in form? Or does gender undecidability here function as a metonym for ontological undecidability? Jesus of Nazareth, spectacularly risen from dehumanizing death on a Roman cross ("I was dead, and see, I am alive forever and ever," Rev 1:18), has begun to become that for which there is no place within the prevailing regime of truth, first of all the Roman regime, and he has received an inassimilable body to match that inassimilable position. The outmaneuvering of the regime staged in and through the reanimation and divinization of this crucified peasant nonentity has something momentous to do with justice. It is not so much that Jesus of Nazareth regains the personhood of which he was stripped through his torture and execution as that he becomes the primal instance of a reenvisioned personhood, an as yet barely apprehensible personhood ("one *like* a human being"). He—or, rather, s/he—becomes a person manifested in and though a body that disrupts the binary gender logic, the "coherent" gender system that hitherto has been the ground of "coherent" personhood. The intersexed body of Jesus Christ in Revelation's inaugural vision may be read a call to consequential decisions about what a person is or may be, and hence a call for justice of the most fundamental kind—but a call that Revelation itself fails to heed consistently, as we are about to see.

If Rev 1:12–16 is the epiphany of the risen Christ as celestial androgyn, Rev 19:11–21 is his epiphany as celestial superwarrior. This gendered apotheosis is accomplished through violence on a spectacular scale (19:17–

21). What manner of foe must be annihilated in order that the apotheosis be achieved? It is a foe that is simultaneously dehumanized, because bestialized—become *the* beast, indeed—and feminized—become the "Mother of Whores"—to the extent that dehumanization and feminization can be cleanly distinguished within Revelation's symbolic universe. One cannot say where the beast ends and the woman, Babylon, begins, since they each symbolize Rome. If *Thea Rhōmē* represents Rome as the attainment of phallic masculinity through the overcoming of primal femininity (one possible interpretation of the Roma iconography, as we saw earlier), the whore in combination with the beast represents Rome as the collapse of masculinity back into the morass of femininity and animality.

"The mark of gender appears to 'qualify' bodies as human bodies," writes Butler. "Those bodily figures who do not fit into either gender fall outside the human, indeed, constitute the domain of the dehumanized and the abject against which the human itself is constituted" (1990, 111). Yet it is not only imperial Rome, Revelation's villain, that is figured in terms that elide the distinction between the animal and the human. Revelation's hero, its messianic protagonist, also slips in and out of humanity as the narrative unfolds. Stripping off his/her ankle-length gown and golden bra, s/he blurs the boundaries not only between male and female, masculine and feminine, but also between human and animal; for if in her first epiphany in the book s/he comes with a pair of mammary glands, in his second epiphany s/he has become a four-footed mammal: "a lamb [*arnion*] standing as though it had been slaughtered" (5:6).[56] In the hyperqueered figure of Jesus in Revelation, all hierarchical binaries dissolve—not only male over female and masculine over feminine but even human over animal. They are digested in his ruminant, ovine stomach and leak as milk from her human, female breasts.

"Gender is an 'act,'" writes Butler, "that is open to splittings, self-parody, self-criticism" (1990, 147). It is the enactment of gender in this self-subverting sense that we have been examining in the book of Revelation. Revelation is a Roman book (see Frilingos 2004, 78). As much as anything else, it is a Roman book on gender. But it is not a book on, or of, hegemonic Roman gender ideology in any pure or simple sense. Rather,

---

56. This theriomorphic epiphany is the principal theme of the final two essays in this volume.

Revelation is a limit-discourse of Roman gender. It is the self-deconstruction of Roman gender.

Epilogue

Earlier I pondered the profound irony that, in the fourth century, Revelation's beast began to mutate into the one monstrosity that the seer seemed incapable of imagining: an empire that was Roman and Christian at one and the same time.[57] And what befell *Dea Roma*, so ill-used by John, even as Jesus became a new Romulus, the founder of a new Rome? The cult of Roma was a provincial creation, as noted above. Provincial emperors were late in appearing, but two of them, Trajan and especially Hadrian, exhibited a special devotion to *Dea Roma*. It was Hadrian who was instrumental in installing her cult in the heart of the metropolis, in the Forum Romanum itself. Erection of the temple of Venus Felix and Roma Aeterna began in 121 CE, and it was dedicated by Hadrian in 135. By all accounts, the temple was the largest and most magnificent the city had ever seen (Curran 2000, 57), and it went on to become the most enduring of its pagan monuments. Under Hadrian and his successors, *Dea Roma* became the preeminent divinity of Roman state religion, even the imperial cult being subordinated to her worship (Mellor 1975, 201; Hope 2000, 81). As Roma Aeterna she remained the primary symbol of the Roman Empire even after the Constantinian watershed and the movement of effective government to Constantinople in 330 CE. She was adopted by Christianity and survived through the Middle Ages and into the Renaissance, becoming the emblem of European civilization itself (Mellor 1975, 202; Rowland 2013).

Roma became Christian, then, even as the Roman Empire morphed into Christendom. But although the former no less than the latter eventuality was one that the seer of Revelation could not have foreseen, the text that he authored seems to adumbrate this very development, all unbeknown to him. Revelation's messianic protagonist appears in anthropomorphic form at the beginning and again at the end of the text; otherwise he trots through the text as a lamb almost without interruption.[58] There are clear points of connection between Jesus' inaugural and climactic

---

57. See 35 above.
58. He features as an infant in 12:5, 13, and, possibly, as "one like a son of man" in 14:14–16 (cf. 1:13), unless the latter figure is an angel (see 14:17–20).

epiphanies, notably, the eyes like a flame of fire (1:14; 2:18; 19:12), the sharp sword issuing from his mouth (1:16; 2:12; 19:15), the rod of iron that is his to bestow (2:27) and with which he rules the nations (19:15; cf. 12:5), and the appellation "faithful and true" (3:14; 19:11). Two other signal features of this singular being, as we have seen, are that he has female breasts (1:13) and that he is an invincible warrior who decimates entire armies single-handedly (19:11–21). He resembles the figure of Roma, in other words. So while Revelation is busy shaming Roma by turning her into a prostitute, on the one hand, Revelation is also busy modeling Jesus on Roma, on the other hand, the one hand not knowing what the other is doing.

Just as the figure of Roma may be read, as we saw, as the celebration of a masculinity that constructs itself through the incessant suppression of femininity, as suggested by Roma's own female physique encased in armor and surrounded by evidence of her incomparable military might, so too can the Jesus of Revelation be read as the celebration of a masculinity that constructs itself through the incessant suppression of femininity (those telltale breasts again)—and of animality. For the androgynous Jesus does not remain bipedal, as we noted. Before long he is walking on all fours, having slid over the edge of the Roman gender gradient altogether and plunged beyond femininity into animality. As such, Revelation's Jesus, no less than *Dea Roma*, may be read as an allegory of hegemonic Roman gender ideology, one equally bristling with debilitating contradictions. Roma and John's Jesus are, to an extent, interchangeable figures. Which is why Roma is also capable of resurrection. She survives the ritual annihilation of the whore that John has attempted to turn her into ("they will ravage her and strip her naked, and they will devour her flesh and burn her with fire," 17:16) and walks off into the sunset of the Roman Empire, arm in arm with Jesus, become, like him, an emblem of Christian empire.

7
Retching on Rome*

This essay was the last in this collection to be written. It is fueled by affect theory (see Clough and Halley 2007; Gregg and Seigworth 2010; Blackman 2012), a seething interdisciplinary field that epitomizes post-poststructuralism (if I may be permitted that barbarism). Affect theory emerged in no small part in reaction to, and from dissatisfaction with, the classic poststructuralist preoccupation with language and, by extension, with cognition. Affect theory is obsessed instead with, well, affect, most often understood as the raw, preprocessed sensory encounter with the world prior to conscious cognition, linguistic representation, or even identifiable emotion. Admittedly, this fixation does not translate easily into literary, including biblical, interpretation. Fortunately, however, affect theory is also preoccupied with emotion, and emotion is an astonishingly underanalyzed topic in biblical studies, given that emotions swirl and eddy so powerfully and so patently through most biblical texts—not least the book of Revelation. What makes affect theory particularly relevant to Revelation, moreover, is the former's recurrent fascination with negative emotions and affects, such as loathing, disgust, revulsion, and aversion, all of which I see as elemental forces in Revelation. Adding to the significance of this dark emotional maelstrom is the fact that two of the primary stimuli that Revelation employs to trigger our social gag reflex and cause our moral gorge to

---

* This essay first appeared under the title "Retching on Rome: Vomitous Loathing and Visceral Disgust in Affect Theory and the Apocalypse of John," *BibInt* 22 (2014): 503–26, and is reprinted with the permission of the publisher.

rise are female characters: "that woman Jezebel" (2:20) and "Babylon the great, mother of whores and of earth's abominations" (17:5). Affect and gender are intimately intertwined and explosively combined in Revelation.

Erin Runions was the first to bring affect theory to bear on a biblical text (Runions 2008a), but I only registered that belatedly. My own interest in affect theory and my sense of its potential for biblical studies was ignited by Maia Kotrosits (2010; 2012; 2013; 2014; see also Kotrosits and Taussig 2013) and Alexis Waller (2014), and further fanned by Jennifer Koosed, with whom I had the privilege of coediting a volume on affect theory and the Bible (Koosed and Moore 2014). I am grateful to these younger colleagues for challenging me to remain connected to the ever-changing world of theory in whose shimmering alien light the biblical text always becomes something other than it was, even in the act of ingestion: "Take it, and eat; it will be bitter to your stomach, but sweet as honey in your mouth" (Rev 10:9).

Ancient Emotions

A lumbering "turn to the emotions" has long been underway in the humanities and social sciences, with much grinding of gears and anxious glances in the mirror. Historian Ruth Leys, undertaking to list the fields in which this turn has been most evident, names "history, political theory, human geography, urban and environmental studies, architecture, literary studies, art history and criticism, media theory, and cultural studies" (2011, 434). She might well have added the field of classics to her eclectic list: much work on ancient Mediterranean emotions has also been accomplished in recent decades, epitomized by such titles as *The Emotions of the Ancient Greeks* (Konstan 2006); *Emotion, Restraint, and Community in Ancient Rome* (Kaster 2005); and *Emotion, Genre and Gender in Classical Antiquity* (Munteanu 2011).[1]

---

1. Comparable titles in biblical studies are considerably harder to name. See, however, Voorwinde 2005; 2011; Kuhn 2009; Egger-Wenzel and Corley 2012; and

Particularly relevant for the present study, which has the Apocalypse of John in its sights, are two of the questions that Robert Kaster includes among those that incited his study of ancient Roman emotions: "How … do various Roman forms of fear, dismay, indignation, and revulsion support or constrain different sorts of ethically significant behavior? … How does [the] interaction [of emotions] create *an economy of displeasure*, a system that causes negative feelings to circulate in constructive ways?" (Kaster 2005, 4, emphasis added). What might such questions yield if brought to bear on that most un-Roman of Roman books, the book of Revelation? How do revulsion, indignation, and other associated emotions, but above all disgust, combine in Revelation to move its audience toward ethically significant—or ethically problematic—behavior? How does Revelation's mobilization of emotions create an economy of displeasure, the circulation of negative feelings toward constructive—or destructive—ends?[2] The present study will draw on that heterogeneous body of work now (perhaps too tidily) termed "affect theory" to tackle these and other such questions.[3]

---

deSilva 2009, 175–228 (which I discuss below). An SBL program unit titled "Bible and Emotion" has also been formed.

2. These "negative feelings" would have circulated, in the first instance, among a listening audience. This essay is not an exercise either in performance criticism or rhetorical criticism, however, and so will not attempt to reconstruct that context of reception. Suffice it to note that although ancient evidence exists for the specific emotions that genres such as comedy, tragedy, epic, or satire were expected to arouse, no such evidence exists for the genre of apocalypse. What Dana LaCourse Munteanu has to say about comedies and epigrams, however, would have been even more true of apocalypses: "[they] stirred emotions that were not permitted unbridled expression in society" (2011b, 2).

3. And as such will immediately part company with Kaster's elegant but theory-shy study. It will also part company with David A. deSilva's *Seeing Things John's Way: The Rhetoric of the Book of Revelation* (2009), two chapters of which analyze Revelation's appeal to the emotions of its audience. Titled "The Strategic Arousal of Emotions in the Oracles to the Seven Churches" and "'I Saw a Monster Rising Up!': Appeals to Pathos in the Visions of Revelation," these chapters are exercises in rhetorical criticism that "rel[y] heavily on the near-contemporary discussions of how to evoke emotions in Greek and Latin rhetorical handbooks" (2009, 176). Jezebel and Babylon, the principal foci of my own study, also feature in deSilva's chapters. For deSilva, the primary emotions that John seeks to arouse against these female characters are enmity and indignation. DeSilva is only secondarily interested in disgust and revulsion, which are my main interest. More specifically, I am interested in how disgust and revulsion intersect with sexual violence, and violence generally, in Revelation—intersections

Intense Affects

Where to begin? Perhaps with the (admittedly crude) recognition that there is something of "the ancient," or at least "the premodern," in affect theory's construal of emotion. First, note David Konstan's argument in *The Emotions of the Ancient Greeks* that "the Greeks did not conceive of emotions as internal states of excitation. Rather, the emotions are elicited by our interpretation of the words, acts, and intentions of others" (2006, xii).[4] Cut now to Sara Ahmed's observation in *The Cultural Politics of Emotion* that "the [contemporary] everyday language of emotion is based on the presumption of interiority" (2004, 8). I *feel deeply*, as does every "emotionally adjusted" person in my world, and that *emotional depth* is contained *inside me* until I *express it* or *put it into words* (or less articulate sounds: cries of grief, snarls of rage, gales of laughter, moans of disgust…), after which my *personal feelings* have been *shared* or *made public*.

This interior-exterior model of emotion has been challenged from many angles. One such challenge, not surprisingly, is bound up with philosophical critiques of the Cartesian legacy, here seen as entailing the cultural construction of the modern inward-turned individual and the concomitant privileging of interiority over exteriority in the conceptualization of human subjectivity.[5] How to rethink emotion in a post-Cartesian

---

that contemporary affect theory enables me to explore in ways that the ancient rhetorical handbooks would not.

4. These ancient emotions or passions, consequently, are conceived as ineluctably corporeal. Juvenal, for instance, expects the auditors of his satires to "turn red" with shame, "sweat" in fear of being exposed, and "shed tears" of remorse (Munteanu 2011b, 3). Similar observations about ancient emotions are made by social-scientific analysts of biblical material. John J. Pilch (1999, 11), for example, notes: "In the contemporary West, anger is perceived primarily as an internal experience…. The nonintrospective [ancient] Middle East, in contrast, tends primarily to somatize experiences. The Hebrew words translated as 'anger' are illustrative…. Anger affects both the nose and breathing. Thus a person described literally in Hebrew as 'short of nose' (Prov 14:17) or 'short of breath' (Mic 2:7; Exod 6:9) is impatient or angry. In contrast, a person 'long of nose' (Prov 14:29) or 'long of breath' (Eccl 7:8) is patient and slow to anger. 'The Lord, the Lord, a God merciful and gracious, long of nose …' (Exod 34:6)…. Often the Hebrew literally speaks of a burning nose (Gen 30:2: Jacob's nose burned against Rachel)."

5. See Brennan 2004, 2: "As the notion of the individual gained in strength, it was assumed more and more that emotions and energies are naturally contained, going no farther than the skin."

register?⁶ The term *affect* has, among other things, assumed a strategic role in such rethinking. For Gilles Deleuze and Félix Guattari, whose *A Thousand Plateaus* (1987) is now retrospectively seen as one of the primary headwaters of affect theory,⁷ "affect" does not denote "a personal feeling," but rather "a prepersonal intensity corresponding to the passage from one experiential state of the body to another" (Massumi 1987, xvi).⁸ *Intensity* and *intensities*, meanwhile,

> **["To the relations composing, decomposing, or modifying an individual there correspond intensities that affect it" (Deleuze and Guattari 1987, 256).]**

are Deleuzian and Deleuzoguattarian shorthand for the incessant sensory bombardment of bodily existence—visual, aural, tactile, olfactory, kinetic, rhythmic, chaotic—prior to its processing by language, cognition, reason, or spatiotemporal organization. The editors of *The Affect Theory Reader* remain implicitly within the Deleuzoguattarian orbit when they announce: "Affect … is the name we give to those forces—visceral forces beneath, alongside, or generally *other than* conscious knowing, vital forces insisting beyond emotion—that can serve to drive us toward movement, toward thought and extension" (Seigworth and Gregg 2010a, 1); and again: "affect is found in those intensities that pass body to body (human, nonhuman, part-body, and otherwise)" (2010a, 1). Next to the term "affect" itself,

> **["the exacerbation of colleagues who 'never want to hear the word affect again'" (2010a, 18).]**

"intensities" is, perhaps, the most insistently intoned term of contemporary affect theory.

One might say, then, that affect theory in its Deleuzian tributary is a series of divergent but overlapping discourses about human bodies

> **["We each go through so many bodies in each other" (Deleuze and Guattari 1987, 36).]**

---

6. See especially Terada 2001. The emotions or passions already constitute a fracturing force in Descartes's philosophy of the *cogito*, as Terada notes (2001, 3, 8–9, 22–24), following Derrida.

7. *A Thousand Plateaus* here serves, however, as a convenient stand-in for the two-volume work of which it is a part (see also Deleuze and Guattari 1983); other Deleuze-Guattari collaborations (e.g., 1986); and Deleuze's extensive solo oeuvre (e.g., 1986; 1989; 1990; 1993; 2003).

8. See Deleuze 1995, 137: "Affects aren't feelings, they're becomings that spill over beyond whoever lives through them (thereby becoming someone else)."

and bodiliness. The body in this heady version of affect theory is ambiguously bounded because invariably excessive. The body always and everywhere outstrips and undoes any and all discursive attempts to encase and contain it, whether regimes of knowledge, systems of signification, or contingent interpretations. It ceaselessly flows through them, around them, and beyond them to affect and be affected by other innumerable, equally illimitable bodies, whether human or nonhuman, organic or inorganic. Unbounded intercorporeality is the human body's constitutive medium. This is because the human body is neither fully material nor yet fully mental, to the extent that these two terms can be held in meaningful opposition. It is affectively experienced as at once an interior and an exterior reality.

Our Bodies, Our Bibles

The capacity of affect theory in its body-preoccupied modality to convert into reflection on textual ontology is arresting.[9] For the text, too, is real and material but also incorporeal and virtual, always indeterminately open to be otherwise than it ephemerally is.[10] Even the "it" that "it is" is not ever one thing. *The* biblical text, for instance, is not, and never can be, *a* general text but is always instead a finite number of nonidentical textual bodies. Even if my NRSV seemed indistinguishable from every other NRSV in the pile when I originally purchased it (on January 6, 1990, at the National Bible Society of Ireland bookshop in Dawson Street, Dublin, as its flyleaf

---

9. But hardly accidental. Affect theory's rejection of semiotic hypertextualism and discursive constructivism in favor of a (chastened) biologism and materialism sets it apart from "classic" poststructuralist theory (see especially Massumi 2002, 1–4; Sedgwick 2003, 93–94, 108–14). Yet there are profound continuities as well as discontinuities between the two theoretical corpora—which is why affect theory on the body is often eerily reminiscent of poststructuralist theory on the text. In particular, affect theory on the unrepresentability of the body regularly echoes poststructuralist theory on the unrepresentability of, say, the real (Lacan), or the semiotic (Kristeva), or *différance* and its many nonsynonymous synonyms (Derrida).

10. I am culling terms here from Brian Massumi's *Parables for the Virtual* (2002), another bravura and already "classic" exercise in affect theory. Massumi himself harnesses a term long associated with theoretical discourse on textual meaning to speak about the human body: the body, a perpetual motion and sensation machine, is always in "a real relation … to its own *indeterminacy* (its openness to an elsewhere and otherwise than it is, in any here and now)" (2002, 5).

reminds me), its material individuality is now transparently apparent in its battered blue cardboard binding, held together with layers of tape; its once glittering gilt title, *Holy Bible*, now threatening to fade altogether out of existence, as though terminally dispirited by the unholy uses to which the book has repeatedly been put; its grubby, dog-eared pages, scored with innumerable scribbles, many now illegible even to me. A parallel tale could be told of my still more decrepit UBS Greek New Testament, purchased at the same Dublin bookshop on February 4, 1980, and hence, embarrassingly, several editions out of date, but essentially irreplaceable because of the still greater profusion of minute scrawls that adorn its pages, many of these scribbles of sentimental value, as they present me with now long-dead versions of myself, handwritten wraiths, with different passions and better Greek.

My relationship to my Bibles, however, is more intimate still. For their pages are also indelibly marked by the secretions and excretions of my sebaceous and sudoriferous glands. My bodily fats, waxes, and acids have leaked copiously into those delicate, unprotesting pages and bonded chemically with them. My NRSV and UBS contain my DNA, while the limbic system of my brain contains much of my Bibles. Our relationship is intercorporeal, intercellular. The words of my Bibles are encoded in their biologically enriched pages, now seething with microscopic life,

> **["The trauma of physiological change is odd. Fractured, bizarre, confounding. What is the agent here? Not who, what" (Gambs 2007, 107).]**

and those words are also encoded, however muddily and murkily, in my medial temporal lobe. They are at once external and internal, material and mental, physical and psychic, visceral and virtual; for are not literary texts in particular, and sacred texts above all, *human* objects,

> **["the human contains all manner of objects within its envelope" (Thrift 2010, 293).]**

or, more precisely, objects that thoroughly "dissipate the onto-theological binaries" of persons and things, "vibrant life" and "dull matter" (Bennett 2010, vii, x)?

My relationship with "the" Bible is a bodily affair, then—a tactile-textual, material-ideational, ethical-emotional affair of two unbounded bodies—and that relationship can only be crudely captured in language, if at all. It is a sensate relationship characterized by intensities, and those intensities are only part-conscious at best, and hence admit only of oblique articulation at most. In Deleuzian-Massumian theory, intensities are what catalyze the making real of the virtual. The unrepresentable intensities of my affective communion with the pre-interpreted, predigested, pre-incorporated textual bodies of my Bible

> ["So I went to the angel and told him to give me the little scroll; and he said to me, 'Take it, and eat'" (Rev 10:9).]

—in unpredictable interaction with the unrepresentable intensities of *your* affective communion with the pre-interpreted, predigested, pre-incorporated textual bodies of *your* Bible (and yours and yours and yours …)—are what catalyze the virtuality of those inherently senseless, yet sense-intensive, texts

> ["it was sweet as honey in my mouth, but when I had eaten it, my stomach was made bitter" (Rev 10:10).]

as disciplined discourse and orderly meaning.[11] In the discipline of biblical studies, affective intensity is what is disciplined most of all.

Ahmed and Affect

How might all or any of this translate into literary interpretation or biblical exegesis?[12] The editors of *The Affect Theory Reader* leave us guessing, and so too, by and large, do all its contributors. Most affect theory tends to huddle under the umbrella of "cultural studies" rather than that of "literary studies."[13] Analyses of literary texts are the exception rather than the rule in affect theory.[14] Sara Ahmed, who, she says, "offer[s] close

---

11. What Massumi (2002, 28) has to say about emotion also holds true for scholarly discourse: "Emotion is qualified intensity, the conventional, consensual point of insertion of intensity into semantically and semiotically formed progressions, into narrativizable action-reaction circuits, into function and meaning. It is intensity owned and recognized."

12. Aside from Deleuze's, or Deleuze and Guattari's, own daunting models for literary interpretation, such as their *Kafka: Toward a Minor Literature* (1986). B. H. McLean makes exciting use of Deleuze and Guattari, especially *A Thousand Plateaus* (1987), to rethink the fundamentals of biblical hermeneutics in his *Biblical Interpretation and Philosophical Hermeneutics* (2012, 268–301), but does not demonstrate, except in passing, the transformed exegetical practice that would result from such rethinking.

13. Emblematic of the cultural studies brand of affect theory is Kathleen Stewart's *Ordinary Affects* (2007), an ethnographic analysis of the affective dynamics of a set of everyday scenes from contemporary U.S. culture.

14. One notable exception is Eve Sedgwick's *Touching Feeling* (2003, see 35–91).

readings of texts" in *The Cultural Politics of Emotion*, seems to feel a need to apologize for it: "For a book on emotions, which argues that emotions cannot be separated from bodily sensations, this book may seem very oriented towards texts" (2004, 12). Yet Ahmed's texts are not literary texts, but texts "that ... circulate in the public domain, and include web sites, government reports, political speeches and newspaper articles" (2004, 14).[15] I will be using Ahmed's enormously suggestive book to tackle an ancient text, which, while literary, also circulates in the public domain. My affect-attuned reading of Revelation is facilitated by the fact that Ahmed refuses the strict Deleuzian-Massumian separation of affect and emotion, opting instead for a version of affect that, while sensate, intense, and transpersonal, is also indissociable from emotion: "Whilst sensation [read: affect] and emotion are irreducible, they cannot simply be separated at the level of lived experience" (Ahmed 2004, 25).[16] The intricate movements of emotion in and around Revelation are tricky enough to track, as we shall see. Reading apocalyptic affect with Deleuze and his disciples

["what do you know about me, given that I believe in secrecy?" (Deleuze 1995, 11).]

will have to await another day.[17]

Telling, nonetheless, is Sedgwick's omission of *texts* from her list of things to which affects are attached: "Affects can be, and are, attached to things, people, ideas, sensations, relations, activities, ambitions, institutions, and any number of other things, including other affects" (2003, 19).

15. They are cultural studies texts, in other words.

16. Ahmed continues: "I am hence departing from the recent tendency to separate sensation or affect and emotion, which is clear in the work of Massumi.... Certainly, the experience of 'having' an emotion may be distinct from sensations and impressions, which may burn the skin before any conscious moment of recognition. But this model ... negates how that which is not consciously experienced may itself be mediated by past experiences.... Even seemingly direct responses actually evoke past histories, and ... this process bypasses consciousness, through bodily memories" (2004, 40 n. 4; cf. Massumi, 2002, 27–28). Ahmed develops this argument further in her *The Promise of Happiness* (2010b, 230–31 n. 1). My privileging of Ahmed over Massumi in what follows, however, rests on pragmatic rather than philosophical grounds.

17. Deleuze himself wrote on the Apocalypse (or, more precisely, on D. H. Lawrence's diatribe against it) in "Nietzsche and Saint Paul, Lawrence and John of Patmos" (1998).

Pure Disgust

The Apocalypse of John is affect-intensive. What Revelation reveals—uncovers, unclothes, indecently exposes—
> ["the veil lifted from about the thing: first of all, if we can say this, man's or woman's sex" (Derrida 1992b, 27).[18]]

is an abyssal loathing. And what Revelation aims to effect, through affect, is to infect its audience with that infinite abhorrence. Like the God-controlled (zombie?) cannibals of 17:16–17,
> ["It is the book of Zombies" (Deleuze 1998, 37).]

the apocalyptic audience must "loathe the whore," both because she is the source of all that is filthy, detestable, and obscene ("mother of whores and of earth's abominations," 17:5) and because entire nations are perversely drunk with lust for her as a result (14:8; 17:2; 18:3, 9; 19:2), eager to wallow in her filth.

How best to take the measure of this immeasurable disgust? Before turning to "The Performativity of Disgust," "The Organization of Hate," and the other relevant chapters of Ahmed's *The Cultural Politics of Emotion*,[19] I want to propose preliminarily that Revelation's apocalyptic disgust is a hyperbolic expression and reciprocal effect of its immense obsession with purity. I have become increasingly convinced by those who argue that purity is a central preoccupation in Revelation.[20] The symbolic lines of separation in Revelation run consistently between the pure and the impure. Those who follow the Lamb have washed their robes white in his cleansing blood (7:14; cf. 3:4–5, 18; 22:14). When they follow him to war,

---

18. Derrida (1992b, 27) continues: "apocalyptic unveiling, … the disclosure that lets be seen what till then remained enveloped, withdrawn, held back, reserved, for example, the body when the clothes are removed or the glans when the foreskin is removed in circumcision."

19. If I turn repeatedly to Ahmed, it is again to make my topic manageable. The literature on disgust, in particular, is becoming colossal and complex. As Daniel Kelly (2011, 2) notes, "disgust has become relevant to discussions across the humanities, especially those engaging the cognitive sciences and those in the midst of the 'affective turn.' It is … already prominent in debates that take place where psychology intersects with philosophy and moral theory."

20. See especially Stenström 2009; Marshall 2009; and Hood 2010. See also Frankfurter 2001, esp. 410–12; Marshall 2001, esp. 155–62; 2007, esp. 249–54. Mayo's book-length study of Revelation and Judaism, in contrast (but in common with most traditional scholarship), accords no major role to purity or impurity (Mayo 2006).

they will wear "fine linen, white and pure/clean [*katharon*]" (19:14), and they will wear it again when they assume their corporate role as the pure bride of the Lamb (19:8). They are a community of priests (1:6; 5:10; 20:6; cf. 3:12a), and as such custodians of purity. They "have not been defiled/made dirty [*ouk emolynthēsan*] with women" (14:4), and as such they are "unblemished" (*amōmoi*, 14:5). "Jezebel" and her followers, in contrast, perform impure acts, eating meat that has been sacrificed to idols and engaging in illicit sex (unless the sex is a metaphor for the idolatry; 2:20–21; cf. 2:14). The woman "Babylon," for her part, not only abandons herself to defiling sexual activity—her cup contains "the impurities/pollutions of her prostitution/fornication [*ta akatharta tēs porneias autēs*]" (17:4)—but even drinks blood (17:6; cf. 16:6; 18:6), "the ultimate impurity."[21] When she mutates into a city, Babylon is home to demons and "every unclean [*akathartou*] spirit" and "every unclean and loathsome [*akathartou kai memisēmenou*] bird" (and animal, in some ancient manuscripts, 18:2). In contrast, nothing "unclean/profane" (*koinon*) shall enter the "holy city" (21:27; cf. 22:15).[22] Its conspicuous cubic structure (21:16) evokes the inner sanctuary of Solomon's temple (1 Kgs 6:20),

["A temple, mobile, fragile, or destroyed: the ark is no more than a little portable packet of signs" (Deleuze and Guattari 1987, 122).]
the "most holy place" (1 Kgs 6:16), and hence the place most inimical to impurity.

In short, as Hanna Stenström concludes, "Revelation provides a textbook example of how 'purity' can structure a symbolic universe" (2009, 49). What affect theory enables us to see in addition is that purity in Revelation is structured in turn by the push-pull movements of emotion. Even as Revelation elaborates a cultural politics of purity, it also elaborates a cultural politics of emotion (to cite the title of Ahmed's book). Or as Ahmed (also) puts it, "emotions work by working through signs and on bodies to materialize the surfaces and boundaries that are lived as worlds" (2004, 191).

---

21. As Marshall notes (2009, 29).
22. Of the formula "nothing *koinon* will enter," Frankfurter remarks: "[it] clearly implies the type of impurity that might pollute a sanctuary" (2001, 410–11).

Sticky Sex Work

What work does "the great whore" (Rev 17:1) perform in the construction of the Apocalypse's affect-intensive world? Ahmed impels us to reconceive the whore as a circulating object that is "sticky"

> ["I begin ... with the messiness of the experiential" (Ahmed 2010b, 22).]

or saturated with affect. "Stickiness," for Ahmed, "involves a transference of affect" (2004, 91). Affect, indeed, is precisely "what sticks, or what sustains or preserves the connection between ideas, values, and emotions" (2010b, 230 n. 1). How does the stickiness of affect come about? Stickiness is "an effect of a history of articulation" (2004, 92), and in and through such layered, laden histories, certain objects accumulate certain affects. The affect *disgust*, for example, "stick[s] more to some bodies than others" (2004, 92). The denigrative terms used to denote such bodies—for instance, the term *pornē* ("whore," "streetwalker," "brothel slave" ...) in Revelation—also stick to them by means of felt associations with other contiguous terms that are not, and do not need to be, spoken: "impure," "unclean," "abominable," "detestable," "filthy," "shamed," and so on (see Ahmed 2004, 92).[23] John, however, as though not altogether trusting in the stickiness of *pornē* to relay its associative baggage safely to the audience, spells out one of these terms: "and on her forehead was written a name, a mystery: 'Babylon the great, mother of *pornōn* and of earth's *bdelygmatōn*'" (Rev 17:5; cf. 17:4; 21:27).

To track the *bdel*-lemma through LSJ, BAGD, and *TLG*

> ["How was disgust discussed before the word disgust?" (Miller 1997, 143).]

(... *bdelygma, bdelygmia, bdelygmos, bdelyktos, bdelyros, bdelyssomai* ...) is to pick one's way gingerly and queasily through a sticky, slimy, and altogether unsavory set of terms while trying hard not to breathe through one's nose: *nausea, sickness, filth, nastiness, defilement, abomination; disgusting, abhorrent, detestable, loathsome; feel a loathing for food, feel a loathing at, cause to stink, make loathsome or abominable....* In Revelation, Babylon is

---

23. For an incisive excavation of the themes of shame and shaming in the Apocalypse, but especially in its representations of the woman Babylon, see Burrus 2008, 14–19. Burrus's book frequently draws on the psychological affect theory of Silvan Tomkins.

the source of all this putrescence and execration and more. She is, after all, the mother of earth's *bdelygmatōn*,

> ["Sperm, saliva, glair, curdled drool, tears of milk, gel of vomit—all these heavy and white substances are going to glide into one another, be agglutinated" (Derrida 1986, 139–40b).]

the fount of all uncleanness. *Bdelygma* sticks to Babylon's forehead like a leech, writhes in the ornate cup she lifts to her lips ("holding in her hand a golden cup full of *bdelygmatōn* and the unclean things [*ta akatharta*] of her fornication," 17:4), and slithers down her throat. She is a hyperbolic figure of disgust.

Yet Jezebel provides an even better example than Babylon of the ways in which stickiness is "*an effect of the histories of contact between bodies, objects, and signs*" (Ahmed 2004, 90). Or to phrase it slightly differently, "stickiness depends on histories of contact that have already impressed upon the surface of the object" (2004, 90). For John, the true followers of Jesus Messiah are "the remnant of Israel [see Rev 7:4–8; 14:1; 21:12], and thus … Israel in its totality; the Others are in John's eyes no longer part of Israel, whatever they may think about themselves [see 2:9; 3:9]" (Stenström 2009, 39; cf. 47). The rival female prophet vilified by John, whom he scathingly renames "Jezebel," epitomizes these intimate others. In the case of "Jezebel," histories of contact between Israel and not-Israel, reimagined in tradition and literature,

> ["the four hundred and fifty prophets of Baal and the four hundred prophets of Asherah, who eat at Jezebel's table" (1 Kgs 18:19).]

have not only "impressed upon the surface of the object" but reshaped the object entirely. Jezebel is one of the stickiest signs in Revelation, and what Ahmed has to say about how stickiness becomes disgusting perfectly fits her or it: "stickiness becomes disgusting *only when the skin surface is at stake such that what is sticky threatens to stick to us*" (2004, 90). In other words, Jezebel is a figure of contamination.[24] She puts filthy, defiling flesh in her body, and the practice is contagious (2:20; cf. 2:14). She pollutes, and her proximity makes that pollution intolerable.

---

24. "If we ask what disgust makes us averse to, then the short answer is *contact*. It … make[s] us averse to … *contamination* … in the sense of feeling oneself to be invaded, violated, made unclean" (McGinn 2011, 41).

Figures of Hate

Jezebel, the Nicolaitan prophet,[25] is not only a figure of contamination. She is also a *figure of hate* ("you hate [*miseis*] the works of the Nicolaitans, which I also hate," 2:6), as is the whore ("they ... will hate [*misēsousin*] the whore," 17:16). Stickiness in Revelation is not merely a matter of how *individual* characters, apprehended in isolation, are saturated with accumulated affect. Revelation's affective economy also operates "by sticking 'figures of hate' together" (Ahmed 2004, 15; cf. 43–44), which are also figures of disgust. This is the answer Ahmed enables to the question I quoted from Kaster (2005, 4) at the outset of this essay: "How does [the] interaction [of certain emotions] create an economy of displeasure ... that causes negative feelings to circulate in constructive ways?"[26] In and through the adhesive and accumulative operations of hate, Jezebel is actively stuck to the whore, while the whore in turn is actively stuck to the beast, and so on. More precisely, the "lesser whore," Jezebel, shares a sex-slimed bed with the "great whore," Babylon;

["I am throwing her on a bed" (Rev 1:22).]

the whore is physically intimate with the beast, her thighs wrapped around it (17:3, 7), and she is destined to be incorporated into the beast ("they and the beast ... will devour her flesh," 17:16), to take the long, twisting trail from its mouth to its anus;

["For in the end the anus also expresses an intensity" (Deleuze and Guattari 1987, 32).]

this monstrous sea beast is aligned with a monstrous land beast, and both beasts are aligned with the great red dragon, itself a hyperbolic figure of loathing ("that ancient serpent, who is called the Devil and Satan, the deceiver of the whole world"—12:8; cf. 20:2),

["When we turn to snakes, here the associated disgust seems to be still more intermixed with anxiety" (Kolnai 2004, 57).]

---

25. I follow the scholarly majority in drawing no meaningful distinction between "those who hold to the teaching of the Nicolaitans" (Rev 2:15; cf. 2:6), "those who hold to the teaching of Balaam" who caused Israel to "eat food sacrificed to idols and practice fornication" (2:14), or those "beguiled" by "that woman Jezebel ... to practice fornication and to eat food sacrificed to idols" (2:20). They are all the same anathematized group.

26. I understand "constructive" here to refer to the construction of symbolic universes.

while to compound the visceral disgust they are designed to evoke, all three figures emit from their mouths "foul spirits like frogs" (16:13); the whore or abominable woman mutates into a no less abominable city (18:1–8); and so on.

The relentless "metonymic slide" from one figure of hate to the next "constructs a relationship of resemblance" between them (Ahmed 2004, 44). What makes them all, in effect, *a single group character* is their common unlikeness from the figures of identification in the Apocalypse, similarly stuck together in its viscous affective economy: the son of humanity (1:13) who is also a lamb (5:6); the woman clothed with the sun who gives birth to a son (12:1–2, 5) who grows up to be the son of humanity who is also a lamb; the 144,000 male virgins who are led by the lamb to the slaughter (14:1–5; cf. 7:13–14) and morph into the bride who is the wife of the lamb (21:9); and so on.[27] Affect's sticky work in the Apocalypse is that of separating its characters into bodies that should be loved and that must be hated (see Ahmed 2004, 52). And ultimately there are only two multimembered bodies in this narrative,[28] a "good" body and a "bad" body—or, more precisely, two multifaceted objects of emotion.[29]

Abject Objects

If the Apocalypse, stripped to its affective essentials, reduces to two character groups and two primary emotions, its plot also reduces to a simple movement. "The replacement of one word for an emotion with another word produces a narrative" (Ahmed 2004, 13).[30] Revelation's narrative, or more accurately its plot, is produced through the successive displacement of one intense emotional state by another. The movement from bitter grief to triumphant joy in 5:1–14, or from maternal anguish to diabolic rage in 12:1–17, or from fascinated revulsion to savage satisfaction

---

27. Revelation's transspecies transformations and interspecies intimacies are the topic of the last two essays in this volume.

28. A narrative explicitly punctuated with multimembered bodies (e.g., 5:6; 9:19; 12:3; 13:1; 17:3, 7, 9), which then stand as visible symbols for the narrative's invisible operations.

29. See Ahmed 2010a, 35: "Objects are sticky because they are already attributed as being good or bad, as being the cause of happiness or unhappiness."

30. In effect, this is Ahmed's affective restatement of E. M. Forster's famous protonarratological pronouncement: "'The king died and then the queen died' is a story. 'The king died, and then the queen died of grief' is a plot" (1927, 86).

in 17:1–19:5 are but exemplary moments in a more pervasive pattern. But the movement of emotion in Revelation is not simply linear. Emotion also moves in rippling circles both within Revelation and between Revelation and its audience.

The woman Babylon emblematizes the circulation of emotion in and around the Apocalypse. She is, as we saw earlier, a circulating object that is sticky or saturated with affect. In this regard, however, the woman Babylon is but the quintessential prostitute (as the text indeed indicates: "Babylon the great, mother of whores," 17:5). For what is a prostitute, a sex worker, but a circulating object—a "sex object," as we say—one that circulates from hand to hand. And that circulation is productive of affect. As Melissa Ditmore observes, "The world's oldest profession is also the world's oldest form of affective labor" (2007, 170). Affective labor is "work that aims to evoke specific behaviors and sentiments…, rather than … being merely about the production of a consumable product" (2007, 171).[31] The work assigned to Revelation's whore by (her) John—paradoxical work for a sex worker, to be sure—is that of inculcating boundary-respecting behavior and boundary-squeamish sentiments in the book's audience, for Revelation is, above all, a border document designed to police the frontier between pure and impure, clean and unclean, as we have seen.

Revelation is a boundary book; but is the boundary, the border, between purity and impurity in Revelation itself clean or dirty? Ahmed, retro-reviving Julia Kristeva's *Powers of Horror* (1982), enables us to think through this odd-sounding but significant question. For at issue here is *the abject*: everything the human subject must exclude or expel

["perhaps we cannot easily live with too much vivid awareness of the fact that we are made of sticky and oozy substances that will all too soon decay" (Nussbaum 2004, 14).]

to establish and maintain its proper borders. In abjection, however, the border itself

> is transformed into an object…. The object that makes us "sick to the stomach" is a substitute for the border…. On the one hand, it is the

---

31. See also Wendy Chapkis, *Live Sex Acts: Women Performing Erotic Labor* (1997), who describes sex work both as "erotic labor" and "emotional labor." Avaren Ipsen reads biblical texts with contemporary sex workers, finding some texts to be sex-worker positive, but not the Apocalypse: "The whore metaphor is just all around bad news to prostitutes" (2009, 170).

transformation of borders into objects that is sickening (like the skin that forms on milk). On the other hand, the border is transformed into an object precisely as an effect of disgust (spitting/vomiting).... Border objects are hence disgusting, while disgust engenders border objects. (Ahmed 2004, 86–87)[32]

The border in Revelation, then, is and must be dirty, as filthy and loathsome as everything it is designed to keep out
> ["Loathing an item of food, a piece of filth, waste, or dung. The spasms and vomiting that protect me" (Kristeva 1982, 2).]

or hold at bay. It transforms, coagulates, or congeals into metonymic objects, two in particular. The city Babylon is *a border town* in Revelation, a refuge for everything illicit and unsavory (demons, unclean birds and animals, spillers of innocent blood, 18:2, 24). Border towns are also intimately associated with sex workers, and the metonym for the city Babylon, itself a metonym for the border, is the *pornē* Babylon. The "whore" must be the ultimate object of disgust in Revelation's affective economy because she stands in for the unclean city, which itself stands in for everything that Revelation's borders are designed keep at a distance, to hold separate, to abject.

Eating and Vomiting

The whore's illicit food ("in her hand a golden cup full of *bdelygmatōn*.... [She] was drunk with the blood of the saints," 17:4, 6), no less than her illicit sexual activity, makes her a hyperbolic object of disgust, just as the same combination of illicit food and (or as) illicit sex earlier made Jezebel an object of disgust. Nauseous food taken into the (social) body
> ["Food loathing is perhaps the most elementary and most archaic form of abjection" (Kristeva 1982, 2).]

figures the fear of contamination in Revelation. Revelation retches on
> ["I experience a gagging sensation and, still farther down, spasms in the stomach ... ; and all the organs shrivel up ... , provoke tears and bile, increase heartbeat, cause forehead and hands to perspire" (1982, 2).]

Rome.

---

32. "Like the skin that forms on milk" is an allusion to Kristeva 1982, 3–4: "When the eyes see or the lips touch that skin on the surface of milk—harmless, thin as a sheet of cigarette paper, pitiful as a nail paring—I experience a gagging sensation."

"To abject something is literally to cast something out, or to expel something" (Ahmed 2004, 94). One important way in which such expulsion is effected is through the performativity of speech acts.[33] The exclamatory speech act, "That's disgusting!," for instance, can function "as a form of vomiting, as an attempt to expel something whose proximity is felt to be threatening or contaminating" (2004, 94). In the woman Babylon's case, the abjecting speech act is emblazoned permanently on her forehead: "Babylon the great, mother of whores and of earth's *bdelygmatōn*" (17:5). The adhesive label "generates the object that it names" (2004, 93), and generates it as an object of revulsion. Does the sticky sign of abomination permanently affixed to Babylon signify that she is perpetually in the process of being vomited, John spewing up the sweet and sour prophetic scroll he had earlier been compelled to eat (10:8–11)? The situation

> ["I give birth to myself amid the violence of sobs, of vomit" (Kristeva 1982, 3).]

is more complex than that.

First, consider the simpler case of Jezebel. "Vomiting involves expelling something that has already been digested, and hence incorporated into the body" (Ahmed 2004, 94). Jezebel, more than Babylon, is the figure in Revelation of an object of disgust already lodged in the body—the social body of Christ followers in the cities of Roman Asia—that needs to be vomited, expelled, abjected. And since the social body has been contaminated, a communal vomiting is what is in view. For "Jezebel," too, is a production of speech acts that, beginning with her scorn-saturated name, amount to the exclamation, "That's disgusting!," and these speech acts are also designed to create "a community ... bound together through the shared condemnation of a disgusting object" (2004, 94),

> ["the word disgust ... does not designate the repugnant or the negative in general. It refers precisely to what makes one desire to vomit" (Derrida 1981, 23).]

a *community of disgust* that signifies its unity by eating together but also by vomiting together.

But if the woman "Jezebel" is the nauseating object that has somehow slid inside, that has been swallowed but must now be ejected,

---

33. "Emotions are performative ... and they involve speech acts" (Ahmed 2004, 13).

["disgust, as when spitting out bad-tasting food, recognizes the difference between inside and outside the body and what should and should not be let in" (Sedgwick 2003, 116).]

what is the woman "Babylon"? To a not insignificant extent, "Jezebel" and "Babylon" are but two alternative names for the same alien entity,[34] whose other (unspoken) name is *Roma* or *Rhōmē*. While Jezebel, however, is a figure for what contaminates from within, Babylon is a figure for what contaminates from without. Babylon *is* the outside, what envelops and threatens to engulf, to contaminate absolutely through the asphyxiating proximity of its infinitely bloated bulk, while Jezebel is a slithering sliver of that alien outside that has surreptitiously slipped inside. But the simultaneous inside-outside nature of the perceived threat against which the Apocalypse apotropaically organizes itself makes its project Promethean. The Apocalypse's impossible project is the vomitous ejection of a loathsome alien entity that has somehow gotten inside one, but in whose monstrous body one is also somehow contained. Jezebel writhes in John's belly, but John writ(h)es in Babylon's belly.

Revelation's response to this uncomfortable predicament accords strictly with the logic enunciated in 18:6: "Render to her as she herself has rendered." The expulsion or annihilation of the intolerable other in Revelation is, in an appearance of contradiction, accomplished through its incorporation or ingestion. John eats the thing that is eating him. More precisely, John stages a scene in which he has the beast, a figure for Rome, devour the whore, equally a figure for Rome (17:16). (Does Rome now retch on Rome, gnawing and choking down its own quivering flesh, digit by digit, limb by limb, organ by organ?) And then John stages a further ghastly but no less necessary scene in which the followers of the beast are themselves devoured in the "great supper of God": "Come gather ... to eat the flesh of kings, the flesh of captains, the flesh of the mighty, the flesh of horses and their riders" (19:17–18, 21). The eaters of John's people—the unclean others who threaten to incorporate or assimilate it—are themselves eaten,

["certain kinds of particularly difficult eating originate not in the search for pleasure but in exploring the meaning of extreme and difficult emotions" (Korsmeyer 2011, 11).]

---

34. On which see especially Duff 2001, 83–96.

and the beast itself is condemned to be eternally devoured by fire (19:20; cf. 20:10).

Is John's eliminative strategy successful? On the surface it would seem to be. The whore vanishes, never to appear again, after being devoured by the beast with the horny horns. The whore disappears into the beast, and the beast disappears into the flames. Jezebel, meanwhile, has, by reason of the horrid history that sticks to her name, always already been devoured by dogs,

> ["when they went to bury her, they found no more of her than the skull and the feet and the palms of her hands.... When they came back and told [Jehu], he said, 'This is the word of the Lord, which he spoke by his servant Elijah..., "The dogs shall eat the flesh of Jezebel; the corpse of Jezebel shall be like dung on the field..., so that one can say, 'This is Jezebel' " ' " (2 Kgs 9:35–37).]

and with impeccable logic the dogs are later denied admittance to the heavenly city, refused incorporation into it: "Outside are the dogs" (22:15).

Equivocal (E)motions

The mechanics of disgust in Revelation, however, are complicated by the fact that the disgusting attracts as well as repels, and so cannot be entirely vomited or excreted from the system. The "great whore," for example, cannot be an absolute object of disgust—at least as Ahmed construes disgust—because "disgust is deeply ambivalent, involving desire for, or an attraction towards, the very objects that are felt to be repellent" (2004, 84).[35] Etymologically, "emotion" is *motion*; it comes from Latin *emovere*, "move" or "move out" (2004, 11). To emote is to be a body in motion, and in motion in relation to an object. "Disgust pulls us away from the object, a pulling that feels almost involuntary, as if our bodies were thinking for us, on behalf of us" (2004, 84). "Come out of her, my people, ... so that you do not share in her plagues," the Apocalypse urges (18:4), seeking to instill primal repulsion in the bodies of its audience. But the countervailing force is desire,

---

35. Many of us might well feel that we could come up with personal exceptions to Ahmed's generalization—recalling or imagining objects so unequivocally repulsive, so abysmally awful, as to shut down any temptation to sneak a second look, touch, taste, or sniff—but her pronouncement holds true for Revelation, as we shall see.

["The 'no' of disgust ... leaves room for various dimensions—ranging from secret to overt—of affirmation" (Menninghaus 2003, 10).]
which "pulls us towards objects, and opens us up to the bodies of others" (Ahmed 2004, 84)—perhaps to recline at table with them and partake of the same food as them, that separately ingested food symbolically uniting us with them. But what if the food in question were itself an object of disgust for still other bodies, "food sacrificed to idols," say, and hence tainted by abomination,

["You shall not eat any abhorrent thing" (Deut 14:3).]
such as the food that Jezebel and her disciples are accused of eating (Rev 2:14, 20; cf. 9:20)? The body recoils from such horrid objects; "it pulls away with an intense movement that registers in the pit of the stomach. The movement is the work of disgust" (Ahmed 2004, 85). The movement is also the predominant (e)motion that Revelation seeks to induce in its audience, a heaving of the stomach, obtained by rhetorically thrusting its fingers down the audience's throat. Revelation is a work of disgust. It wants its audience to

["Depart, depart, go out from there! Touch no unclean thing; go out from the midst of it, purify yourselves" (Isa 52:11).]
retch on Rome.

But Revelation also flirts with the object of disgust even while performing the work of disgust. "Disgust brings the body perilously close to an object only then to pull away from the object in the registering of the proximity as an offence" (Ahmed 2004, 85).

First the proximity: "When I saw her, *ethaumasa … thauma mega*," John concedes (Rev 17:6b). He was amazed with a great amazement, astonished with a mighty astonishment, fascinated with an immense fascination. This is the only vision among all his visions to elicit this reaction from the seer—the same reaction, as it happens, that "the entire earth" has to the beast, and the reaction that teeters over into outright worship: "The entire earth was amazed [*ethaumasthē*] by [lit., "after"—*opisō*] the beast.... And they worshiped the beast" (13:3–4).

Now the distance: "Why are you so amazed?," the interpreting angel asks John, and continues in effect, "I will tell you this woman's disgusting secret" (17:7). As Ahmed notes, "[the] distancing requiring proximity is crucial to the intercorporeality of the disgust encounter" (2004, 85). More precisely, "the object becomes disgusting, in a way that allows the subject to recoil, only after an intimate contact," whether real or imagined, "that is felt on the surface of the skin" (2004, 88). Hence the necessity that Rome, the intimately encroaching yet ultimately alien culture, be a prostitute in Revelation. In other words, Rome is represented in intensely sexualized terms in Revelation, terms that engender "an intensity of affect" (2004, 89), because the intolerable cultural closeness of Rome requires corporeal metaphors that evoke intimate contact, desired but unwanted, felt on the surface of the skin. Disgust "'disturbs' the skin with the possibility of desire" (2004, 88), and the disturbance does not always

> ["even to talk about affect virtually amounts to cutaneous contact" (Sedgwick 2003, 17).]

happen at a distance. The skin of the unclean other may be pressed too close to easily move away from it, to push away or push off from it, though the (e)motion of disgust. Babylon brushes up against John. Babylon touches John in an unexpected place. The "whore" has followed John to Patmos. He feels her hot breath on his neck as he pens his apocalypse and plots her destruction.

Revelation's ambivalence toward Rome, then,[36] can be affectively refigured as a simultaneous pulling toward and pulling away

> ["desire and disgust are dialectically conjoined" (Ngai 2005, 332-3).]

from the sickening object of loathing. Pulling back, pulling out, entreating *coitus interruptus*. "Come out of her, my people," pleads John (18:4), but through teeth clenched in rage. "Bodies that are disgusted," observes Ahmed, "are also bodies that feel a certain rage, a rage that the object has got close enough to sicken, and to be taken over or taken in. To be disgusted is after all *to be affected by what one has rejected*" (2004, 86).

Just Feelings

John's rage has always been contagious, as has his revulsion
> ["In fixing its objects as 'intolerable,' disgust undeniably has been and will continue to be instrumentalized in oppressive and violent ways" (Ngai 2005, 340).]

and attraction. Yet emotion as such (rage, revulsion, attraction, hate, fear, love, desire, shame …) does not circulate socially, Ahmed argues, whether in and around the Apocalypse or elsewhere. It is only objects of emotion (the one on the throne, the lamb that was slain, the beast from the abyss,

---

36. See 30–31 above.

the great whore, "that woman Jezebel" …) that circulate.³⁷ And in circulating, "such objects become sticky, or saturated with affect, as sites of personal and social tension" (Ahmed 2004, 11). Ahmed's use of the word "tension" here is also apt in relation to Revelation. The conflicting, contradictory affects that stick to the "whore" in particular, and by extension to Rome, evoke a stretched or strained state, a condition of being rent asunder, torn apart, that is felt uncomfortably in the body of the text and, by extension, in the social body this tense text conjures up—the divided body of Christ-followers in the cities of Roman Asia. From the ancient textual remains of this sundered social body

["Write in a book what you see" (Rev 1:11).]

the ghostly voices of these warring Christ-followers still echo faintly,

["I, John, am the one who heard and saw these things" (22:8).]

triangulated with the no less spectral, but also no less visceral, voice of Babylon, these distorted voices densely layered with, and continually reconfigured by, the viscously adhesive affects they have accumulated through the centuries:

Babylon: "In Tierra del Fuego a native touched with his finger some cold preserved meat which I was eating … and plainly showed disgust at its softness; whilst I felt utter disgust at my food being touched by a naked savage, though his hands did not appear dirty."³⁸

Jezebel: "Of course, it is significant that this cross-cultural encounter takes place over food, partly because the politics of 'what gets eaten' or consumed is bound up with histories of imperialism. … Through emotions, the past persists on the surface of bodies. Emotions show us how histories stay alive, even when they are not consciously remembered; how histories of colonialism, slavery, and violence shape lives and worlds in the present. The time of emotion is not always about the past, and how it sticks. Emotions also open up futures, in the ways they involve different orientations to others."³⁹

---

37. It is not that the objects of emotion are therefore static in Ahmed's model. Rather, the objects of emotion themselves "take shape as effects of circulation" (2004, 10). Ahmed is nuancing the common model of emotional contagion that derives in part from Silvan S. Tomkins and that he termed "affective resonance." For a brief introduction to the concept, see Nathanson 2008, xv–xvi.

38. Darwin 1872, 257, quoted in Ahmed 2004, 82.

39. Ahmed 2004, 83, 202.

John: "Food is significant … because food is 'taken into' the body. The fear of contamination that provokes the nausea of disgust reactions hence makes food the very 'stuff' of disgust."[40]

Jezebel: "Of course, we must eat to survive. So the very project of survival requires that we take something other into our bodies. Survival makes us vulnerable in that it requires we let what is 'not us' in; to survive we open ourselves up, and we *keep the orifices of the body open.*"[41]

John: "I feel sick, you have sickened me, you are sickening."[42]

Jezebel: "Of course, we are not just talking about emotions when we talk about emotions. The objects of emotions slide and stick and they join the intimate histories of bodies, with the public domain of justice and injustice. Justice is not simply a feeling. And feelings are not always just. But justice involves feelings, which move us across the surfaces of the world, creating ripples in the intimate contours of our lives. Where we go, with these feelings, remains an open question."[43]

---

40. Ahmed 2004, 83.
41. Ahmed 2004, 83.
42. Ahmed 2004, 85.
43. Ahmed 2004, 202.

8
Derridapocalypse*

Co-authored with Catherine Keller

Jacques Derrida once remarked that postmodernism and religion were two things that were foreign to him (Malabou and Derrida 2004, 95). In 2002, however, two years before his death, Derrida somehow found himself at the joint annual meeting of the American Academy of Religion and the Society of Biblical Literature, condemned to sit through paper after paper on Derrida and religion or Derrida and the Bible, mainly delivered by card-carrying postmodernists. Theologian Catherine Keller and I jointly presented one of those papers. Yvonne Sherwood, who had invited the paper, had asked that it engage with "later Derrida." In other words, we were not being asked to pull yet another deconstructive rabbit out of the increasingly tattered hat, or to demonstrate how biblical and theological texts burst apart at the seams when touched by the deconstructive wand to release slithering handfuls of slippery signifiers into the world. Instead, our brief was to engage exegetically and theologically with the Derrida who, it seemed, had found religion, or at least become intensely interested in it. For the later writings of Derrida teem with what Yvonne and I dub "big, flabby, old-fashioned words" (Moore and Sherwood 2011, 122–31) such as *justice, forgiveness, friendship, gift, hospitality, faith,* and *the messianic*. Several of these terms, not least justice and the messianic, evoked intimate

---

* This essay first appeared in *Derrida and Religion: Other Testaments* (ed. Yvonne Sherwood and Kevin Hart; New York: Routledge, 2005), 189–207, and is reprinted in lightly revised form with the permission of the co-author and the publisher.

points of interconnection with Revelation, but so did still further terms from Derrida's mid- to late oeuvre: *testimony/ witnessing, the absolute secret,* even *apocalypse/apokalypsis* itself along with the apocalyptic invocation, "Come!" (Rev 22:17, 20). In short, Derrida and John of Patmos were already in dialogue, sometimes explicitly, more often implicitly, and Catherine and I only needed to interject, to interrupt, to divert the conversation into the areas that most intrigued us, whether the dread-inducing monstrosity of Revelation's Messiah, or the curious propensity of Derrida's own messianic eschatology, for all its apophatic austerity, to echo orthodox Christian theology—perhaps the most covert secret of all.

Catherine Keller: As nothing like a philosopher or a biblical scholar, but something like a theologian, I perch at this table with fear and trembling. But then theology is always trembling. It oscillates between Bible and philosophy, between a ghostly apocalypse of conjurations and the discipline of the reasonable doubt. Theologians have been embarrassed by the oscillations; we have (unlike biblical scholars) tended to disavow the apocalypse and the doubt. So no wonder some of us are grateful for the mysterious resonances of deconstruction with our own lost irony, with our haunting uncertainty, and, more recently, with our politico-messianic hopes. But beyond this table, among most theologians, such appreciation of Derrida sounds at best like gratitude for crumbs—crumbs from the banquet of high theory for the hungry dogs. (Not that there is any shame in the posture of the Syrophoenician woman, the grief-stricken mother who, for that moment, healed Jesus of his Abrahamic chauvinism [Mark 7:24–30].)

In the light of Derrida's coming, it would at any rate be inhospitable not to risk admitting this gratitude. But the risk is double-edged, like the Messiah's tongue. Gratitude in the present context may be the inhospitable itself. As it has been said: "When a gift is given, first of all, no gratitude can be proportionate to it.... As soon as I say 'thank you' for a gift, I start canceling the gift, I start destroying the gift, by proposing an equivalence, that is, a circle which encircles the gift in a movement of reappropriation" (Derrida 1997b, 18). So without saying thank you, without fantasizing equality or proportionality, without preaching a Sunday school poststructuralism—shall we risk an appropriation in order to avert a destruction? Shall we risk apocalypse in order to defer it? Doesn't *he*?

Stephen Moore: As it *is* a Sunday,[1] and as "poststructuralism" (however we define the term or be ourselves defined by it) is, let's face it, the only thing, other than "religion" (no more amenable to definition, no less amenable to deformation), that brings us together around this table, the notion of a Sunday school poststructuralism might not, after all, be such a fanciful conceit. What we preach on Sunday, indeed (those of us who do), we practice throughout the week—or so Jacques Derrida has recently been teaching us.

Derrida, who has confessed, indeed circumfessed, that he "quite rightly pass[es] for an atheist" (1993a, 155–56), has, paradoxically, also declared himself *for faith*, albeit a faith that is "not religious" per se but is instead "absolutely universal" (1997b, 22). Faith is what enables any and every address to the other, for to address the other, any other, is always to ask to be believed. This request, this *demand*, for faith—utterly quotidian and ordinarily implicit—is, as such, the structural a priori of any address whatsoever. (The elucidation *of* Derrida *to* Derrida, in the *presence* of Derrida[2]—whatever that expression might mean *after* Derrida—is a somewhat bizarre public ritual, a ritual of torture at times, no doubt, to which Derrida has repeatedly been subjected over the years. And it is not without a certain dismay—but, if I may say so, also not without a certain pleasure—that I now find myself charged with turning the screws.)

If the demand for faith is to be regarded as the structural a priori of any address whatsoever, what then are we to say about the extended epistolary address that is the Apocalypse of John? Several decades ago, Derrida, in the course of a dual analysis of the Apocalypse and an antiapocalyptic essay by Kant, argued that the former reveals, in exemplary fashion, "a transcendental condition of all discourse, of all experience even" (1992b, 57).[3]

---

1. Or was, at any rate, an auspicious day on which to tackle the Apocalypse of John: "I was in the spirit on the Lord's day" (Rev 1:10).

2. Modestly seated in the second row of the audience as this paper unfolded.

3. Derrida arrives at this conclusion through reflection on the inaugural moments of the Apocalypse in particular, in which the revelation passes from God to the seven churches by way of a circuitous series of relays: Jesus, an angel, John, John's written testimony … (Rev 1:1–2). To this convoluted structure of relays—perpetually in danger of derailing or arriving at an unintended destination—Derrida perversely (re)attaches the term *apocalypse*: "as soon as one no longer knows who speaks or who writes, the text becomes apocalyptic" (1992b, 57). What apocalypse, thus reconceived, reveals is

It seems to me, however, that Derrida has succeeded in making a better case for that contention in his more recent work on faith—even though the Apocalypse is not, so far as I can see, mentioned by name in that work. (Derrida's most incisive commentary on biblical texts, however, often occurs when his attention is directed elsewhere.) "In testimony, truth is promised beyond all proof," Derrida contends near the end of his extended meditation on faith and knowledge. And again: "The act of faith demanded in bearing witness exceeds, through its structure, all intuition and all proof" (2002c, 98).

On the one hand (to employ a formulation long familiar to readers of Derrida, the corresponding "on the other hand" characteristically being deferred so long as to lull the reader into forgetting that there ever was an "on the one hand" in the first place), the Apocalypse promises, indeed provides, "proof" of the truth to which it testifies, announcing: "These words are trustworthy and true [*houtoi hoi logoi pistoi kai alēthinoi*], for the Lord ... has sent his angel to show his servants what must soon take place" (Rev 22:6; cf. 1:1–2). Is this *visible* proof? Faith, for Derrida, is inextricably bound up with blindness, and as such with an eclipsing of the ordinarily privileged sense of sight and the entire attendant epistemology of vision (see Derrida 1993b, 12). Indeed, "faith, in the moment proper to it, *is* blind. It sacrifices sight, even if it does so with an eye to seeing at last" (1993b, 30, emphasis added). The Apocalypse is faithless in this sense. It does not sacrifice sight. It is an affair, not even so much of seeing at last, but of seeing from first to last ("John, who testified ... to all that he saw," 1:2; "I, John, am the one who ... saw these things," 22:8). It testifies not to the unseen but to the seen. Being "blind" in the Apocalypse, in consequence, is equated with being "wretched," "pitiable," "poor," and "naked" (3:17).[4]

On the other hand, the Apocalypse cannot show, cannot make present the revelatory radiance that is its theme. It can only bear witness to

---

not, however, no one saying nothing to nobody, but rather "a transcendental condition of all discourse, of all experience even, of every mark or every trace." As such, apocalypse is "an *exemplary* revelation of this transcendental structure" (1992b, 57). The essay was originally published as *D'un ton apocalyptique adopté naguère en philosophie* (1983)—unless we count the English translation of it that somehow managed to precede the French original by a year: see Derrida 1982b.

4. "You do not realize that you are wretched, pitiable, poor, blind [*typhlos*], and naked," the Son of Man harangues the Laodicean church. "Therefore I counsel you to buy from me ... salve to anoint your eyes so that you may see [*hina blepēs*]" (Rev 3:17–18).

that luminescence. The Apocalypse testifies to what it claims to have seen in order to elicit faith from the other. And yet the Apocalypse is no ordinary demand for faith. The testamentary structure of everyday speech acts amounts, Derrida suggests, to declaring: "Believe what I say as one believes in a miracle" (2002c, 98).[5] But the testamentary structure of the Apocalypse amounts to a still more audacious declaration (and demand): "Believe what I say as one believes in a miracle—precisely because I am testifying to the truly invisible as that which I have truly seen." As such, the Apocalypse might indeed be said to exemplify the quasi-transcendental structure of any and every speech act: it makes manifest (or *reveals*, to use the Apocalypse's own idiom) the structural conditions of the speech act as such (even while chafing at the operational restrictions of those conditions, as we have seen). The Apocalypse, any apocalypse, would thus be a privileged instance of what Derrida has termed "pure attestation" ("if there is such a thing") (2002c, 99), which is precisely attestation to the unseen as seen, demanding a response of blind faith. "I am telling you this truth, believe me, believe what I believe, there, where you will never be able to see nor know the irreplaceable yet universalizable, exemplary place from which I speak to you" (2002c, 98).

Now, faith, in the Derridean sense (or, perhaps, in any sense), bears a privileged relationship to *the secret*. The secret subtends my address to the other insofar as that address, as testimony and appeal for blind faith, ordinarily gestures to that which is veiled from the other. The secret that most preoccupies Derrida, however—what he has termed *the absolute secret* (1995c, 59)—is not something subject to provisional concealment, and that could, in consequence and in principle, be made manifest under different conditions. The absolute secret—which, extrapolating a little from Derrida's own reflections on it, might be said to be the structural prerequisite of faith itself, and, hence, by extension, of each and every address to any and every other—does not admit of manifestation, revelation, apocalyptic uncovering, unveiling, or denuding. The absolute secret is absolutely closed, absolutely clothed, but as such infinitely open because undecidable.

Can an apocalypse contain an absolute secret, or does the apocalyptic genre necessarily make open secrets of all secrets? Is the apocalyptic gesture always and only one of unveiling? Certainly, Revelation's Son of Man

---

5. He continues: "Even the slightest testimony concerning the most plausible, ordinary or everyday thing cannot do otherwise: it must still appeal to faith as would a miracle" (2002c, 98).

seems to have trouble keeping a secret ("As for the secret [*to mystērion*] of the seven stars that you saw in my right hand, and the seven golden lampstands: the seven stars are the angels of the seven churches, and the seven lampstands are the seven churches," 1:20), and so do Revelation's *angeles interpres* ("I will tell you the secret [*to mystērion*] of the woman, and of the beast with seven heads and ten horns that carries her," 17:7). But let us begin at the beginning. A secret is evoked in the Apocalypse's opening words; whether it is the *absolute* secret remains to be seen. "The unsealed secret of Jesus Messiah, God's gift to him," begins the text (in my admittedly customized rendering of it: *apokalypsis Iēsou Christou, hēn edōken autō ho theos*). God's gift, then; but given when, given where? In answer, the text enjoins us to gaze, to gawk, to gawp through the gaping door of heaven itself, seductively left ajar by the divine doorkeeper (4:1)—the same one, no doubt, who earlier identified himself as he "who opens and no one shuts, who shuts and no one opens" (3:7; cf. 3:8). Thus it is that we become openly covert witnesses to the gift of the sealed scroll, the secret scroll—or perhaps it suffices to say: the secret. "Then I saw in the right hand of the one seated on the throne a scroll [*biblion*]," John testifies, "written on the inside and on the back, sealed with seven seals" (5:1). The only anthropomorphic physical trait attributed to "the one seated on the throne" (other than the implied backside doing the sitting—the same derriere formerly paraded before Moses in response to his plea for a vision of the divine glory, Exod 33:17–23) is this hand, and the only purpose attributed to the hand is the clutching of the scroll. Thus encircled by the divine fingers, this mystified cylindrical object looks and acts suspiciously like a phallus, and not just any phallus, but the Lacanian phallus that, as "the signifier that has no signified" (Lacan 1982b, 152), can only perform its function when veiled (Lacan 2006, 581). For it appears that the sealed scroll, the secret scroll—again, it will suffice simply to say: the secret—is indeed absolute at first, indecipherable because inaccessible, and hence unpossessable and impossible: "no one in heaven or on earth or under the earth was able to open the scroll or look into it" (5:4)—that is, until the mortally wounded Lamb, who, up until this moment, has been bleeding quietly and unnoticed nearby (5:6), working earnestly but unsuccessfully, it seems, at accepting his own castration,[6] precipitously steps forward to

---

6. Castration, in Lacanian terms, being the recognition that "the phallus, even the real phallus, is a *ghost*" (Lacan 1982a, 50).

claim the scroll as his own, with all the phantasmatic power and pomp that possession of it apparently confers: "Worthy is the cut Lamb [*to arnion to esphagmenon*] to receive power and wealth and wisdom and might and honor and glory and blessing!" (5:12). But enough of this Lacanian digression, or regression.[7]

When the Lamb unzips the very first seal (6:2), the secret threatens to leap whole and entire out of the scroll—or so it seemed, at any rate, to certain patristic expositors in particular, beginning with Irenaeus, who, taking their cue from the messianic cut of the rider on the white horse thereby let loose (cf. 19:11), imagined that the Parousia was already underway.[8] But the denuding of the secret has barely begun. And even when the seventh seal has been broken, and heaven itself has been plunged into suspenseful silence ("When the Lamb opened the seventh seal, there was silence in heaven for about half an hour," 8:1), all that ensues is another series of seven—seven further deferrals of climactic disclosure. Seven trumpets are distributed to seven angels, who proceed to blow them in turn. When the sixth trumpet is blown, a further angel—anxious, perhaps, at the prospect of yet another nail-biting half hour of heavenly silence, issuing in yet another stupendous anticlimax—blusters that "there will be no further delay, but in the days when the seventh angel is to blow his trumpet, God's secret [*to mystērion tou theou*] will be fulfilled" (10:6–7).

Immediately before this portentous announcement, however, the "seven thunders" (*hai hepta brontai*) have sounded—or spoken (*elalēsan*), rather—and John, pen poised as always to spill every secret, is unexpectedly instructed instead to "seal up what the seven thunders have said, and do not commit it to writing" (10:4). All of which raises the question: What if the real secret, the absolute secret, in the Apocalypse, were the secret revealed, unveiled, uncovered by the seven thunders—and then immediately reveiled, covered over, closed up again; in which case the absolute secret would, once again, have slipped surreptitiously through our grasp? The secret announced by the seven thunders remains secret in the Apocalypse even after all else has been laid bare. It is not covered (nor is it uncovered) by the closing injunction to the seer, "Do not seal up [*mē sphragisēs*] the words of the prophecy of this book" (22:10). But does the text dismiss the absolute secret even as it demarcates it? As if in refusal of the very

---

7. Especially as further pursuit of it would necessitate difficult passage through Derrida's own reading of Lacan (Derrida 1987, 411–96).

8. See Aune 1998a, 393; Beale 1999, 375; Weinrich 2005, 82–83.

concept of an unspillable secret, the Apocalypse, following the sounding of the seven thunders, conjures up an impatient angel, as we have seen, who, raising his right hand to heaven for dramatic effect, swears that "in the days when the seventh angel is to blow his trumpet, God's secret will be fulfilled" (10:5–7).

So what *is* the secret that is fulfilled, or rather leaked, when the seventh angel finally blows his trumpet? First and foremost, it is *a secret empire*: "Then the seventh angel blew his trumpet, and there were loud voices in heaven, saying, 'The empire [*hē basileia*] of this world has become the empire of our Lord and his Messiah, and he will reign forever and ever'" (11:15). A secret empire, then, that is also a global empire, and as such always already an open secret, administered from a heavenly throne room that, the more we peer through the door left ajar for our edification and instruction, seems to resemble an oval office—except when it resembles a CIA debriefing room instead, or a Pentagon war room.[9] But if this is the secret intelligence that the Apocalypse is only too eager to leak, indeed to flaunt, what might be the secret that it would prefer to keep under wraps, first and foremost from itself? Here is where Derrida's earlier reading of the Apocalypse, aided and abetted by his more recent reflections on justice, proves especially illuminating, enabling us to read the Apocalypse against the Apocalypse and thereby decrypt its internal communications.

The testimony of the Apocalypse, of any apocalypse, to a secret conceived, not as a closed body of content but an open space of possibility, is for Derrida encapsulated in the apocalyptic injunction, "Come!" (as in Rev 20:17: "The Spirit and the bride say, 'Come [*erchou*].' And let everyone who hears say, 'Come.' And let everyone who is thirsty come"; and again in 20:20: "The one who testifies to these things says, 'Surely I am coming soon.' Amen. Come, Lord Jesus!"). By the time Derrida has finished with it, indeed, the apocalyptic "Come" shows all the signs of having become yet another nonsynonymous synonym for *différance*: "'Come' … could not become an object, a theme, a representation" (1992b, 64). But "Come" also beckons us beyond *différance*. As a radical, irruptive opening to and for the other, otherness, the future, "Come" is also inextricably intertwined with certain later Derridean themes or nonthemes, not least *justice*

---

9. Rev 5 describes the opening of a top-secret file and the breaking of its code. Chapters 6–9 and 14–16 display a series of spectacular, shock-and-awe-inducing air strikes. Chapter 19 reports on the last-resort, boots-on-the-ground operation, as the armies of heaven invade the kingdom of the beast.

beyond the law; *hospitality beyond reciprocity*; *the gift beyond debt* (up to and including the gift of death); *democracy without sovereignty* (which is to say, the democracy to come); but most of all *the messianic* (more precisely, the messianic without messianism).[10]

Derrida is enamored of a particular anecdote about the Messiah that Maurice Blanchot relates (Blanchot 1986, 141–42; cf. Derrida 1997a, 46 n. 14, 173–74; 1997b, 24–25), in which the Messiah appears one day at the gates of Rome, but disguised as a beggar or leper—a dissimulation designed to *defer* his advent, as it turns out. One of those who lays eyes on this ragged Messiah does see through his disguise—but tellingly elects to reveil rather than reveal him, putting the denegating question to him: "When will you come?" For the Messiah, in order to be the Messiah, can never actually be present, can never actually have arrived, any more than justice—justice beyond the law, that is—or hospitality—hospitality beyond reciprocity—can ever simply be assumed to be present, to have arrived. To assume their arrival would be to evade their demands. Derrida's "messianicity *without* messianism" would be "a waiting *without* waiting," which is to say "a waiting for an event, for someone or something that, in order to happen or 'arrive,' must exceed and surprise every determinant anticipation. No future, no time-to-come [*à-venir*], no other, otherwise; no event worthy of the name, no revolution. And no justice" (1999b, 250–51).

Appropriately enough, therefore, when the Messiah does finally show up in the Apocalypse (19:11–16)—and at the shattered gates of Rome, no less (see 18:1–24)—the indiscretion, the inappropriateness, the scandal of the event is duly, if obliquely, marked in the text that announces his advent. His name is secret: "he has a name inscribed that no one knows but himself [*echōn onoma gegrammenon ho oudeis oiden ei mē autos*]" (19:12). He is incognito, then, in disguise. But it is a pitifully thin disguise. It is not as a beggar or a leper that he comes, although it might well have been (see Matt 25:35–45: "I was hungry and you fed me.... I was naked and you clothed me, I was sick ... and in prison and you visited me"). And that, perhaps, is the problem. We dread his appearance, appropriately enough, but for all the wrong reasons.

First, the dread. The Messiah, in order to be the Messiah, is, and must be, a figure of foreboding, as Derrida compellingly argues:

---

10. See Derrida 2002a, 230–98 (on justice); 2000 and 2002a, 358–420 (on hospitality); 1992a and 1995b (on the gift); 1994b, 73–83; 1997a; and 2005 (on democracy); and 1994b, 166–69 and 1999b, 250–56 (on the messianic).

who has ever been sure that the expectation of the Messiah is not, from the start, by destination and invincibly, a fear, an unbearable terror—hence the hatred of what is thus awaited? And whose coming one would wish both to quicken and infinitely to retard, as the end of the future? ... How could I desire [the] coming without simultaneously fearing it, without going to all ends to prevent it from taking place? Without going to all ends to skip such a meeting? ... The messianic sentence carries within it an irresistible disavowal. In the sentence, a structural contradiction converts a priori the called into the repressed, the desired into the undesired, the friend into the enemy. (1997a, 174)

The Messiah of the Apocalypse, too, is a figure of dread no less than desire—but less because his parousia marks the impossible arrival of an altogether unanticipatable future, oriented to justice beyond the law and hospitality beyond reciprocity, than because the Apocalypse's "Come," which impatiently holds the door open for the imminent advent of the Messiah ("I am coming soon!," 3:11; cf. 16:15; 22:7, 12, 20), is an implementation of justice as slaughter on a horrific scale,[11] and an implementation of hospitality as a horrid invitation to feast on the mangled mountain of the slain. The invitation to the dreadful banquet also opens with "Come,"[12] as it happens: "Come, assemble for God's great banquet, to devour the flesh of kings, of captains, of the mighty, of horses and of riders—the flesh of all, whether free or slave, small or great" (19:17–18). That which the Messiah establishes through the cataclysm of his coming (in a word, empire: "the empire [*basileia*] of our Lord and his Messiah," 11:15) is also that which the Messiah has come to destroy (in a word, empire: "I rule as an empress [*basilissa*]; ... I will never see grief," 18:7). Because its Messiah can build only by destroying—and by destroying on a stupendous scale—the Apocalypse converts the desired into the undesired, the friend into the foe, the Christ into the Antichrist. We have long been conditioned to regard the Antichrist as a *monster*: "And I saw a beast [*thērion*] rising out of the sea, with ten horns and seven heads" (Rev 13:1). But what if the Messiah, the Christ,

---

11. The locus classicus of this theme is Rev 6:9–11, wherein "the souls of those who had been slaughtered for the word of God and for the testimony they had given [cry] out with a loud voice, 'Sovereign Lord, holy and true, how long will it be before you judge and avenge/exact justice for our blood [*Heōs pote ... ou krineis kai ekdikeis to haima hēmōn*] on the inhabitants of the earth?'" They don't have long to wait, as it turns out (see 16:5–7; 19:1–2).

12. Although as *deute* rather than *erchou*.

were the true apocalyptic monster, the emblem and revealer of a monstrous truth? Do we dread the coming of the Messiah precisely because he *is* a monster, *the* monster, the very form of monstrosity itself?

The future, when it is absolutely unanticipated and unanticipatable, assumes monstrous form, Derrida insists (indeed, it is one of his oldest themes).[13] To embrace such a future—one not simply reducible to "a predictable, calculable, and programmable tomorrow"—would be "to welcome the monstrous *arrivant*, … to accord hospitality to that which is absolutely foreign or strange" (Derrida 1995d, 387). The future is always a monster at the door. Is this the monster—the monster Messiah—that we have begun to identify in the Apocalypse ("Behold, I stand at the door and knock," 3:20)? Yes and no.

The future that the Apocalypse, whipped on by its God and his Messiah, so frenziedly rushes forward to embrace—a war-ridden, famine-ridden, utterly ecocidal, altogether cataclysmic future—is far from unfamiliar. This insufficiently unfamiliar, inadequately unanticipatable future can always, and all to easily, be regarded simultaneously as our present (which, of course, is how the Apocalypse has managed to live on—to live an improbably long life—impossibly surviving the demise of Rome that, on its own account, should have ushered in the end of history). But might not the intolerability, the unacceptability, of an all too familiar present, or an all too easily anticipated future, be far more *monstrous*, in the end—more unsettlingly strange in its absolute familiarity, more disturbingly alien in its absolute intimacy—than a wholly unanticipatable future? Why pretend to cage the monster in the secret structurally destined to remain forever sealed—the absolute secret that the seven thunders have sounded—when it is an open secret that the monster is, and was, and is still to come (Rev 1:4, 8; 17:8)—and then to come yet again?

Catherine Keller: Well, frankly, I would prefer not to know what is coming, because of all those all too predictable processes, like the U.S.

---

13. His 1966 manifesto, "Structure, Sign, and Play in the Discourse of the Human Sciences," ended with "a glance toward those who … turn their eyes away when faced by the as yet unnamable which is proclaiming itself and which can do so … only under the species of the nonspecies, in the formless, mute, infant, and terrifying form of monstrosity" (Derrida 1978, 293). And again a year later in *Of Grammatology*: "The future can only be anticipated in the form of an absolute danger. It is that which breaks absolutely with constituted normality and can only be proclaimed, *presented*, as a sort of monstrosity" (Derrida 1976, 5).

push for a war in the neighborhood of Israel. As Derrida said already in 1992, "the war for the appropriation of Jerusalem is today the world war."[14] Like the boundless reach of the newly revealed American Empire; like the boundless "blowback" of terrorism; like the boundless filling of the globe and the exhaustion of the gift of the bounded earth. The finite future of the infinite drive to profit requires no prophet. Where I come from, the four horsemen star in movies, they have fans on every street. You can rap, dance, or tap your fingers to their familiar hoofbeat. It is, as Stephen Moore suggests, the *anticipatable* future that sends us back into the hard arms of John of Patmos. And as Derrida insists, it is in the *unknown* coming, the *avenir* in uncertainty, that hope would lie. So we turn (again) to the Derrida of what Gayatri Spivak (somewhat self-justifyingly) calls his "ethical turn" (1999, 431).[15] Turning is already apocalypse: "then I turned to see whose voice it was that spoke to me" (Rev 1:12).

So how would we read, with Derrida's help, the open secrets of apocalypse? How would we see its voices (see Keller 1996, 36–83)?[16] With eyes wide shut? "And I began to weep bitterly because no one in heaven or on earth or under the earth was able to open the scroll or to look into it" (5:4). But it is the Messiah with seven eyes, the gory lamb, the first and last, whom John (in his "prayers and tears") inscribes as the ultimate reader, who is worthy to read the scroll of seven seals. Or might we mistake Derrida for the hyperreader (after all his first and last names both have seven letters)? John's lamb who comes displays monstrosity from the start, with "seven horns and seven eyes, which are the seven spirits of God" (5:6). So each eye *is* a spirit, an optical specter. The seven-lensed spectacles fit

---

14. "It is happening everywhere, it is the world, it is today the singular figure of its being 'out of joint'" (Derrida 1994b, 58). If Derrida's presciently expansive "today" seems all too empirically correct, let us remember he is rereading its disjointedness by way of *Hamlet*.

15. In a posture not unfamiliar among certain theologians, Spivak is straining toward an activist appropriation of Derrida, while distancing herself from the taint of a merely academic deconstruction. Hence the last sentence of this hefty book: "The scholarship on Derrida's ethical turn and his relationship to Heidegger as well as on postcolonialism and deconstruction, when in the rare case it risks setting itself to work by breaking its frame, is still not identical with the setting to work of deconstruction outside the formalizing calculus specific to the academic institution" (1999, 431).

16. I did once imagine a certain begrudging pneumatological kinship with John's anti-imperial vision/audition. For a Derridean afterthought to this counterapocalypse, see Keller 2001.

the Lamb-Messiah to read the spectacular predictions hidden behind the seven seals. No blind faith, this, but true supervision.

Read under the supervision of these ghost-glasses, what is the book of Revelation but a book of specters? Its angels of terror, its ghosts under the altar, its ghost riders—not to mention John as the ultimate ghostwriter for God, or is it for the spooky Messiah, head and hair "white as white wool, white as snow," with red burning eyes (1:14). Does it not anticipate Derrida's *Specters of Marx*? Derrida invokes the dread not-quite-dead ghost of Marxism (which haunts the triumph of capitalism) but also the host of ghosts that haunted Marx himself. But Derrida shows that Marx failed to develop patience for ghosts—including the specters of the Jewish messianism that energizes all political eschatology. So Derrida proposes his eerily hospitable spectropoetics. In the interest not of exorcising but of discerning these spirits, he writes some of his most theologically important prose: "If there is a spirit of Marxism I will never be able to renounce, it is not only the critical idea or questioning stance.... It is even more a certain emancipatory and messianic affirmation, a certain experience of the promise that one can try to liberate from any dogmatics, from any metaphysico-religious determination, from any messianism" (1994b, 89).

Is it also the Messiah of the Apocalypse from whom a messianic deconstruction would liberate us? Is Derrida's democracy to come, promise, gift that which in its vulnerability must be protected from the apocalyptic Coming? The *avenir* from the *futur*? But if we look into it, isn't such a binary too oppositional, indeed too apocalyptic, for deconstruction? It would make the Messiah into the Antichrist, and Derrida's messianicity into the *true* Coming.

Yet once one reads the scroll with the spectral lenses of deconstruction one recognizes how closely the monster and the Messiah mimic and mock one another—right down to their display of wounds, their surplus of horns and of eyes, their coupling with an urban femininity: the whore of Babylon, as Rome, in one instance, the new Jerusalem as the bride in the other. Such a politically charged mimicry: for the Messiah has always signified the antiempire, and the whore the beast empire itself.[17] But as it

---

17. I will not rehearse the case here, which is not hard to make, for the anti-imperial intentions of the book, for the millennium-long history of politically revolutionary deployments of John's Apocalypse, and the twentieth-century liberation apocalypse among Christians of the so-called developing world. These are recapitulated in *Apocalypse Now and Then* (Keller 1996). Let me point only to Ernst Bloch's *Principle*

turns out they both stand under the banner of the "coming": the Lamb-Messiah is "the one who is and who was and who is coming"—*ho erchomenos* (1:4). But according to John, the beast also was and is to come (17:8).

Yet the leading eschatological thinker in twentieth-century theology, Jürgen Moltmann, has theology depend on the "coming" as the distinguishing mark of the Jewish and Christian Messiah. This politically progressive Protestant translation of eschatology into "hope" rather than "end things," and hope as the *Zukunft*, the to-come of *adventus* rather than the calculable linearity of *futurum*, parallels Derrida significantly—if not, as we will see, unproblematically.[18] But it is not just the messianic that comes! People "will be amazed when they see the beast, because it was, and is not, and is to come" (17:8). Amazed, perhaps, because the beast iterates, it parodies, the temporal structure of the messianic hope—but with a *différance*: one recognizes that the difference between the Messiah and the monster comes, indeed comes down, to the copula. Both were, both will come. But only the Messiah *is*.

This is the infinitesimal but infinite gap: the beast is only as an is-not, as a present of absence, whereas the Messiah is the subject of a tense presence, the present tense of a "to be" that conjugates the entire alphabet of salvation history. But the copulating beast-whore couple mocks the copula itself, it haunts the alpha-omegic order of "is" with monstrous writing; the beast is "full of blasphemous names" (13:1; 17:3). We will have been alerted (by a certain critique of the metaphysics of presence) to the totalizing potential of this revelation of a pure present. But how would this messianic wisp of minimally hellenized ontology, in this abysmally unphilosophical text, written in an inelegantly hebrewized Greek, have caught the *ousia* virus? Or does it rather carry the precondition for the subsequent ontotheologizing of Christianity—that which Derrida calls, in distinction

---

of Hope (1986) and, less enthusiastically, Norman Cohn's *Pursuit of the Millennium* (1970) as pivotal accounts of the political *Wirkungsgeschichte* of the text.

18. Moltmann's *Theology of Hope* (1967), a twentieth-century theological classic, made the key transitions: "The more Christianity became an organization for discipleship under the auspices of the Roman state religion and persistently upheld the claims of that religion, the more eschatology and its mobilizing, revolutionizing and critical effects upon history as it has now to be lived were left to fanatical sects and revolutionary groups." But once we read the biblical testimonies as "full to the brim with future hope of a messianic kind for the world," we realize that "the eschatological is not one element of Christianity, but it is the medium of Christian faith as such" (15–16). His specific enunciation of the "coming" as *adventus/Zukunft* will be discussed below.

from and as the ghostly precondition of ontology, *hauntology*? Might this precondition lie in the hope, all too human among subjects of imperial injustice then and now, for the cessation of brutality? But more, for its reciprocation, and finally—oh surely, so deservedly—for the gift of a life without suffering, without death?

The Apocalypse wants an end to mourning (at least for its own people). Yet Derrida is teaching, if I am not mistaken, that any politics that would eschew brutality, that would not replicate empire even as it revolts against empire, can never eschew the work of mourning.[19] John's ecclesia, with its ghost-martyrs crying for vengeance, its strange angels bringing justice by way of global terror and mass death, wants no more tears. No more death. No more sea (21:1-4). For tears condense out of the chaos of the primal salt waters. So immediately after 9/11 the Bush regime forced mourning toward violence; one hundred times that many children die daily, daily, from avoidable causes: and as Derrida has noted, we let them.[20] Nor can we grieve for all whom the peoples of the book daily make into ghosts.

No wonder: at the end of the *biblos*, an entire *bibliotheque* of texts rich with grief, mourning got shut down once and for all. Along with messianic comfort, the apocalypse offers a merciless preemption of history: dis/closure as closure. A closing of the very space of disclosure (final revela-

---

19. Derrida's "topology of mourning" as the "spectral spiritualization that is at work in any *techne*" may be as interminable as mourning itself, and so extends of course indefinitely beyond, if it can, the contours of the specifically political: "A mourning in fact and by right interminable, without possible normality, without reliable limit, in its reality or in its concept, between introjection and incorporation. But the same logic ... responds to the injunction of a justice which, beyond right or law, rises up in the very respect owed to whoever is not, no longer or not yet, living, presently living" (1994b, 97).

20. See Derrida's Kierkegaardian meditation on responsibility, suggesting of course no ethical fix to a paradox that perhaps the invocation of "Bush" flattens—but also tests (for at what point does the inevitability of "sacrifice" enable the most vulgar collusion with brutality)? "As soon as I enter into a relation with the other, with the gaze, look, request, love, command, or call of the other, I know that I can respond only by sacrificing ethics, that is, by sacrificing whatever obliges me to also respond, in the same way, in the same instant, to all the others. I offer a gift of death, I betray, I don't need to raise my knife over my son on Mount Moriah for that.... I am sacrificing and betraying at every moment all my other obligations: my obligations to the others whom I know or don't know, the billions of my fellows (without mentioning the animals that are even more other others than my fellows), my fellows who are dying of starvation or sickness" (1995b, 68–69).

tion). We who dwell in the land of the doctrine of preemption, the land of *Ghostbusters*, must now newly grieve and resist our beastly messianism. Among us the tearless white warrior of they-are-evil-we-are-good have-a-nice-day righteousness comes hybridized with the drag queen of Babylon/Romanhattan, she who said "I will never see grief" (Rev 18:7). How else can we read the peculiar production of a born-again Christian (who says his favorite philosopher is Jesus) as Roman-style emperor? Not that Dubya Caesar is performing an original—there is a long history of copulation between messianism and colonialism. It was born in Western form as the Crusade to Jerusalem. But now this hybrid Messiah-Caesar complex metabolizes in the high speed global media of what Hardt and Negri have dubbed the "postmodern Empire" (2000).

Yet the medium of John's Apocalypse already seems spectropoetic. In its scrolling bombardment of images, blunt bits of the poetry of prophets kaleidoscope at an oneiric speed.[21] It prefigures what Derrida calls "techno-tele-iconicity": the medium of the media, the "techno-tele-discursivity" he says "determines the spacing of public space, the very possibility of the *res publica* and the phenomenality of the political. ... This element itself is neither living nor dead, present nor absent: it spectralizes" (1994b, 51). Indeed the *res publica* is now *res privata*—so notes Néstor Míguez, who like most liberation theologians (all knowingly haunted by the ghost of Marx) loves John's Apocalypse for its denunciation of the globalizing greed then and now.[22] "Public things" are being privatized for profit, while what was private appears in televised public spectacles, pubic impeachments, talk shows.

Indeed, Derrida's "Faith and Knowledge" (2002c) tracks the alliance of religion with "tele-technoscience," which he calls globalization itself. But "on the other hand," it declares war against this power that dislodges religion from "all its proper places, *in truth from place itself*, from the *taking-place* of its truth"—hence the "auto-immune reaction" within religion: "the auto-immunitary haunts the community ... like the hyperbole of its own possibility" (Derrida 2002c, 82). Intriguingly,

---

21. I discuss the breathless compression of Hebrew poetry, especially Ezekiel, in this flashing proto-MTV vision, in "Eyeing the Apocalypse" (2001) as well as *Apocalypse Now and Then* (1996).

22. For this invaluable formulation, I thank Néstor O. Míguez for his unpublished keynote address for the Oxford Institute, "The Old Creation in the New, the New Creation in the Old" (2002).

this ghostly global techno-tele-iconicity, which is so effectively deployed among the apocalyptic hyperboles of Abrahamism (the so-called fundamentalisms), is specifically what provokes in *Specters* the announcement of a hauntology.

So is it too much of a stretch to suggest that the global spatiality (and what place is more global than apocalyptic space?) of the dissolution of the public and private into each other sheds light on Derrida's "taste for the secret" (Derrida and Ferraris 2001)? Instead of reading the latter as a symptom of his crypto-bourgeois individualism, we could recognize its protection of a space of alterity, of nonbelonging. That space characterizes not only one Franco-Algerian Jew but also the ever more migratory masses of the globe: "the demand," he writes, "that everything be paraded in the public square ... is a glaring sign of the totalitarianization of democracy." (We who mourn the possibility of U.S. democracy will be needing this phrase.) "In terms of political ethics: if a right to the secret is not maintained, we are in a totalitarian space" (Derrida and Ferraris 2001, 59). In the space of the apocalyptic utopia, the displacement of space itself, darkness, ocean, and death have been eliminated. "God is the light" of the new Jerusalem, "and its lamp is the Lamb" (Rev 21:23). A ghost-white transparency of goodness and security rule: a neon panopticon, shining through the lamb-lamp. For the seven spectral eyes do not just see but shine.

On the other hand: these city streets "transparent as glass" (21:21) are lined with trees leafing "for the healing of the nations" (22:2): they encode the oppression of those often inhabiting filthy streets. They yearn for "the water of life as a gift" (22:17) as the desert spreads, water wars loom, and the empire privatizes every public good. The book concludes with an entire riff on coming: "I am coming soon.... The spirit and the bride say, 'Come.' ... Let everyone who hears say, 'Come.' ... Let everyone who is thirsty come" (22:12a, 17). To this water, always at least literal, every other is invited along with the invitation of the utterly other. *Tout autre est tout autre* (Derrida 1995b, 82–115).

Does John's Apocalypse dis/close—or only close—what Derrida has here and there—delicately empirical—referred to as the chaos of the gaping mouth, of thirst and hunger as well as speech? Having looked into it for all too long, I still see no closure of this undecidable scroll with the End always already in its sights. Its empire and antiempire continue to conjugate history, separated only by the negation of a presence too pure to recognize its own irony: its own *is/is not*. No end in sight of apocalypse or of empire, of the autoimmune violence of our bloody Abrahamic purities.

If this text won't close, don't we need an opening within the space of its haunted iconicity? But within the terms of the sibling rivals of the Abrahamic patrilineage how would that space open—except as more desert? Derrida finds a promising chaos in that very desert, a deconstructive kenosis. But what of the rivalrous women, Sarah and Hagar, unsisterly, divided but never quite conquered? Would their ghosts settle now, after so much movement of women, for these desert patrimonies—for the messianic masculinities? For their crumbs? Unexpectedly Hagar survived in the desert, as did the anonymous goddess of the apocalypse chased there by the first beast.[23] Then the earth opened its mouth: the very maw of chaos nonviolently swallows the vomit of the dragon, the effluvium with which he had sought to drown her. But now—would these desert women, practiced in a wide variety of open mouths, not also (re)open the watery chaos, *thalassa*, the mythic sea, the salty birth waters, the bottomless flux or *tehom* that apocalypse nihilates along with death, night, and tears?

Not a pure femininity (goddess forbid), not a feminist apocalypse, and certainly not a pure origin, a patristic *ex nihilo*—but something more like a Joycean "chaosmos of Alle"? Might *tehom* (in some dream of a divine woman) lend another "nonsynonymous substitution" to what Derrida calls, in a chaosmically clarifying paradox, the "heterogeneity of origin"? "Heterogeneity opens things up," he says (1994b, 33). Is this the very dis/closure that opens the apocalypse up, precisely there where it would shut everything down, a counterapocalypse that is no mere pro- or antiapocalypse?

Here my question becomes confessedly, though not circumfessedly, theological. If the "heterogeneity of origin" deconstructs (as I believe it must) the *ex nihilo* of an orthodox origin, doesn't it also call for a heterogeneity of the eschaton? But wouldn't such a heterogeneous future upset the purity, the absoluteness, the unilateral gift, of the coming? At times Derri-

---

23. "But the woman was given the two wings of the great eagle, so that she could fly from the serpent into the wilderness, to her place where she is nourished for a time, and times, and half a time. [Time out of joint indeed!] Then from his mouth the serpent poured water like a river after the woman, to sweep her away with the flood. But the earth came to the help of the woman: it opened its mouth and swallowed the river that the dragon had poured from his mouth" (Rev 12:14–16). Amid the many graphic oralities of the Apocalypse, the nurturing desert and the vomiting yet voracious beast invoke the scene of a burning and many-orificed desire (see Keller 1996, 70–73).

da's messianicity seems to invoke such a purity: when he calls for the "absolutely undetermined messianic hope" (1994b, 65), or, with Kierkegaard, for the "absolute secret," *ab-solutum*, absolved from any bond, detached, out of joint (Derrida 1995b, 57). Then it is as though any moment of joining, any connectivity, would deny the time out of joint; as though one is either detached or fused, as though attachment entails determination, confinement, closure; as though we might disavow the chaosmic fluidities of our interrelations for the sake of a deconstructive absolute, purified even of the possible. I realize that Derrida—at these present-transcendent moments—means to save the undetermined future from any (theological) foreclosure: "As soon a determinate outline is given to the future, to the promise, even to the Messiah, the messianic loses its purity, and the same is true of the eschatological" (Derrida and Ferraris 2001, 20).

Still: is messianic purity the only alternative to determination? This question is posed within a tradition in which the omnipotent One, himself the essence of origin and end, routinely determines outcomes; in the name of opening up a transcendent future he closes down history. (Oh please, whoever comes fresh to religion, "turn and see" the force, the violence, the homogenizing Presence of every unhistoricized enunciation of this "he": please do not casually erase the grammatology of a few decades of fragile, feminist theology.) From this "Nobodaddy" (Blake) the indeterminate certainly needs messianic salvation.

Derrida proposes therefore a "messianic eschatology so desertic that no religion and no ontology could identify themselves with it" (Derrida and Ferraris 2001, 21). This is an intriguing tactic: to dry the ontotheology out of eschatology, to bake the religious out of the messianic. Of course, I love its negation of dogma itself, and ipso facto of all the dogmatisms that keep women in the role of God-dogs, licking the leavings of the religious masters. This desert eschatology answers to Derrida's "faith without religion." But here is my worry (and let me state it without frisson of feminist fury, without for the moment the distraction of symbolic sex, without apocalyptic ambush): might this very strategy not be echoing—so inadvertently, indeed with such gentle intent—the foundation of orthodox theology? For "in the beginning"—not of Genesis but of Christian orthodoxy—the *ex nihilo* had evaporated the *tehom*, the watery abyss whose traces remain in Scripture until they are vaporized in the apocalypse. The *ex nihilo* purged the Jewish and mythic residues of chaos, and at the same time established a divine sovereignty of pure power—which determines through grace the purity of faith. After all, wasn't Protestant neo-orthodoxy founded on such an opposition between the purity of faith and the heterogeneity of religion—Karl Barth's "Christian faith" versus any, including Christian, "religion"? (Naturally enough, Derrida's Christian interlocutors are understandably, but massively, Roman Catholic: John Caputo, Kevin Hart, Jean-Luc Marion, even David Tracy—so I am aware that a certain problematic within Protestantism, involving the totalizing effects of Protestant versions of transcendence *sola scriptura, sola gratia, sola fides*, may for Derrida lack comparable mediation, except by way of Kierkegaard, Barth's inspiration, or Heidegger oddly, and fundamentalism repugnantly.)

Nonetheless some of us within and between the religions depend on Derrida to help release the infinite indeterminacy—khoric and tehomic— from the anxious grip of every orthodoxy, even the most progressive. Indeed, for this bottomless indeterminism—in its democratically cosmopolitan justice as well as its meditative apophasis—some of us have come to depend on his mysterious overflow into theology, his divine surplus. So then one does not want some spectral afterimage, some theological ghost, of Derrida to be reinforcing the kind of paternalist dichotomy that invests even the socially responsible messianisms of theology. (Perhaps it is not he who is responsible for such Derridean specters, but those of us who interpret him theologically.)

As I noted earlier, Derrida's assertion of the pure coming, the *avenir*, over against the determinate *futur* structurally nearly parallels Moltmann's binary of a pure and promised *Zukunft* versus the emergent future: coming versus becoming (not only against being but also against any Nietzschean or Whiteheadian immanence of becoming). But Moltmann criticizes the Parmenidean eternal presence only to yield to a *theologia gloriae* of "lasting being in the coming presence of God": the *parousia* yields total *ousia*, in the end—after death and transience have been overcome.[24] The One who comes arrives in his (*sic*) glory, never again to suffer the *zimzum* of nonbeing. Of course, even if it comes dangerously close to mirroring the two-kingdom structure of law and grace, Derrida's own binary of determinate history and absolute promise heralds no total

---

24. "God's Being is in his coming, not in his becoming. If it were in his becoming, then it would also be in his passing away. But as the Coming One (*ho erchomenos*), through his promises and his Spirit (which precede his coming and announce it) God now already sets present and past in the light of his eschatological arrival.... The coming of God means the coming of a being that no longer dies and a time that no longer passes away" (Moltmann 1996, 23–24).

or final coming. Au contraire. He presumes—with *theologically* crucial insight—that the hardening of the messianic into a Messiah will produce such totalizing effects. But if the only alternatives are a determinist appropriation, on the one hand, and the gift of a separative absolute, on the other, might Derrida's own "gift" not harmonize, hauntingly, with the triumphant chorale of God's absolutely free and transcendent gift, *charis*, grace, *sola gratia*—a unilateral, pure omnipotence, whether coming from above or from the future?

This would be a spooky surplus indeed—at least for those theologies, including most feminist and ecological varieties, which for the sake of a sustainable justice and a credible faith resist the imaginary of omnipotence, indeed for those heterodoxies in which the divine morphs into the *ruach*, *Geist*, spirit of infinite indeterminacy. Can Derrida's "messianic performative" work within Christianity to gird the loins of a *deus absconditus* who absconds once again with all agency, leaving humanity enough rope to hang itself with? Or mainly, as I hope, to provoke spirited—indeed sometimes graceful—actualizations of what might not otherwise have been possible?

Instead of reestablishing the dry abyss—between the future, which will come predictably from our efforts, and *ho erchomenos*, that which comes despite all effort: can we not admit the Derridean heterogeneity into the gap itself? Need we understand the agency of our efforts as a linear determinism rather than as a complex, uncertain multicausality, unfolding at the edge—the *eschatos*—of chaos? I find a hint of this alternative flow of agencies enacted in Derrida's notion of the "I": "there is no 'I' that ethically makes room for the other, but rather an 'I' that is structured by the alterity within it, an 'I' that is itself in a state of self-deconstruction, of dislocation." And so "the other is there in any case, it will arrive if it wants, but before me, before I could have foreseen it" (Derrida and Ferraris 2001, 84). The messianic other—*tout autre*—as arriving before "I am" upends the linear determinism of any closed system. It counters apocalypse with dis/closure. At the same time it suggests a momentary "I" always already heterogeneous with—indeed, co-constituted by—the future coming. This "I" comes-to-be as event (to borrow Whitehead's language from the 1920s, loved by Deleuze) only through its prehensions of the others that precede it—indeed, that haunt it!

This impure "I" can never be absolved of its other. So why impose purity onto the other itself? Why not let the *tout autre*, whatever or whoever it will come-to-be, also appear as impure, heterogeneous, already taking account of *its* others (of me), *as* it comes? Then *ruach* is emptied of the dominological structure of *sola gratia*—though perhaps not of her tehomic grace. As to Derrida's so graceful gift to theology: he will not offer us an apocalyptic feast, *dieu merci*, but healing crumbs. In the shared spirit of an indeconstructible justice—as indeconstructible as deconstruction itself (see Derrida 2002a, 243)—he will not cease to haunt Scripture and its interpreters. As we have come—to haunt him.

9
Quadrupedal Christ*

Biblical ecocriticism has featured in my teaching since the late 1990s, but it is only recently that I have begun to write in that mode (Moore 2011b; 2013; 2014) and hence to etch out a modest niche for myself in that ever more important field. I have done so in response to an animal gaze. I would like to be able to attest that I was convicted by a numinous encounter with some animal other in the wild—a mountain gorilla, say, or some other critter that less comfortably mirrored my own creaturely contours. But the arresting animal gaze came instead from the feline eye emblazoned on the cover of Jacques Derrida's *The Animal That Therefore I Am* (2008a), as encountered on Amazon.com, that lush habitat for book covers. This cover was thoughtfully displayed, for my added buying pleasure, with covers of other recent animal books by other leading philosophers and theorists. And immediately I was seized by the hope that here, coiled under this accusatory eye, lay resources that would add intellectual bite to my often toothless classroom ruminations on ecology and the biblical texts.

This essay and the next one emerged from a series of marginal scribbles in *The Animal That Therefore I Am* and its littermates, the two volumes of *The Beast and the Sovereign* (Derrida 2009; 2011). What the essays attempt to do is bring the field of posthuman animality studies, a field catalyzed

---

* This essay first appeared under the title "Ruminations on Revelation's Ruminant, Quadrupedal Christ; or, the Even-Toed Ungulate That Therefore I Am," in *The Bible and Posthumanism* (ed. Jennifer L. Koosed; Semeia Studies 74; Atlanta: Society of Biblical Literature, 2014), 301–26, and is reprinted with the permission of the publisher.

by Derrida's animality theory, into dialogue with Revelation, itself an extraordinary animal book. If the earlier essays in this volume focused variously on the one seated on the throne, the one like a Son of Man, the celestial superwarrior on the white horse, and the wicked stepsisters Jezebel and Babylon, these concluding essays finally focus on Revelation's central character, who, rather astonishingly, happens to be a ruminant, quadrupedal, even-toed ungulate—a sheep, in short, although one with irregular ocular and cornual features. Most importantly, however—and also most mundanely—this is a butchered sheep: it has met the fate of sheep everywhere. And this, for me, is the logical place to begin ecological reflection on Revelation.

All the things that a sheep has inside it and that he has inside him too. (Coetzee 1997, 98)

Anomanimality

The Lamb has long been the elephant in the room of Revelation scholarship. What does it mean—theologically, philosophically, ecologically—that the figure introduced as "like a Son of Man" (*homoion huion anthrōpou*) in Revelation's inaugural vision (see 1:13) has ceased to be anthropomorphic by the time we reach Revelation's throne room scene ("I saw ... a Lamb [*arnion*]," 5:6)? What does it mean that Revelation's Christ moves through most of the subsequent narrative not on two legs but on four? By and large, the burgeoning body of ecocritical and ecotheological work on Revelation[1] is oddly silent on this highly conspicuous spectacle and on the no less obvious fact that Revelation in general is a bizarre bestiary,[2] more thickly populated with nonhuman animals than any other early Christian

---

1. See, e.g., Rossing 1999b, 2002, 2005a, 2008; Reid 2000; Maier 2002b; Hawkin 2003; Bauckham 2010, 174–78; 2011, 163–84; Bredin 2010, 165–80; Cate 2010; Horrell, 2010, 98–101.

2. As are other ancient Jewish apocalyptic works or sections of such works, most notably the "Animal Apocalypse" of *1 Enoch* 85–90.

text.[3] Such work has tended to grapple instead with the ecocidal excesses of Revelation 8 and 16, seize on the fleeting moment of agency accorded the earth in 12:16 ("But the earth [hē gē] came to the help of the woman"), and contentedly come to rest in the city park of 22:1–2.[4] But even in the latter locale, it is the water flowing "through the middle of the street of the city" and the tree on either side of the stream that has tended to capture the ecological imagination, not the nonhuman animal that also features in the vision—an altogether anomalous animal, as we shall see, enthroned, not encaged, in the city park (22:1, 3).[5] This anomalous animality—*anomanimality*, if you will—is the principal focus of the present essay.

BEFORE THE ANIMAL

This essay will have recourse to Jacques Derrida's three posthumously published animal books, *The Animal That Therefore I Am* (2008a) and the two volumes of *The Beast and the Sovereign* (2009; 2011), to analyze and defamiliarize Revelation's animal Christology.[6] The first of these books—

---

3. Theologian Catherine Keller comes closest, perhaps, to being the exception to the rule in the Revelation chapters of her *God and Power* (2005, 34–95). Her ecofeminist, poststructuralist, postcolonial reading of Revelation frequently engages with its "cosmic bestiary" (72)—although less with the animality of the Lamb, ultimately, than of the four living creatures around the throne (see 67–95), which were also the focus of her earlier "Eyeing the Apocalypse" (2001). While not explicitly ecological in thrust, Ingvild Saelid Gilhus's brief survey of the animals of Revelation in her *Animals, Gods and Humans* (2006, 176–80) is also worth consulting, as is James Resseguie's treatment of Revelation's animals as literary characters in his *Revelation Unsealed* (1998, 117–36). Revelation is all but absent, however (and oddly so), from Robert Grant's *Early Christians and Animals* (1999).

4. Barbara Rossing's position is typical: "Revelation emphasizes that our future dwelling will be with God on earth, in a radiant, thriving city landscape" (2005a, 171). For less typical, more cautious treatments of the heavenly city and the representation of nature within it, see Martin 2009; Horrell 2010, 100–101. For my own extended reflections on the new Jerusalem, see 235–43 below.

5. The spectacle is not to be confused with the Central Park Zoo, then.

6. With the appearance of the second volume of *The Beast and the Sovereign* (see also Derrida 1995a; 2004), more than eight hundred pages single-mindedly devoted to the animal had appeared in Derrida's name—even apart from all of the less sustained engagements with the animal that had marked his writing from the beginning ("the innumerable critters that … overpopulate my texts" [2008a, 37–38])—making animality one of his central and most enduring philosophical themes. For book-length

or, more precisely, its first chapter[7]—has been a crucial catalyst (one of several) for a heterogeneous academic field that has attracted various (nonsynonymous) names, notably, *animal studies*, *critical animal studies*, *animality studies*, and *posthuman animality studies*.[8] The term *posthuman* in this context is frequently a synonym for "post-Cartesian."[9] Descartes is, indeed, something of a bête noire for animal studies. The Cartesian elevation of individual subjectivity, it is now commonly asserted, was obtained by reconceiving the relations between human and nonhuman animals in terms that were absolutely oppositional and hierarchical.[10] But the term *animal(s)* is perhaps not the best one in this context. Prior to the Cartesian revolution in philosophy there were no "animals" in the modern sense. There were "creatures," "beasts," and "living things," a bionomic arrangement reflected in, and reinforced by, the early vernacular Bibles. As Laurie Shannon notes (2009, 476), "*animal* never appears in the benchmark English of the Great Bible (1539), the Geneva Bible (1560), or the

---

discussions of Derrida's animal work, see Badmington 2007; Lawlor 2007; Berger and Segarra 2011; and Krell 2013; and for biblically and/or theologically oriented discussions, see Chrulew 2008 and several of the essays in Koosed 2014 and Moore 2014.

7. Originally published as "L'animal que donc je suis (à suivre)" (Derrida 1999a); ET: "The Animal That Therefore I Am (More to Follow)" (Derrida 2002b).

8. As Marianne DeKoven (2009, 368 n. 3) notes, "There is disagreement in the field over terminology. In general, those primarily motivated by animal advocacy and by the human-animal relation favor *animal studies*, while theorists of the posthuman, who want to move beyond the human-animal distinction, often prefer *animality studies*." See further Lundblad 2009.

9. For a more comprehensive treatment of the posthuman than is possible in this essay, see Wolfe 2009b; Braidotti 2013.

10. Descartes was radicalizing philosophical and theological views of the animal with deep roots in antiquity. Greco-Roman philosophy was characterized by a broad and complex range of positions on human-animal relations, certain of which anticipated those of Descartes. Aristotle in his voluminous writings on animals distinguished them from humans by their alleged lack of reason, speech, and upright posture. The Stoics built on and extended Aristotle's ideas on animals, and their ideas in turn were adapted by Jews such as Philo and Christians such as Augustine. More nuanced views of human-animal relations stemmed from the Platonic and Pythagorean traditions, and received extended expression in the works of such philosophers as Plutarch and Porphyry. For overviews of these ancient debates, see Gilhus 2006, esp. 37–63; and Spittler 2008, 15–26. For a magisterial discussion of Descartes's ideas on animals in relation to those of Aristotle, Augustine, Aquinas, and other seminal philosophers and theologians, see Steiner 2005, 132–52, together with 53–131.

King James Version (1611)."[11] More significantly, the continuum evoked by the term *creature* also included angels and demons, so that premodern humans saw themselves as embedded in a multilayered biosphere. Missing was "the fundamentally modern sense of the animal or animals as humanity's persistent, solitary opposite" (Shannon 2009, 476).[12] The term *animal* could be employed without implying stark, dichotomous opposition to *man*. Susan Crane (2013, 1–2) cites an illuminating medieval evocation of human animality or bestial humanity:

> John Trevisa's fourteenth-century translation of Bartholomaeus Anglicus places the human within the animal category: "All that is compounded of flesh and spirit of life, and so of body and soul, is called *animal*, a beast, whether it be of the air like birds, or of the water like fish that swim, or of the earth such as beasts that go on the ground and in fields, *like men* and wild and tame beasts."

Even Trevisa's subsequent theological qualification of "man" does not de-animalize him. Trevisa cites Isidore of Seville's sixth-century *Etymologies*, which "says that a man is a beast that resembles God" (Crane 2013, 2).[13]

Descartes was the prime creator of the animal in the peculiarly modern sense of the term. What Descartes did was cull the human creature, conceived as the only one "equipped with a rational soul, from the entire spectrum of creatures," all others being consigned to "the mechanistic limits of purely instinctual behavior" (Shannon 2009, 476). This radical reconception of the nonhuman animal is commonly termed the *bête-machine* ("animal-machine") doctrine for its equation of animals with clocks and other

---

11. The title of Shannon's article, "The Eight Animals in Shakespeare; or, Before the Human," refers to the fact that the term *animal* occurs only eight times in Shakespeare's entire oeuvre, while the terms *beast* and *creature* occur hundreds of times. "As the *OED* confirms, *animal* hardly appears in English before the end of the sixteenth century" (Shannon 2009, 474). She has since developed these ideas in more detail (Shannon 2013, esp. 1–19).

12. Donna Haraway pointedly uses the term *critters* for both human and nonhuman animals. She writes: "Critters are always relationally entangled rather than taxonomically neat" (2007, 330 n. 33).

13. Work on medieval views on human-animal relations began in earnest with Salisbury 1994. Not all medievalists would agree, however, that medieval culture at large was comfortable with soft human-animal distinctions. For a sharp contestation of that position, see Steel 2011.

mechanisms with automatic moving parts.[14] The Cartesian human/animal antithesis has powerfully catalyzed both a philosophical and physical erasure of the animal, one whose effects are manifested with unprecedented starkness in our own time. As Shannon (2009, 477) observes:

> The disappearance of the more protean *creatures* into the abstract nominalizations of *animal, the animal,* and *animals* parallels livestock's banishment to a clandestine, dystopian world of industrial food production, where the unspeakable conditions of life depend on invisibility. It mirrors, too, the increasing confinement of wildlife in preserves as wild spaces disappear with alarming speed.

Shannon's article was one of fourteen on human-animal relations that appeared in the March 2009 issue of *PMLA*, the flagship journal of the Modern Language Association. "Why Animals Now?" is the title of the lead article in the collection (DeKoven 2009).[15] One possible answer to the question of why human-animal relations have become a locus of intellectual energy and ethical investment in the humanities is that prominent theorists and philosophers have been writing on them and thereby providing models for other ecologically minded academics who also want to write on them. The most influential of these theoretical/philosophical writings (to return to the claim with which this section began) has been Derrida's "The Animal That Therefore I Am" (2002b) and the posthumously published book of the same name (2008a).[16] Derrida's title is a riposte to

---

14. For the doctrine, see Descartes 2006 (French orig. 1637), 35–49; and 2000, 275–76, 292–96 (two letters from 1646 and 1649 respectively).

15. Compare Kari Weil's *Thinking Animals* (2012), the subtitle of which is *Why Animal Studies Now?*

16. Cary Wolfe (2009a, 570), in his magisterial survey of the field of animal studies, contends that Derrida's article "([along with the] book that shares its title) is arguably the single most important event in the brief history of animal studies." The work of Donna Haraway, especially her *When Species Meet* (2007; see also Haraway 1990, 2003), has also been highly influential. Giorgio Agamben's *The Open* (2004) has been another prominent contribution, as also (from the analytic side of the analytic/continental philosophical divide) has been Stanley Cavell et al.'s *Philosophy and Animal Life* (2008). For introductions to the major philosophical/theoretical work on animality, see Calarco and Atterton 2004; Calarco 2008; Oliver 2009. For the intersection of animal studies and postcolonial studies, see Huggan and Tiffin 2010, which includes a biblically oriented chapter, "Christianity, Cannibalism and Carnivory" (162–84); and DeLoughrey and Handley 2011. For more general reflections on the complex relations

Descartes's "I think, therefore I am"—"a summons issued to Descartes," as he himself puts it (2008a, 75).

To summarize, then, the preeminent modern philosophical category—*the human*—has been based on a conceptual subjection of the animal, and the material corollary of that conceptual subjugation has been an actual subjugation, even annihilation, of the animal on an unprecedented scale. "A war against the animal," is how Derrida phrases this phenomenon, a "war to the death" that threatens to "end in a world without animals, without any animal worthy of the name," living for anything other than as a means for the human (2008a, 101–2; cf. 2009, 302–3). Derrida writes of the "sacrificialist current" that animates the Cartesian *cogito* and other influential philosophical discourses on the animal (Kantian, Heideggerian, etc.)—not "sacrificial," however, in the sense of a "ritual sacrifice of the animal" but rather in the sense of a "founding sacrifice" enacted "within a human space where … exercising power over the animal to the point of being able to put it to death when necessary is not forbidden" (2008a, 90–91). As we are about to see, Revelation both affirms and disturbs this sacrificial logic, at once age-old and peculiarly modern.

The Hyphen between God and Sheep

On the one hand (hoof, paw, claw …), instead of the asymmetrical, antithetical human/animal dyad endemic to post-Cartesian modernity, Revelation presents us with a divine/human/animal triad, each of the three terms bleeding profusely into the other two. Revelation opens with a vision of one *homoion huion anthrōpou* (1:13–16)—one like a Son of Man, a Son of Humanity, a Human Being. Although labeled as human, however, this numinous figure bears the marks of divinity on his physical person: most conspicuously, the wool-like whiteness of his hair (*hōs erion leukon*, 1:14a) evokes the wool-like whiteness of the Ancient One's hair in Daniel 7:9 (LXX: *hōsei erion leukon*).[17] The Human Being is also a Divine Being.[18]

---

between animal advocacy and advocacy for oppressed human groups, see DeKoven and Lundblad 2012.

17. The "Son of Man" designation spills into Rev 1:13 from Dan 7:13, where it is used to differentiate the human from the animal. For preliminary reflections on human-animal relations in Dan 7, see Moore 2011, 87–88, and for extended reflections on Daniel's animal apocalypse, see Koosed and Seesengood 2014.

18. As well as an androgyne: the "Son of Man" sports a pair of female breasts

The wool metaphor, however, also conjoins this Human proleptically with the Animal, and with one animal in particular. For when the figure next appears it has undergone a theriomorphic metamorphosis. It shimmers uncertainly for a moment, taking the form of a Lion (5:5), but resolves into the form of a Lamb (5:6).[19]

This is not the only metaphoric lamb in early Christian literature (see also, e.g., Luke 10:3; John 1:29, 36; 21:15; Acts 8:32; 1 Cor 5:7; 1 Pet 1:19; Justin Martyr, *Dialogue with Trypho* 40, 72; Melito of Sardis, *On the Passover* 7–8, 71; *Gospel of Philip* 58, 14–15), but it may be the only four-legged one. When John the Baptist, for instance, on "[seeing] Jesus coming toward him" in John 1:29 exclaims, "Here is the Lamb of God [*ho amnos tou theou*] who takes away the sin of the world!" (cf. 1:36), few if any readers or hearers have visualized a quadrupedal, ruminant mammal of the *Ovis aries* species trotting up to John. But a four-legged lamb is precisely what the Christian imagination has tended overwhelmingly to visualize in Revelation's throne room,[20] albeit an anomalous specimen of lambhood, multihorned and many-eyed (5:6).[21] In the terms associated with conceptual metaphor theory, more characteristics of the source domain (*lamb*) are mapped onto the target domain (*Jesus*) in Rev 5:6ff. than in John 1:29, 36.[22] The result is a theriomorphic Messiah or quadrupedal Christ, a Jesus

---

(*mastoi*, 1:13), as we saw earlier (149–51). Last but not least, the Human Being is also an angelic being: most of the details of his/her head-to-toe description are copied from Dan 10:5–6, where they describe an angel, probably Gabriel (see Carrell 1997, 129–74). Densely imbricated in this category-defying figure, then, are animal, angelic, human-female, divine, and human-male elements, and in no discernible hierarchical order.

19. In which guise it then trots through most of the remaining narrative. As Loren Johns (2003, 22) notes, "Not limited to one or two scenes, the term [*arnion*, "lamb"] appears in fully half of the 22 chapters of the Apocalypse.... [It] is by far the most frequent designation for Christ in the Apocalypse. It appears more than twice as often as any other name or image for Christ—even more than the simple name *Iēsous*, the title *Christos*, or variations thereof."

20. See Kovacs and Rowland 2004, 74–75, for a brief review of some of the better-known artistic representations of Revelation's Lamb.

21. Might the horns even disqualify it from being regarded as a lamb at all? Might we be looking at a ram instead? Apparently not. "The idea that lambs could have horns was not unknown in the ancient world. According to one tradition, some lambs immediately begin to develop horns at birth (cf. Homer, *Odyssey* 4.85; Aristotle, *Historia Animalium* 7.19)" (Johns 2003, 24 n. 11).

22. The classic exposition of conceptual metaphor theory is Lakoff and Johnson

who now adds species crossing to the other border-crossing activities regularly attributed to him.

With the exception of the human animal, as Derrida remarks, "no animal has ever thought to dress itself" (2008a, 5). Clothing or its absence is yet another means by which the Christ of Revelation shuttles in and out of humanity. In his initial appearance as "a Son of Man" he is clothed ("clothed with a long robe and with a golden sash across his chest," 1:13) and seen only by the seer. In his second appearance as the slain Lamb he is unclothed, presumably, even though the object of a mass gaze (5:6–14); yet he is not naked, because animal. When he resumes his human form following the demise of Babylon,[23] he is clothed once more ("clothed in a robe dipped in blood," 19:13; cf. 19:16). Yet his robe bears a residual mark (if not ineradicable stains) of the animal identity that it conceals: it is inscribed with a name, "King of kings and Lord of lords," which was earlier attributed to the Lamb (17:14), and the blood in which it has been dipped has been interpreted most often as issuing from the Lamb's slaughter.[24] Clothes do not make the man or Son of Man in Revelation so much as remind us that he is always liable to be unmade and remade as animal. Jesus' humanity flickers indecisively in Revelation, and is ultimately eclipsed by his animality.

For the Lamb, not the Lion (cf. 5:5) or even the (Son of) Man, is the king of beasts in Revelation, including human beasts. If the anthropomorphic warrior on the white horse is "King of kings and Lord of lords" (19:16), his inverted image, the Lamb, is "Lord of lords and King of kings" (17:14); but while the warrior only has followers (19:14), the Lamb has followers (14:4) and adorers. It is the Lamb, not the Man, that is the object of mass adulation, mass adoration, for "every creature in heaven and on earth and under the earth" (5:13; cf. 5:8–14; 7:9–10). And even if the Lamb is ambiguously positioned *en mesō tou thronou* in its initial appearance (5:6; cf. 7:17)—"on the throne"? "in the inner court area around the throne"?[25]—by the time we eventually arrive at the heavenly

---

1981. Lynn Huber applies the theory to Revelation's images of the bride (2007) and the 144,000 male virgins (2008), while Ingvild Saelid Gilhus (briefly) applies it to the dove and lamb images of the New Testament (2006, 173–74).

23. Setting the ambiguous 14:14–16 aside for now.

24. The interpretation may be traced back to the early centuries of the church (Weinrich 2005, 311).

25. G. K. Beale is among those who favor the latter rendering, arguing that "in

city, "God's dwelling place among human beings" (*hē skēnē tou theou meta tōn anthrōpōn*, 21:3), the throne has become "the throne of God and of the Lamb" (*ho thronos tou theou kai tou arniou*, 22:1, 3; cf. 3:21), the Lamb now lording it with God over humans, who have now become its slaves (*douloi*, 22:3) even as it has become unequivocally divine. Revelation's Lamb, then, is at once a human-animal hybrid and a divine-animal hybrid. And for now, at least, its sharp little horns seem proleptically to be ripping the Cartesian human/animal hierarchy to shreds.

But the Lamb is not the only animal whose habitat is Revelation's throne room. Surrounding the throne are the four "living creatures" (*zōa*), encrusted with eyes in front and behind and fitted with multiple wings, one creature lionlike, another calflike, a third "with a face like a human face" (*echōn to prosōpon hōs anthrōpou*), and a fourth "like a flying eagle" (4:6b–8). These four creatures may represent the entire created order of animate beings, as has sometimes been suggested (see especially Brütsch 1970, 230–33). More significant, however, for our topic is the fact that the human does not represent the apex of creation in this bestial tableau.[26] Its placement as the third item in the series of four is decidedly nonemphatic. The humanoid face is briefly glimpsed among the (other) animal visages, but it does not rise above them, look down on them, or see beyond them. This creature has exactly the same number of wings and eyes as its fellows and the exact same lines to utter in the eschatological script (4:8b; cf. 5:14; 6:1, 3, 5, 7; 19:4).

What Derrida has to say in a different context, then, seems eminently applicable to Revelation's initial throne-room scene: "there are gods and there are beasts, there is, there is only, the theo-zoological, and in the theo-anthropo-zoological, man is caught, evanescent, disappearing, at the very most a simple mediation, a hyphen between the sovereign and the beast, between God and cattle" (2009, 13). Or between God and lions, God and eagles, or God and sheep, as is also the case in our throne-room tableau.

5:6 it appears that the Lamb is near the throne, preparing to make his approach to be enthroned" (1999, 350). Matthias Reinhard Hoffmann, however, prompted by the spectacle of the Lamb taking the scroll from "the right hand" of the one seated on the throne (5:7), proposes a third alternative: "the Lamb is placed at the right hand side of God [the position of exaltation] after (or when) he takes the scroll from God" (2005, 138).

26. As Catherine Keller (2005, 68) insightfully notes in her reflections on the four living creatures.

Derrida defines the "ahuman," which he also names "divinanimality," as "the excluded, foreclosed, disavowed, tamed, and sacrificed foundation of … the human order, law and justice."[27] Prior to that exclusion, that foreclosure—which, most of all, is a Cartesian exclusion—the divine is both theriomorphic and anthropomorphic, and such anthropomorphic divinanimality comes to sublime expression in Revelation.

SLAVES OF THE SHEEP

On the other hand (hoof, paw, claw …), Revelation's divine/human/animal triad—the divine intimately conjoined to the human and the animal, and the human consequently conjoined to the animal and the divine—can hardly be said to be symmetrical. Hierarchy continues to rear its ugly head in Revelation, and the head is frequently that of a young sheep—paradoxically, a rather petite young sheep, if the diminutive form of *arnion*, the term in Revelation ordinarily translated as "lamb," is to be accorded its full (if meager) weight.[28]

Arguably, however (and this should be said before we surrender fully to the imperious grip of *the other hand*), even the form taken by hierarchy in Revelation where it pertains to the paradoxical figure of the Lamb is significant for ecotheology. For Revelation, however inadvertently, inverts the Aristotelian-Stoic species hierarchy that elevated the human over the

---

27. Derrida 2008a, 132, in the course of his critique of Lacan's conception of the animal (see also Derrida 2009, 127). Earlier Derrida writes of "the ahuman combining god and animal according to all the theo-zoomorphic possibilities that properly constitute the myths, religions, idolatries, and even sacrificial practices within the monotheisms that claim to break with idolatry" (2008a, 131; also 2009, 126).

28. See also Derrida 2009, 258: "There is no more reason to call a superterrestrial God great ('God is great') than small.… In certain religions the manifestation of divine presence or sovereignty passes through the small, the smallest: the weakness and smallness of the baby Jesus for example, or the lamb." But how little is Revelation's Lamb? Technically, *arnion* is the diminutive form of *arēn* ("young sheep"). Loren Johns (2003, 26), however, echoes the views of many when he writes: "Although diminutives normally express either smallness ('small lamb') or endearment ('Lämmlein,' 'lambkin,' or 'lamby'), the historical linguistic evidence suggests that neither of these can be pressed in New Testament times apart from corroborating contextual evidence, which is certainly lacking in this case." Other scholars are less certain. David Aune (1997, 368), for instance, writes: "it is extremely difficult to argue that *arnion* was consistently used as a faded diminutive [by the first century CE]."

animal.²⁹ As noted above, the scene in Revelation in which the Lamb first makes its entrance has every creature in heaven and on earth—angelic creatures and human creatures included—worshiping the Lamb (5:11–13); while the final scene in Revelation in which the Lamb appears characterizes the human inhabitants of the heavenly city as "slaves" (*douloi*, 22:3)—apparently of God and the Lamb, humanity in thrall to a~~human~~ divin~~animal~~ity.³⁰ Revelation's new Eden, then, appears to overturn the order established in the old Eden that, as Jewish and Christian tradition has most often understood it, accorded humankind dominion over all nonhuman creatures (Gen 1:28; 2:18–20; cf. Ps 8:3–8).³¹ This species hierarchy is unceremoniously toppled head over hoof in Revelation. The animal domesticated to serve human beings

> ["When did a sheep last die of old age? Sheep do not own themselves, do not own their lives. They exist to be used, every last ounce of them, their flesh to be eaten, their bones to be crushed and fed to poultry. Nothing escapes, except perhaps the gall bladder, which no one will eat. Descartes should have thought of that. The soul, suspended in the dark, bitter gall, hiding" (Coetzee 1999a, 123).³²]

now rules over every human being, including every human ruler: the Lamb is "Lord of lords and King of kings" (17:14), as we recall. The human subject is subjected to the animal for all eternity. That, however, is but the

---

29. Contrast John's near-contemporary Philo of Alexandria, for example, who, channeling Stoic doctrine, declared: "To raise animals to the level of the human race and grant equality to unequals [*anisoi*] is the epitome of injustice" (*On Animals* 100, my trans.).

30. G. K. Beale argues: "That 'they will serve *him* [*latreusousin autō*, 22:3b]' likely does not refer only to God or only to the Lamb. The two are conceived so much as a unity that the singular pronoun can refer to both" (1999, 1113). Thomas Slater changes the "him" to a "them" in his paraphrase of the passage—"God and the Lamb ... will provide the highest quality of life possible and the servants of God will worship them" (1999, 200)—and cites the commentaries of J. P. M. Sweet, Gerhard A. Krodel, Leon Morris, Robert H. Mounce, and George Eldon Ladd in support of his interpretation.

31. Derrida, echoing this tradition, parses out the combined effect of the two Genesis creation accounts as follows: God "has created man in his likeness *so that* man will *subject, tame, dominate, train,* or *domesticate* the animals born before him and assert his authority over them" (2008a, 16).

32. Cf. Cicero, *The Nature of the Gods* 2.63: "What other use have sheep, save that their fleeces are dressed and woven into clothing for men?" (LCL trans.).

outer layer of the paradox that, like a wooly fleece, envelops Revelation's Lamb.

Murder in the Sheepfold

The necessary precondition for the subjection of humans to the anomalous animal of Revelation is that the animal first had to be subjected to slaughter by humans: "Worthy is the Lamb that was slaughtered [*to arnion to esphagmenon*] to receive power" (5:12; cf. 5:6, 9; 13:8). The Lamb suffers, then. The standing-as-though-slaughtered (*hōs esphagmenon*, 5:6)—or slaughtered-but-still-standing—Lamb is, indeed, the privileged metaphor (and not only in Revelation) for the salvific suffering of the god-man. The god-man suffers like a god-man-animal, a theo-therio-anthropmorph. And suffers in silence. As animal, as *arnion*, the god-man does not—and perhaps cannot—speak in Revelation (see Isa 53:7; Acts 8:32), if by "speaking" we mean the utterance of human language. Not a single line, nor even a single word, is accorded to the Lamb in John's talking animal book (contrast 4:7–8; 5:13–14; 6:1, 3, 5, 7; 8:13; 19:4).[33]

"Man alone among the animals has speech," Aristotle declared (*Politics* 1253a 10).[34] In Aristotelian terms, then, the Lamb is inherently inferior to the man, even the (speaking) Son of Man with whom it is, yet is not, identical. And not just in Aristotelian terms: as Derrida observes, philosophers otherwise as different as Aristotle, Descartes, Kant, Heidegger, Levinas, and Lacan "all … say the same thing: the animal is deprived of language. Or, more precisely, of response, of a response that could precisely and rigorously be distinguished from a reaction" (2008a, 32). But even if the Lamb is deprived of speech, it is hardly deprived of response, beginning with its decisive claiming of the sealed scroll ("It went and took the scroll from the right hand of the one who was seated on the throne," 5:7), the action that sets the entire ensuing narrative in motion.[35]

---

33. What of the phrase *tēn ōdēn tou arniou* in Rev 15:3a? Most contemporary commentators (e.g., Aune 1998a, 873) translate the phrase as "the song about the lamb" (objective genitive) rather than as "the song of [i.e., sung by] the lamb" (subjective genitive), not least because the Lamb itself is not the singer, as the context makes clear.

34. Derrida's *The Beast and the Sovereign*, vol. 1, concludes with an analysis of this declaration and the larger passage in which it is embedded (2009, 343–49).

35. And thus is revealed the mystery of how the Lamb "took [or: "has taken"

But perhaps speech, or even response, is not the crucial issue. Channeling Jeremy Bentham's late eighteenth-century plea on behalf of the animal, Derrida remarks: "the question is not to know whether the animal can think, reason, or speak.... The *first* and *decisive* question would rather be to know whether animals *can suffer*" (2008a, 27).[36] If the crucial question is not whether animals can speak but rather whether animals can suffer, the Lamb both answers and complicates the question. On the one hand, the Lamb suffers without speaking, which is to say that it suffers as an animal suffers.[37] In Revelation, then, the torturous death of Jesus of Nazareth is figured as animal suffering. Crucifixion is implicitly represented through the figure of the butchered Lamb as an altogether abject death, an utterly dehumanizing death, a death more fitting to an animal than a human—a theme to which we shall later return.[38] To that extent, the

---

(*eilēphen*)] the scroll." With its mouth? With its hoof? No, with its hand. For even if the Lamb as a quadrupedal mammal of the *Ovis* genus (albeit a metaphorical mammal with irregular ocular features and an abnormal number of horns) does not and cannot have a hand, its epochal action of taking and subsequently unsealing the apocalyptic scroll shows that it does have a hand in the Heideggerian sense. In "Heidegger's Hand," Derrida takes Heidegger to task for denying a hand to the animal (Derrida 2008b; cf. 2011, 83). Only *Dasein*, the human existent, can have a hand, according to Heidegger. Only *Dasein* is capable of the kind of thought and action that merits the term *hand*, while the animal (even the ape) has no hand, properly speaking, but only a prehensile grasping organ at best. "The hand is infinitely different from all grasping organs—paws, claws, or fangs—different by an abyss of essence" (Heidegger 1968, 16). Revelation's Lamb, a Heideggerian impossibility, can be said to emblematize hand-endowed animality. Hand over hand, it clambers out of Heidegger's abyss.

36. Bentham's plea for animal rights, epitomized in his pronouncement "the question is not, Can they *reason*? nor, Can they *talk*? but, Can they *suffer*?," occurs in his *Principles of Morals and Legislation* (1879 [1st ed. 1780], 310–11).

37. Although the perceived "silence" of animals depends on a rigidly narrow conception of "speech." Derrida notes in an interview that the structural elements that make human language possible (the elements that his early work isolated and that he here itemizes as the mark, the trace, iterability, and *différance*) "are themselves not only human" (1995a, 285).

38. Suffice it for now to note that as a slain domestic animal, the Lamb is always already about to be eaten, and as such its fate curiously mirrors that of the woman Babylon, annihilated by being savagely devoured (17:16), and the enemies of the rider on the white horse, also obliterated through ingestion (19:17–18, 21), and even Jezebel (2:20), whose symbolic name also connotes the fate of being ignominiously consumed (see 1 Kgs 21:23; 2 Kgs 9:10, 30–37). What Derrida (2011, 55) has to say about *Robinson Crusoe* is only slightly less true of Revelation: "the great gesture, the great phantas-

image of the slaughtered Lamb reinscribes the hierarchical human/animal divide, writes it in blood. And yet the slaughtered-but-still-standing Lamb also represents a leveling of the human in relation to the animal. Forever bearing the marks of death,[39] the human-animal amalgam that is the Lamb figures the finitude that humans share with other animals. At the center of the throne room that is the locus of absolute power in Revelation is a curious nonpower, an abject inability, whose emblem is a butchered animal. Mortality stands in the place of eternity in Revelation's central theophany.[40]

The Lamb is also singular in that it is also, as we are about to discover, an emblematic challenge to the logic that sacrificing an animal, exploiting it to death, does not constitute murder, a logic as ancient as Genesis 4, God's preference for the firstlings of Abel's flock over Cain's fruits of the earth, and as recent as factory farming.[41] In our own era, the scale of

---

matic *gesta* of [this] book, which rules its whole vocabulary, its speech, its mouth, its tongue and its teeth, is that of eating and devouring, eating the other."

39. Johns (2003, 111 n. 9) cautions that the phrase *hōs esphagmenon* in 5:6 "should not be translated 'as if [slaughtered],' suggesting that the marks of slaughter are ambiguous. The lamb of the Apocalypse is clearly a slain lamb."

40. See Derrida 2008a, 28, which, although not about Revelation or Christian soteriology, has impelled these reflections.

41. The traditional view of Revelation's Lamb as sacrificial animal (reflected, for example, in Aune 1997, 371–73, and most other commentaries) has been contested, notably by Loren Johns (2003, 22–39), who observes that *amnos*, not *arnion*, is the standard term used for lambs as burnt offerings in the Septuagint, *arnion* being used instead in what Johns sees as nonsacrificial contexts. This pattern continues in the earliest Christian literature, Johns argues, *amnos* being used of Christ when the sacrificial cult is evoked (John 1:29; 1 Pet 1:18–19), whereas John 21:15, the only occurrence of *arnion* in the New Testament writings outside of Revelation, is a nonsacrificial usage. Johns doubts in particular that Revelation's *arnion* is a Passover lamb, because although Revelation alludes extensively to Exodus, it never explicitly evokes the Passover (contrast 1 Cor 5:7). George Heyman (2007, 137–45) counters Johns's technical discussion with one of his own and arrives at a more nuanced conclusion: "'Sacrifice' can be both expiatory *and* communion-oriented. The rhetorical effect created by the image of a 'slaughtered lamb who conquers' was precisely to engender a sense of identity and community among the early Christians," themselves perpetually vulnerable to summary slaughter. "Thus, while Revelation might not have directly emphasized that the actual slaughter of the lamb effected the expiation of human sin and impurity, it did effect a bond of solidarity among believers" (139–40). I'm not sure that either Johns or Heyman accord sufficient weight to Rev 7:14 ("they have washed their robes and made them white in the blood of the Lamb"; see also 1:5b; 5:9; 12:11). All that matters, however, for my own analysis is that the Lamb has been slaughtered to ben-

this mass nonmurder has achieved grotesque proportions, necessitating a prodigious disavowal and dissimulation of the colossal cruelty it entails. Like other writers on these matters, Derrida has recourse to the figure of genocide to

> ["'They went like sheep to the slaughter.' 'They died like animals.' 'The Nazi butchers killed them.' Denunciation of the camps reverberates so fully with the language of the stockyards and slaughterhouses that it is barely necessary for me to prepare the ground for the comparison I am about to make. The crime of the Third Reich, says the voice of accusation, was to treat people like animals....
>
> "It was and is inconceivable that people who did not know (in that special sense) about the camps can be fully human. In our chosen metaphorics, it was they and not their victims who were the beasts. By treating fellow human beings, beings created in the image of God, like beasts, they had themselves become beasts.
>
> "I was taken on a drive around Waltham this morning. It seems a pleasant enough town. I saw no horrors, no drug-testing laboratories, no factory farms, no abattoirs. Yet I am sure they are here. They must be. They simply do not advertise themselves. They are all around us as I speak, only we do not, in a certain sense, know about them.
>
> "Let me say it openly: we are surrounded by an enterprise of degradation, cruelty, and killing which rivals anything that the Third Reich was capable of, indeed dwarfs it, in that ours is an enterprise without end, self-regenerating, bringing rabbits, rats, poultry, livestock ceaselessly into the world for the purpose of killing them" (Coetzee 1999b, 20–21).[42]]

---

efit human beings. The term *sacrifice* as I employ it is shorthand for that simple, yet pivotal, plot element. (I am grateful to Maia Kotrosits for pressing me on this issue.)

42. This is an excerpt from one of two invited lectures that fictional novelist Elizabeth Costello, protagonist of J. M. Coetzee's novella "The Lives of Animals," delivers at fictional Appleton College in nonfictional Waltham, Massachusetts. Posthumanist critic Cary Wolfe utters a similar condemnation in his *Animal Rites* (2003, 190): "I think it entirely possible, if not likely, that a hundred years from now we will look back on our current mechanized and systematized practices of factory farming, product testing, and much else that undeniably involves animal exploitation and suffering ... with much the same horror and disbelief with which we now regard slavery or the genocide of the Second World War."

express his own revulsion at "the *unprecedented* proportions of this subjection of the animal" (Derrida 2008a, 25). "One should neither abuse the figure of genocide," Derrida states,

> nor too quickly consider it explained away. It gets more complicated: the annihilation of certain species is indeed in process, but it is occurring through the organization and exploitation of an artificial, infernal, virtually interminable survival, in conditions that previous generations would have judged monstrous, outside of every presumed norm of a life proper to animals that are thus exterminated by means of their continued existence or even their overpopulation. (2008a, 26)[43]

These present-day abominations far exceed the animal sacrifices of the Bible (see Derrida 2008a, 25), even at their most extravagant ("Solomon offered as sacrifices ... to the Lord [at the dedication of the temple] twenty-two thousand oxen and one hundred twenty thousand sheep," 1 Kgs 8:63 [= 2 Chron 7:5]; cf. 1 Kgs 8:5). The unprecedented proportions of our current subjection of the animal intensifies Revelation's paradox of a butchered animal bearing the bloody marks of subjection unto death, yet to which all human beings are now subjected and to whose vengeance they are now subject. "Hide us from ... the wrath of the Lamb [*tēs orgēs tou arniou*]," the human inhabitants of the earth, "everyone slave and free," cry out in panic as they themselves scamper like frightened animals, hiding in caves and among rocks (6:15–16).[44] The slaughtered sacrificial victim has returned to life, causing the priest to drop his knife and flee from the altar in terror. But there are also more subtle significations encrypted in the figure of the slain-but-standing Lamb than these lurid dramas of reversal and revenge.

Derrida dissects Heidegger's argument that only man, as *Dasein*, "has an experiential relation to death, ... to his *own* death, his own being-able-to die, to its possibility, ... whereas the animal ... perishes but never dies, has no relation worthy of the name to death" (Derrida 2009, 307–8; 2011, 115–17, 290).[45] As Heidegger himself aphoristically puts it, "Only man

---

43. This statement occurs as part of a lengthy passionate protest (Derrida 2008a, 25–27) that erupts rather abruptly in what has up to then been a somewhat cerebral meditation on human-animal relations.

44. Note, too, the species inversion of Rev 7:17: "for the Lamb ... will be their shepherd [*to arnion ... poimanei autous*]" (cf. 14:4b).

45. See also Derrida 1994a, 35–38, 74–76, one of several earlier texts in which

dies. The animal perishes" (Heidegger 1971, 176). Derrida parses out the implications of this claim: if the animal is indeed incapable of an "authentic" relation to death, then the animal is a living creature that can only live, that can never die, and as such is an "immortal" being (Derrida 2008a, 129). If this were all there was to the matter, we would now have explained in full why and in what sense the slain Lamb of Revelation is immortal. It lives, it lives on—eternally—precisely as an animal, which, although slaughtered, cannot truly die. It has simply perished. There is, however, more to the Lamb that makes its animality anomalous—which is to say that Revelation is not simply reducible to what Derrida imagines the Heideggerian corpus to be: "a text given over to the gnawing, ruminant, and silent voracity of … an animal-machine" (1989, 134).

Elsewhere Derrida takes Levinas to task for his explicit hesitation to ascribe a "face" to the animal and hence the ethical obligation that is due to the human (Derrida 2008a, 105–18; 2009, 237–39). Derrida recounts that when Levinas was challenged by a questioner at a 1986 symposium, "Does the animal have a face? Can one read 'Thou shalt not kill' in the eyes of the animal?," Levinas vacillated: "I cannot say at what moment you have the right to be called 'face.' The human face is completely different and only afterwards do we discover the face of an animal. I don't know if a snake has a face" (Derrida 2008a, 107–8).[46] Levinas's recourse to the example of the snake is telling, as Derrida notes. Many more "disturbing examples" might have been adduced—"for example, the cat, the dog, the horse, the monkey, the orangutan, the chimpanzee—whom it would be difficult to refuse a

---

Derrida previously mused on this Heideggerian theme. In effect, Heidegger epitomizes, for Derrida, the post-Cartesian absolutization of the human/animal divide: "The distinction between the animal … and man has nowhere been more radical nor more rigorous than in Heidegger" (2005, 268). Consequently, it is mainly in relation to Heideggerian thought that Derrida engages in sustained fashion with "the question of the animal" prior to "L'animal que donc je suis" (1999a).

46. A slightly different version of the exchange is presented in Derrida 2009, 237, after which Derrida takes up the specific example of the snake at some length. Many centuries earlier, we find Augustine also pondering the question of whether "Thou shalt not kill" applies to animals. Lining up behind Aristotle and the Stoics, Augustine declares that it cannot apply to "the irrational animals that fly, swim, walk, or creep, since they are dissociated from us by their want of reason, and are therefore by the just appointment of the Creator subjected to us to kill or keep alive for our own uses; if so, then it remains that we understand that commandment simply of man" (*The City of God* 1.20, *NPNF* 1/2:15).

face and a gaze. And hence to refuse the 'Thou shalt not kill' that Levinas reserves for the face" (2008a, 110). What the exchange with Levinas impels is the introduction of the category of *murder* into our consideration of the slain Lamb. If the Lamb does not possess a face in the Levinasian sense, is not a candidate for murder, then a Levinasian reflection on the Lamb as animal takes us no further than a Heideggerian reflection on it, and to a death that is not worthy of the name.[47] Inasmuch as it is categorically incapable of being a murder victim, the Lamb still cannot die. It lives on forever as the quintessential sacrificial animal.

But this is not what Revelation implies, hence its interest and relevance for contemporary ecotheology. The slaughter of its singular animal was a heinous crime, so much so that when this creature returns "with the clouds"—whether as theriomorph, anthropomorph, or therioanthropomorph—"every eye will see him, even those who pierced him; and on his account all the tribes of the earth will wail" (1:7; cf. 6:15–17), implicitly because of the unspeakable injustice done to him. Does the Lamb have a face? Yes, it would seem, to the extent that killing the Lamb was

> ["A Kent head teacher at the centre of a row about the slaughter of a school lamb has resigned. Andrea Charman will step down as head of Lydd Primary School in Romney Marsh at the end of the week, Kent County Council has said.
>
> "Mrs. Charman was criticised in September after sending Marcus the lamb—who had been hand-reared by pupils—to slaughter, despite calls to save him....
>
> "But Mrs. Charman went ahead with sending the animal to slaughter, which was part of a project to teach children about the food cycle" ("Slaughtered Lamb Head Teacher Resigns" 2010).]

a culpable act. The Lamb is that anomalous animal in whose (seven) eyes "Thou shalt not kill" can be read. This, then, is yet another way in which Revelation problematizes in advance the Cartesian conception of the human as categorically distinct from the animal and hence the sole object of ethical obligation. But it is not just the Cartesian conception that is called into question. Far more ancient is the logic that declares

---

47. On the question of the animal, Derrida finds Levinas to be "profoundly Heideggerian" (2008a, 110). For Levinas's own most profound meditation on human-animal relations, see his "The Name of a Dog, or Natural Rights" (1990). For theological reflection on this essay, see Gross 2009, and for biblical-critical reflection on it, see Stone 2014.

that sacrificing an animal, slaughtering it for food, or otherwise exploiting it to death, does not—indeed cannot—constitute murder (see Derrida 2008a, 110).[48] Revelation presents us with the ethical paradox of a sacrificial animal whose slaughter constitutes unlawful killing, which is to say manslaughter or murder.

Yet Revelation also relies on the sacrificial logic it deconstructs. That the slaughter of the Lamb was a culpable act, an unjust killing, does not render it an unproductive act, an ineffective sacrifice. On the contrary, Revelation represents this judicial murder as the most spectacularly efficacious sacrifice ever performed. "You were slaughtered and by your blood you ransomed [or: purchased (*ēgorasas*)] for God saints from every tribe and language and people and nation," exults the heavenly chorus (5:9; cf. 1:5; 7:14; 12:11; 19:13), including the four living creatures, themselves more animal than human; and before long "every creature in heaven and on earth and under the earth and in the sea" (5:13) has joined in. A vast chorus of creatures, including every nonhuman animal, rejoices in the sacrifice of the god-man who died a death so ostensibly ignoble, so unbefitting of an honorable man, much less a god, that it elicits representation as an animal death: "I saw ... a slaughtered Lamb standing" (5:6). Do they exult because this unique sacrifice has made all further animal sacrifices unnecessary? Is the explicit argument of the letter to the Hebrews (see especially 10:1–14) implicit in the book of Revelation?

To settle for such a solution would be to domesticate Revelation's wildly anomalous Lamb, a sacrificial victim that is also a murder victim.

---

48. This was not a uniform logic in antiquity, however, a complication of which Derrida seems unaware. The ideas of Pythagoras and his disciple Empedocles regarding the transmigration of souls were revived during the Roman principate, and together with the Orphic tradition formed the basis for ethical arguments for vegetarianism. Attributed to Orpheus was the view that slaughtering animals was murder—equivalent, indeed, to killing one's own kin. Also relevant here is the "contractual" view of animal sacrifice common in antiquity, the notion that animals led to the altar were expected to consent to their own slaughter, even to the point of nodding their assent before the knife or ax descended. This nod was regularly produced by pouring water, flour, or some other substance over the animal's head; yet many of the human participants in the rite seem to have deemed the nod significant nonetheless. Plutarch, for example, remarks: "people are very careful not to kill the animal till a drink-offering is poured over him and he shakes his head in assent. Such precautions they [take] to avoid any unjust act" (*Table Talk* 729F, LCL trans.). Further on all of these topics, see Gilhus 2006, especially 25–26, 35–38, 87, 119–21, 141–47.

In effect, the slain Lamb is the sacrifice of Cain—not Cain's "offering of the fruit of the earth," however, for which God "had no regard" (Gen 4:3–5), but the slaughter of his brother Abel that occasions divine horror: "What have you done? Listen, your brother's blood is crying out to me from the ground!" (4:10). For now at least, God much prefers the slaughtered "firstlings of the flock" that is Abel's offering (4:4). In Revelation's throne room, however, sheep and man, sacrificial victim and murder victim, become one. In order to effect divine remission of human sin, the slaughter of the sacrificial victim must itself be a sin, a crime. In order for sacrifice to be fully and eternally efficacious, the sacrificial victim must have a face, must die a human death—but that death must also be so abject,

> ["To slaughter you will need one sharp butchering knife, a small skinning knife, and a steel to keep the knives sharp. If you are butchering only one or two lambs, you can work outdoors under a tree that has an overhanging limb.... If you are working indoors, you should have a solid beam to hang the lamb on....
> However, if you are going to slaughter many animals, a sawbuck rack large enough to hold a lamb placed on its back with its head hanging off the end is a convenience that will allow you to, in effect, guillotine the lamb.... To use the guillotine method, strap the lamb to the sawbuck or have someone hold it there. Grab the lamb's muzzle, bend the head back a bit and, with one clean stroke of a sharp butcher's knife cutting down toward the backbone, sever the jugulars, carotids, gullet, and windpipe. Twist the head and with the knife disjoint the head from the body where the backbone joins the skull....
> Cutting from the inside out, open the skin on the neck down to where you cut the throat. Using your clenched fist instead of a knife to separate the skin from the body, 'punch' or 'fist' the hide loose over the brisket as far back as the navel.... Fist the hide loose over the shoulders and back and as far up as the tail.... Using the knife, skin around the tail and anus.... Cut around the bung, deep into the pelvis, and tie the rectum off so manure will not spill out.... To do this it will be necessary to pull the bung out of the pelvis; it is easier to have a second person tie the string" (Mettler 2003, 73–74, 76–78).[49]]

---

49. See Wesley Bergen's *Reading Ritual*, which has recourse to the operations

so awful, as to compel metaphorization as a death only befitting an animal. Had Jesus of Nazareth expired of old age, quietly passed away in his sleep, there would be no butchered animal bleeding all over Revelation's throne room. What remains undisturbed in Revelation is the notion that certain forms of death potentially reduce the human being to animal status. Equally undisturbed by extension, therefore, is the notion that animals, in death as in life, are inherently inferior to humans.

Also undisturbed, finally, are the operations of the ancient sacrificial machine. Far from declaring the machine obsolete, Revelation's Lamb ransoming saints by its blood shows that the machine still works—that it is, indeed, spectacularly effective (see especially 7:9, 13–14: "After this I looked, and there was a great multitude that no one could count, ... robed in white.... 'Who are these, robed in white?' ... 'These are they who have come out of the great ordeal; they have washed their robes and made them white in the blood of the Lamb'"; cf. 1:5b; 5:9; 12:11). There is no *explicit* critique of animal sacrifice in Revelation, then, no intimation that exploiting an animal to death for human benefit is unethical, even though such critiques were not unknown in Revelation's world.[50] And yet, as we have seen, Revelation's Lamb also presents us with the ethical paradox of a sacrificial animal whose slaughter constitutes unlawful killing—which is to say that there is, nonetheless, in Revelation (and irrespective of whether its author intended it or not), an *implicit* critique of animal sacrifice, and hence, by extension, of our continuing

> ["**On November 24th, 2011 a video from Live Leaks surfaced on YouTube showing a group of US soldiers dragging a sheep into a crowded room and laughing with delight as one of their officers repeatedly and savagely smashes it in the head. As this is happening, several Afghani children jump up and down with excitement as soldiers clap, cheer and encourage the attacker on until the limp and lifeless body of the animal is dragged across the ground and out of the view of the camera.**

---

of a modern meat-packing plant to defamiliarize the prescriptions on ritual animal slaughter in Lev 1–7. His reflections begin (2005, 14): "I used to work on the killing floor of a modern meat packing plant. This means that I spent my day within sight and hearing of the gun that killed an animal every twenty seconds. Is there any way this experience can serve as a bridge between ourselves and Leviticus 1–7?"

50. See Gilhus 2006, 138–60; Steiner 2005, 47–48, 105–7, on the most prominent ancient critics of animal sacrifice.

To date, the US Army has released only one statement saying that they are investigating the matter but animal activists and concerned citizens have expressed their distrust in the process as in the past only minor disciplinary charges have been given out for similar offences.

According to the Live Leaks website the incident occurred on November 6th, 2011 as part of the holy festival of Eid and that the killing was 'to represent a sacrifice made by Abraham of a ram when the angels told him that he had fulfilled the dream ordering him to sacrifice his young son, at which he laid down the knife and sacrificed the animal instead'" (Williams 2011).]

sacrificial war against the animal.

10
Ecotherology*

As much as anything else, Revelation is a tale of two animals: a many-eyed, multihorned Lamb and a many-headed, multihorned beast. Behind these fabulous creatures stand still other unlikely animals. The beast serves a many-headed, multihorned, great red dragon, while the Lamb serves a God who appears ever more beastlike the longer we gaze at him or it, as we shall see. But Revelation is also a tale of the women who are involved with these animals. Intimately related to the beast or wild beast (*to thērion*) is a wild woman, "Babylon the Great, mother of whores," while intimately related to the Lamb is a domesticated woman or domestic goddess, "the bride." The interspecies intimacy of Revelation's butchered sheep with its spotless bride is one of the queerer conceits of this hypersurreal book.

My own book will end in the home that the bride creates for the Lamb, a city often visited but seldom seen. For if Revelation's four-legged Christ, its woolly Messiah, has been oddly underremarked by ecotheologians and ecoexegetes, so has the sheer, stupendous size of the eschatological city in which Revelation and hence the Christian Bible finally come to rest. The bride morphs into a habitat that would be rather bleak for a sheep or any other nonhuman animal: a megalopolis that is a continent-sized shopping mall, as we shall discover, with a single stream and a token tree. I confess to finding this

---

* This essay first appeared in *Divinanimality: Animal Theory, Creaturely Theology*, ed. Stephen D. Moore (Transdisciplinary Theological Colloquia; New York: Fordham University Press, 2014), 196–209, and is reprinted with the permission of the publisher.

celestial megacity singularly ill-designed to serve as a prophetic counterexample to the contemporary paving over of the planet and the annihilation of plant and animal species. The new Jerusalem is better equipped to serve instead as a grim parody of our apocalyptically theriocidal world. And yet the shopping mall with the single stream and token tree also contains a lone animal, and that singular sheep seems to me to possess stunning ecotheological significance. Plucking an ecotherology, however frail, from a megacity is arguably more pertinent at the present time than plucking it from the swiftly vanishing garden planted in the original Eden.

And I saw a beast. (Rev 13:1)

Midway through the first of the thirteen weekly course lectures from 2001–2002 that make up the first volume of *The Beast and the Sovereign*, Jacques Derrida alludes to "all the beasts from John's Revelation, … the reading of which would merit more than one seminar" (Derrida 2009, 24). Whether all or any of these beasts receive even one seminar of the fourteen thousand pages of unpublished seminars that Derrida left behind at his death in 2004, I am not in a position to know.[1] Taking a back-row seat, at any rate, in Derrida's weekly seminar, I attempt once again in this essay to read the two volumes of *The Beast and the Sovereign* (2009; 2011), together with *The Animal That Therefore I Am* (2008a), as incisive if unintended commentary on Revelation's theological bestiary—its theotherology, if you will. In effect, I read the beast of *The Beast and the Sovereign* as the beast of the book of Revelation. Midway through the essay, however, I slip out of the seminar, leaving Derrida to his characteristic preoccupations, in order to extend my analysis into the area of sex and gender in Revelation where it leaks into the area of animality. My focus in this section of the essay is on Revelation's notorious wild woman astride her wild beast ("I saw a woman sitting on a scarlet beast," 17:3). My focus is even more on that other, far queerer instance of interspecies intimacy, the radiant bride who is married

---

1. The material is being published in English translation in a series titled the Seminars of Jacques Derrida, which promises to run to forty-three volumes (so Bennington et al. 2009, ix).

to a slaughtered-but-still-standing sheep ("the marriage of the Lamb has come, and his bride has made herself ready," 19:7). In the final section of the essay, I fall in behind the joyful throng of those who "follow the Lamb wherever he goes" (14:4), and trail them into the continent-sized shopping mall that, as we shall see, is the centerpiece of Revelation's climactic vision. As will become particularly apparent in this final section, the aim of the essay is to relate what Revelation has to say about nonhuman animals—and about creatures that are neither human, animal, nor divine—to the plight of nonhuman animals in our apocalyptically theriocidal world.

Apocalyptic Animetaphors

> I will explain the mysterious symbol of … the beast to you. (Rev 17:7)

To begin to address the theme of animality in Revelation is to run immediately into a problem. On the one hand, Revelation is an animal book extraordinaire. On the other hand, there are almost no nonhuman animals as such represented anywhere in Revelation,[2] only metaphorical animals, chimerical animals, and metaphorical-chimerical animals, beginning with the many-eyed, multihorned Lamb "standing as if it had been slaughtered" (5:6), and extending to the many-eyed, multiwinged "living creatures" ensconced in the heavenly throne room (4:6b–8); the human-faced, lion-toothed, scorpion-tailed, human-torturing locusts that swarm out of the bottomless abyss (9:1–10); the lion-headed, fire-breathing, serpent-tailed horses sent forth "to kill a third of humankind" (9:13–19); the many-headed, multihorned, great red dragon whose tail "[sweeps] down a third of the stars of heaven and [throws] them to the earth" (12:3–4); the many-headed, multihorned, species-blurring beast that rises out of the sea (13:1–2); and the lamb-horned, dragon-voiced beast that rises out of the earth (13:11). The animals that do not automatically fit into the metaphorical or chimerical categories, meanwhile, are, more often than not, philosophical conundrums. For instance, can the Greek word *hippos* ("horse") meaningfully be said to signify a bona fide nonhuman animal if it is being ridden by Conquest, War, Famine, or Death? (6:1–8).

Revelation stands loosely, then, within the fable tradition. "The fictional use of animals for didactic purposes reaches back to … Aesop's

---

2. As Ingvild Saelid Gilhus remarks, "John, the author of Revelation, did not intend to say anything about real animals" (2006, 177).

animal fables," notes Colleen Glenney Boggs. "The fable tradition, however, ... is not interested in animals as such" (2009, 535).³ As Graham Huggan and Helen Tiffin phrase it, "the animal *as animal* becomes invisible" (2010, 173). Metaphorical animals—*animetaphors*⁴—are as thick on the ground in much of Revelation as in Aesop. How best to relate to them? I propose to take my lead from the fact that Revelation presents us with an anthropomorphism of the animal that is qualitatively indistinguishable from its anthropomorphism of the divine. And just as I have found it fruitful elsewhere (Moore 1996, 117–38; 2001, 175–99 passim) to read Revelation's God *as* human—more precisely, to ask what kind of divine-human relations are encoded in this human, all too human deity—so I am attempting here to read Revelation's metaphorical, all too metaphorical animals *as* animals in the interests of deciphering the human-animal relations encrypted in them. In other words, and taking my cue from Rosi Braidotti, I am attempting a "neoliteral" reading of Revelation's animetaphors.⁵

Of God and Other Beasts

Yet I have been the Lord your God
    ever since the land of Egypt....
It was I who fed you in the wilderness....
When I fed them, they were satisfied; ...
    therefore they forgot me.
So I will become like a lion to them,
    like a leopard I will lurk beside the way.
I will fall upon them like a bear robbed of her cubs,
    and will tear open the covering of their hearts;

---

3. Cf. Derrida 2008a, 37: "We know the history of fabulization and how it remains an anthropomorphic taming, a moralizing subjection, a domestication. Always a discourse *of* man, on man."

4. Akira Mizuta Lippit's term (1998; 2000, 162–98).

5. Braidotti cautions against "the metaphoric habit of composing a sort of moral and cognitive bestiary in which animals refer to values, norms, and morals," as in "the nobleness of eagles, the deceit of foxes, or the humility of lambs." Instead, she urges "a neoliteral relation to animals.... The old metaphoric dimension has been overridden by a new mode of relation. Animals are no longer the signifying system that props up humans' self-projections and moral aspirations.... They have, rather, started to be approached literally, as entities framed by code systems of their own" (2009, 527–28; see also Braidotti 2013, 69–70).

there I will devour them like a lion,
  as a wild animal would mangle them. (Hos 13:4–8)

What better place to begin a consideration of Revelation's bestiary than with *the* beast, its best-known figure and most infamous animal? The most popular candidate at present for the beast's secret identity appears to be Barack Obama,[6] a disturbing reality that merits an essay of its own.[7] The small tribe of critical biblical scholars takes a more pedantic view. For us, almost without exception, the beast is a figure for ancient imperial Rome and/or its emperor(s). This unexceptional interpretation, however, immediately takes us to the heart of *The Beast and the Sovereign*.

A recurrent preoccupation of Derrida in *The Beast and the Sovereign* is the rhetorical trope whereby "the essence of the political and, in particular, of the state and sovereignty has often been represented in the formless form of animal monstrosity" (2009, 25).[8] As it happens, Revelation's beast qualifies eminently as a monster: "And the beast that I saw was like a leopard, its feet were like a bear's, and its mouth was like a lion's mouth" (13:2; cf. Dan 7:2–6).[9] Of the chimera of classical antiquity, Derrida observes:

---

6. How does one ascertain such a fact? By utterly unscientific means. Typing "Obama Beast Revelation" into the Google search box on January 12, 2013, yielded around 4,640,000 hits, while typing the same words into the YouTube search box yielded around 3,670 hits.

7. And has, in fact, received one: see Amarasingam 2011. Especially disturbing for me is a 2009 survey cited by Amarasingam (97), which concluded that 24 percent of young voters in my (now) home state of New Jersey "believed Obama to be the Antichrist."

8. Derrida is in transit from his passing reference to "all the beasts from John's Revelation, which clearly present themselves as political or polemological figures, the reading of which would merit more than one seminar" (2009, 24), to his analysis of Thomas Hobbes's *Leviathan*, that archetypal work of political theory from 1651, whose argument Derrida paraphrases as follows: "So the state is a sort of robot, an animal monster, which ... is stronger ... than natural man. Like a gigantic prosthesis designed to amplify ... the power of ... the living man that it protects, that it serves, but like a dead machine, or even a machine of death" (28).

9. Another animal, too, lurks within this beast. The beast is, more specifically, a sea beast ("And I saw a beast rising out of the sea," Rev 13:1) as distinct from a land beast ("Then I saw another beast that rose out of the earth," 13:11). David Aune remarks of this beastly duo: "These two beasts clearly reflect the Jewish myth of Leviathan, the female monster from the sea, and Behemoth, the male monster from the desert" (1998a, 728; cf. 728–29, 732).

"Its monstrousness derived precisely from the multiplicity of animals … in it (head and chest of a lion, entrails of a goat, tail of a dragon)" (2008a, 41). A monster, then, is that which does not respect the "proper" divisions between animal species—which divisions, however, the collective, catchall noun *animal* itself disregards, "in spite of the infinite space that separates the lizard from the dog, the protozoon from the dolphin, the shark from the lamb, the parrot from the chimpanzee, the camel from the eagle, the squirrel from the tiger, the elephant from the cat, the ant from the silkworm, or the hedgehog from the echidna" (34).[10] As such, the animal is always a monster. Or, if you prefer, a beast.

Bestiality has always been a convenient figure for political despotism. Derrida unpacks the logic of the metaphor. The absolute sovereign possesses the power not only to make the law but also to break the law, to suspend its operations at will. But this godlike power also has a beastly aspect. "This arbitrary suspension or rupture of right … runs the risk of making the sovereign look like the most brutal beast who respects nothing, scorns the law.… Sovereign and beast seem to have in common their being outside-the-law" (Derrida 2009, 17). Is this, at base, why Rome is—must be—a beast in Revelation? Because for the Christ-confessing Jewish author of Revelation, Rome, as a blasphemous aberration (13:1, 5–6; 17:3), operates outside the law of God? But why, then, is *divine* power in Revelation also accorded an animal face, that of a Lamb (see especially 5:6–14)? Is it because the divine sovereign, too, and his messianic agent, are also outside and above the law, even (or especially) the law of God, including the divine command, "Thou shalt not kill"? (Exod 20:13; Deut 5:17). Mountains of corpses, both human and animal, loom over the landscapes of Revelation as the direct result of actions initiated by God or the Lamb.[11] For all who do not acknowledge their sovereignty, God and the Lamb are monstrous agents of terror, beastly objects of horror. "Fall on us," these terrified rebels cry out to the mountains, "and hide us from the face of the one seated on the throne and from the wrath of the Lamb; for the mighty day of their wrath has come and who can stand before it?" (6:16).

---

10. To say "animal" in the singular, then, is to utter "an asininity [*bêtise*]" (Derrida 2008a, 31). This Revelation does and does not do. On the one hand, it presents us with the specificity of the Lamb. On the other hand, it presents us with the nonspecificity of the beast.

11. See 55–56 (left-hand column) and 82 above.

"Beast, criminal, and sovereign have a troubling resemblance," muses Derrida. "There is between [them] a sort of obscure and fascinating complicity, or even a worrying mutual attraction, … an … uncanny reciprocal haunting.… [They] resemble each other while seeming to be situated at … each other's antipodes" (2009, 17). This unsettling family resemblance finds telling expression in the description of Revelation's second beast, the one that rises out of the earth, as having "two horns *like a lamb*," Revelation's master metaphor for Jesus, even while speaking "*like a dragon*" (13:11), Revelation's master metaphor for Satan (see 12:9; 20:2). In Revelation we discern "the face of the beast under the features of the sovereign" and vice versa—not least the divine sovereign—"the one inhabiting or housing the other," the one serving as "the intimate host of the other" (Derrida 2009, 18).

All of this is to say that Revelation, compulsively if surreptitiously hybrid,[12] constantly undercuts its own insistent dualisms. For Revelation's God is a beastly figure in other ways as well. "The one seated on the throne" (Revelation's preferred term for its deity—4:2, 9–10; 5:1, 7, 13; 6:16; 7:10, 15; 19:4; 21:5) is notably unresponsive, almost entirely aphasic, speaking only in 1:8 and 21:5–8.[13] As Derrida remarks of such divine monarchs, "the sovereign's sovereign, God himself, like the beast, does not respond.… And that is indeed the most profound definition of absolute sovereignty" (2009, 57).[14] That the visible face of this sovereign in Revelation is an animal face, an ovine face, is no accident. And although this Lamb is not entirely incapable of response, in general it exhibits the imperial nonresponsiveness of the figure on the throne, even outdoing that figure in aphasic inexpressiveness. The Lamb is not assigned a single word in this book in which it is the central character. The Lamb is mute, precisely as humans have almost always imagined animals to be. The Lamb is as dumb as a beast—or as *the* beast, which likewise has no speaking role in Revelation.

---

12. See 32–34 above.

13. See 81 above.

14. Derrida is glossing Hobbes here, and articulating a stereotype of bestiality that he also wishes to problematize: "this place of nonresponse that is commonly and dogmatically called bestiality, divinity, or death" (2009, 57). Derrida later concludes a lengthy analysis of *Robinson Crusoe* by noting: "What Robinson thinks of his parrot Poll is pretty much what Descartes, Kant, Heidegger, Lacan, and so very many others, think of all animals incapable of a true responsible and responding speech" (2011, 278; cf. 260).

The Lamb and the beast face each other mutely on the same side of the chasm long imagined to separate the human from the animal. In effect, this presumed aphasia is the abyss out of which the beast has crawled ("the beast that ascends from the abyss [*ek tēs abyssou*]," 11:7; see also 17:8), but to which it is still tethered, perhaps by the "mighty chain" mentioned in 20:1 ("Then I saw an angel descending from heaven, holding in his hand the key of the abyss and a mighty chain [*kai halysin megalēn*]"). But the Lamb also seems to be tethered by it and so also perches precariously on the lip of the abyss, unable to walk away from it. In other words, the Lamb and the beast may each be figures of fable, human entities draped in animal skins. But their animality does not sit lightly on either of them. Especially in relation to the shibboleth of speech, they behave as animals have almost always been imagined to behave. They are animetaphors that know how to pass as animals—up to a point, at least. The beast does not altogether succeed in passing its savagery off as animal savagery or predatory ferocity, as we are about to see.

Derrida remarks (2008a, 64) how the animal has traditionally been imagined "in the most contradictory and incompatible generic terms [*espèces*]: absolute (because natural) goodness, absolute innocence, prior to good and evil, the animal without fault or defect (that would be its superiority as inferiority), but also the animal as absolute evil, cruelty, murderous savagery." Both versions of the animal pad their way through Revelation, the domestic animal and the wild animal—the Lamb ostensibly without fault or defect, which makes it the perfect sacrificial victim (5:6, 12; 7:14; 12:11; 13:8), and the beast ostensibly the embodiment of absolute evil (11:7; 13:5, 14–15), its cruelty indicating that it is a figure for human savagery, for "cruelty implies humanity."[15]

The Sheep's Wife

Come, I will show you the bride, the wife of the Lamb. (Rev 21:9)

Revelation's *thērion*, its "beast," is also a "wild animal." Not only does the Greek admit both meanings, but "wild animal" is also what *thērion* most often meant according to most Greek lexicons (and what it means in Rev

---

15. The dictum "cruelty implies humanity" (as opposed to animality) is Lacan's (2006, 120). Derrida discusses (and dissects) this dictum (2009, 97, 102–11). He later returns to the paradoxical humanity of inhumanity (2011, 140–41).

6:8: "they were accorded authority over a fourth of the earth, to kill by sword, famine, and plague, and by the wild animals [*tōn thēriōn*] of the earth"). Human savagery, however, is represented in Revelation not only by the figure of a wild animal but also by the figure of a wild woman, a sexualized woman utterly out of (male) control—"Babylon the great, mother of whores and of earth's abominations" (17:5). Entire nations are drunk with lust for "the great whore" (14:8; 17:2; 18:3, 9: 19:2), but she herself is drunk on violence—"And I saw the woman drunk on the blood of the saints and the blood of Jesus' witnesses" (17:6; cf. 19:2)—a feral feast of blood that is subsequently and hyperbolically extended to "all who have been slaughtered on earth" (18:24). One cannot easily say where the ravenous woman ends and the ravening beast begins, and not only because both are figures for imperial Rome: the woman's ferocious appetite for blood makes her akin to a predatory animal.

Just as the wicked woman and the wild animal are intimately intertwined in Revelation, the woman's thighs wrapped around the beast ("I saw a woman sitting astride a scarlet beast," 17:3), so too are the virtuous woman and the domesticated animal intimately intertwined. But now we fully set foot—or hoof—on what Susan McHugh has termed the "queer spectrum of interspecies intimacies" (2011, 117). As it happens, McHugh is in the midst of a chapter titled "Breeding Narratives of Intimacy: Shaggy Dogs, Shagging Sheep." "Come, I will show you the bride [*nymphē*], the wife of the Lamb," John is told (21:9), having earlier learned that "the marriage of the Lamb has come, and his bride has made herself ready" (19:7). The apocalyptic bestiary is not innocent of bestiality—not that the sheep's marriage is a salacious tale, I hasten to add. If our sheep is involved in any shagging, it occurs discreetly behind the narrative curtain. What we are faced with, nonetheless (at least in our neoliteral construal of it), is what McHugh (2011, 132–33) would term a "nonstandard intimacy" conducted across species lines.

But who or what is "the bride, the wife of the Lamb"? "I will show you the bride," the angelic interpreter informs John, but what he sees instead is "the holy city Jerusalem" descending from heaven (21:9–10; cf. 21:2). The good empire, "the empire [*basileia*] of our Lord and his Messiah" (11:15), is here condensed as a good city, and that good city is figured in turn as a good wife. But the imagistic condensation and displacement does not end there, with the good wife cuddled up with her blood-drenched sheep. The imagery took a detour on its way from the city to the bride chamber. The detailed description given of the city (21:11–21) shows it to be nothing other than an elaborate, multifaceted symbol for the redeemed people of God.[16] The upshot of John's intricately layered symbolism is that all the subjects in his ideal empire are represented by (indeed, reduced to) the figure of an ideal wife. In John's good empire, all other subjects are subsumed in that one subject. The bride *is* the redeemed people of God. In consequence, no other relationship exists within the eschatological empire of empires than the submissive relationship of an obedient, worshipful wife to her husband and lord—a master who, however, happens also to be a domestic animal.[17]

The figure of the Lamb is fraught with paradox in Revelation. The Lamb is a nonhegemonic symbol for a hegemonic entity, a docile (indeed, domesticated) trope for domination. But so too is the bride. This is an all but wordless wife, who, as such, mimics the animal muteness of her four-legged bridegroom. Whereas the wild woman, "the whore," is accorded only one line in Revelation (18:7), the domestic woman, "the bride," is accorded only one word (albeit an enticing one): "And the Spirit and the bride say, 'Come' [*erchou*]!" (22:17).[18] Yet as city, metropolis, and megalopolis, this domestic goddess will lord it over the kings of the earth who must bring tribute (in)to her (21:24, 26; cf. Isa 60:1–16). And just as it is the destiny of the wicked city, Babylon/Rome, represented by the wicked woman, to be delivered back to undomesticated nature and become "a dwelling place of demons, a haunt of every foul and hateful bird, a haunt of every

---

16. On the city's gates are inscribed the names of the twelve tribes of Israel (Rev 21:12), while on the foundations of its wall are inscribed the names of the twelve apostles (12:14; cf. Eph 2:19–20). The twelve jewels that adorn the foundations correspond to the twelve jewels on the breastplate of Israel's high priest (Rev 21:19–20; cf. Exod 28:15–21; 39:8–14). This in turn evokes Revelation's description elsewhere of the redeemed as a company of priests (1:6; 5:10; 20:6). The bride herself likewise appears to be a symbol for the redeemed (cf. 2 Cor 11:2; Eph 5:23–32): the "fine linen, bright and pure" in which she is arrayed is glossed as "the righteous deeds of the saints" (Rev 19:8). Further on this commonplace construal of the city and the bride, see Gundry 2005.

17. Nevertheless, this husband-wife relationship is implicitly a master-slave relationship, because the Lamb's relationship to the redeemed is explicitly that of a slave owner. "Slave(s)" (*doulos/douloi*) is used fourteen times in Revelation (1:1 [twice]; 2:20; 6:11; 7:3; 10:7; 11:18; 15:3; 19:2, 5, 10; 22:3, 6, 9) for those who serve God or Jesus (and each time the term is euphemistically rendered as "servant[s]" in every major English translation).

18. For a resourceful reading of this one word, see Økland 2005, especially 327–32.

foul and hateful beast" (Rev 18:2; cf. Isa 13:21–22),[19] so is it the destiny of the virtuous city, "the new Jerusalem" (Rev 21:2), represented by the virtuous woman, to be the repository of domesticated nature, nature adapted to human needs. "Through the middle of the street of the city" will run "the river of the water of life," a river whose water is a life-giving gift for human beings (22:1–2; cf. 7:17; 21:6; 22:17; Gen 2:10; Ezek 47:1–11). On either side of the river, "the tree of life" will flourish, a tree whose leaves are for healing human beings (Rev 22:2; cf. 2:7; 22:14, 19; Gen 2:9; 3:22, 24; Ezek 47:12; *4 Ezra* 8:52). And at/as the source of the river and hence of the tree will stand the ultimate domestic animal—an always already slaughtered Lamb whose blood confers absolute benefit on human beings, which is to say, eternal life (Rev 22:1; cf. 1:5b; 5:9; 7:14; 12:11). Already one wonders how good this good city is for thinking with ecologically, or ecotheologically, or, especially, ecotherologically. And we have yet to take the measure of the city's biggest drawback: its sheer, stupendous size.

The Sheep in the Shopping Mall

[They] follow the Lamb wherever he goes. (Rev 14:4)

Ecotheological and ecojustice engagement with Revelation has tended to have recourse to the blueprint of the new Jerusalem to extract positive ecological visions from the blighted landscapes of this disaster-ridden book. Such reflection has typically gravitated not to the heavenly city itself so much as its river (*potamos*), which, somewhat peculiarly, courses down the center of the city's main street (*en mesō tēs plateias autēs*)—less a river, then, than a stream or channel?—and to the tree (*xylon zōēs*) that straddles the stream, or, alternatively, lines its banks, if *xylon* is to be read as a collective noun (Rev 22:1–2).[20] The obvious problem, however, is stated remarkably seldom in such reflection.

---

19. On the textual uncertainty of the final clause in Rev 18:2, see Beale 1999, 895, who argues that it is original, and Aune 1998b, 965, who suspects that it is not.

20. The work of Barbara Rossing is especially representative of this approach to Revelation and ecology; see, for example, her "River of Life in God's New Jerusalem" (1999a) and "For the Healing of the World" (2005a). See also Reid 2000, esp. 243–44; Maier 2002, 177–79; Bauckham 2010, 174–78; Bredin 2010, 172–77; Cate 2010, 153–55. Many other such examples could be listed.

The metaphors on which we have been musing (stream, tree, animal) are themselves situated within another metaphoric structure (city) so surreally outsized as to look unsettlingly like a cartoon rendition of what we are so busily turning our planet into anyway, as though too impatient to await the promised arrival of the heavenly megalopolis. "The city lies foursquare," writes John, "its length the same as its width" (Rev 21:16). John's angelic interpreter measures the city with his "measuring rod of gold" and discovers it to be exceedingly large (just how large we will see below).[21] The new Jerusalem is indeed a symbolic structure, as noted earlier. But size does matter here, as consultation of other ancient descriptions of the eschatological Jerusalem suggests. In the Dead Sea Scrolls (4Q554) its dimensions are 140 by 100 stadia (that is, 18.67 by 13.33 miles), "larger than any ancient city" (Wise, Abegg, and Cook 2005, 558);[22] in the *Sibylline Oracles* (5.250–52) its wall will reach to Joppa on the Mediterranean coast, 30 miles distant; while in *Song of Songs Rabbah* (7.5.3) its walls will extend all the way to the gates of Damascus, 135 miles distant, prompting even as serious a scholar as David Aune to quip about "eschatological urban sprawl" (1998b, 1162). Presumably, the eschatological Jerusalem is outsized because in ancient thought colossal size could function as a metaphor for divine transcendence.[23] And Revelation's new Jerusalem is biggest of all, Brobdingnagian in its dimensions. "He measured the city

---

21. Attention to the size of the heavenly city has been all but absent from ecocritical, ecotheological, and ecojustice work on Revelation. The notable exception, so far as I am aware, is Thomas Martin's "The City as Salvific Space" (2009); also see David Horrell's briefer reflections in *The Bible and the Environment* (2010, 100–101). Like the scholars listed in n. 20, however, Martin's perspective on the heavenly city through most of his colorful and intriguing article is overwhelmingly positive. Horrell's analysis of the city is far more critical.

22. The editors remark of the dimensions of the visionary city in this fragmentary text: "This new Jerusalem would have been larger than any ancient city and could only have been built by divine intervention" (Wise, Abegg, and Cook 2005, 558). Or modern technology. At just under 250 square miles, it would have been fractionally smaller than El Paso, Texas, and only one-twentieth the size of Tokyo.

23. For instance, gargantuan stature was frequently imputed to gods, angels, and other heavenly beings in the ancient Near East. As Wesley Williams (2009, 20) phrases it, "while the gods possessed an anthropoid or human-like form, this form was also in a fundamental way unlike that of humans in that it was transcendent, either in size, beauty, the substance of which it was composed, or all three." Gigantic angels are a frequent feature of ancient Jewish apocalyptic literature (and not unknown in Revelation either: see especially 10:1–10).

with his rod, twelve thousand stadia; its length and width and height are equal" (Rev 21:16). That amounts to "1,416–1,566 miles in each direction," depending on whether the city's 12,000 stadia cubed are Attic stadia, Olympic stadia, or Roman stadia (Aune, 1998b, 1161). The new Jerusalem is a sovereign city, not just in the sense that it is seat to a divine sovereign (Rev 22:3–5; cf. 21:22–26), but also in the sense that it partakes of the excess intrinsic to sovereignty. As Derrida observes, "What is essential and proper to sovereignty is … not grandeur or height as geometrically measurable, sensible, or intelligible, but excess, hyperbole … : higher than height, grander than grandeur, etc. It is the *more*, the *more than* that counts, … the absolute supplement that exceeds any comparative toward an absolute superlative" (2009, 257).

Revelation's vision of Edenic restoration, then, is more than a mere diversion of the river that watered the original garden (Gen 2:10) and more than a mere transplantation of the tree of life that was the centerpiece of that garden (2:9). River and tree are now situated in a sovereign city whose literal dimensions (limiting ourselves to length and breadth alone) would encompass more than half the continental United States. The problem for ecotheology (and even more for ecotherology) is that the proportions are horribly wrong—but uncannily right if a dystopian vision is needed of where contemporary urban hyperdevelopment, and not just in the United States, is headed—if climate change does not put all our megalopolises underwater before they have managed to link hands, or rather suburbs, and cover entire continents with housing developments and office developments, industrial parks and business parks, gas stations and fast food restaurants, strip malls and parking lots.

And Eden will have to adapt accordingly. Environmental historian Carolyn Merchant has argued compellingly that the high-end shopping mall may be read as emblematic of the re-creation of Eden through the anthropocentering of nature.

> The modern version of the Garden of Eden is the enclosed shopping mall. Surrounded by a desert of parking lots, malls comprise gardens of shops covered by glass domes, accessed by spiral staircases and escalators reaching upward toward heaven.… The "river that went out of Eden to water the garden" [Gen 2:10] is reclaimed in meandering tree-lined streams and ponds filled with bright orange goldfish.… Within manicured spaces of trees, flowers, and fountains we can shop for nature at the Nature Company, purchase "natural" clothing at Esprit, sample organic foods and "rainforest crunch" in kitchen gardens … and play virtual

reality games in which SimEve is reinvented in cyberspace. The spaces and commodities of the shopping mall epitomize consumer capitalism's vision of the recovery from the Fall of Adam and Eve. (2003, 157–58)

John's "measuring rod of gold" (Rev 21:15) may also be applied to the shopping mall, for size also matters to the latter, most of all when it is a megamall. Merchant describes the surreal dimensions of a megamall: "Canada's West Edmonton Mall, the first of a generation of megamalls, is eight city blocks long by four blocks wide and covers 5.2 million square feet. It sports an indoor surfing beach with adjustable wave heights" (2003, 158). For millions, as she notes, "malls are places of light, hope, and promise—transitions to new worlds. People are reinvented and redeemed by the mall. Said one ecstatic visitor, '… This place is heaven'" (158). And heaven keeps getting bigger. By 2012, the West Edmonton Mall, once the largest mall in the world, had dropped to tenth place in the international megamall rankings. At the time of writing, the nine largest malls on the planet are all located in Asia, the most immense of all being the "New South China Mall, Living City," in Dongguan—which, paradoxically, is also a "dead mall," having been 99 percent vacant since its 2005 opening.[24]

How should we classify the new Jerusalem—as a "living city" or a "dead mall"? Or simply as a megamall, whether living or dead? Is Revelation's heavenly city not all too readily—all too eerily—evocative of this most iconic of postmodern urban spaces,[25] complete with its central fountain, single stream, and token tree? And as such, is it not singularly ill-designed to serve as a prophetic counterexample to the contemporary paving over of the planet and the attendant obliteration of plant and animal species? One animal does survive in the heavenly city, however, and to that singular creature we turn once again.

Derrida writes apocalyptically of an ongoing sacrificial war against the animal, and specifically of an Abrahamic war: "I think that Cartesian-

---

24. "Dead mall" is the technical term for a commercially failing or shut down shopping mall. The New South China Mall was the subject of a brief but notable documentary directed by Sam Green and Carrie Lozano and titled *Utopia, Part 3: The World's Largest Shopping Mall* (Lucky Hat Entertainment, 2009).

25. Fredric Jameson in his now classic *Postmodernism, or, The Cultural Logic of Late Capitalism* describes his exploration of his topic as an attempt to keep hold of an "Ariadne's thread on its way through what may not turn out to be a labyrinth at all, but a gulag or perhaps a shopping mall" (1991, xi).

ism belongs ... to the Judeo-Christiano-Islamic tradition of a war against the animal, of a sacrificial war that is as old as Genesis" (2008a, 101). Revelation's Lamb is ambiguously positioned in relation to this ancient war. As slaughtered sacrifice, the Lamb is a victim of the war and has the war wounds to prove it ("Then I saw ... a Lamb standing as if it had been slaughtered," 5:6). But the Lamb is also a prime perpetrator of the war. The Lamb's incremental opening of the "scroll ... sealed with seven seals" (5:1; cf. 5:7) precipitates a chain of catastrophic events, many of which entail ecocidal devastation on a planetary scale (6:12–14; 8:7–12; 16:3–4, 8a, 10a, 12, 18, 20–21). Not only are "those [humans] who destroy the earth" destroyed (11:18),[26] but also the earth itself is destroyed in the process along with the nonhuman animals who depend on it for sustenance—yet another knotty contradiction in which the fraught figure of the Lamb is enmeshed.

Derrida warns that our sacrificial war against the animal—now characterized by exploitation and annihilation of the animal on an unprecedented scale—threatens to "end in a world without animals, without any animal worthy of the name living for something other than to become a means for [the human]" (2008a, 101–2), whether as source of meat, source of dairy products, source of clothing, domestic pet, model for children's toys or cartoon characters, hunter's trophy, hunter's aide, zoo specimen, or experimental life form. Is this the world with which Revelation presents us in the end, and as the end? What becomes of animals—animals other than the anomalous animal that is the Lamb—in Revelation's blueprint for the outsized urban enclosure that will form the center of the "new heaven and ... new earth" (21:1)?

Explicitly named in Revelation's climactic vision are "the dogs" (*hoi kynes*, 22:15)—metaphoric canines, assumedly, even if not certainly.[27] They are metonymically associated with human iniquity ("dogs and sorcerers

---

26. Rev 11:18 reads: "The nations raged, but your wrath has come, and the time for judging the dead, for rewarding your slaves, the prophets and saints and all who fear your name, ... and for destroying those who destroy the earth." Rossing plausibly identifies the destroyers of the earth as imperial Rome (2002; see also Maier 2002, 176).

27. At least one scholar has argued that literal, not metaphorical, dogs are in view here: "Because dogs were regarded as purificatory animals among both Greeks and Romans, and because Revelation 22:14 refers to purification by washing, Revelation 22:14–15 might well have been heard by Asian audiences as a polemic against pagan purificatory rites, especially those related to the cult of Hecate" (Strelan 2003, 148).

and fornicators and murderers and idolaters, and everyone who loves and practices falsehood"), and, most likely, are themselves metaphoric for a further form of activity that the author deems subhuman (homoeroticism is a common guess).[28] As a result, the dogs are banned from the heavenly city: "Outside [*exō*] are the dogs."

But if John casts the first stone at these dogs, the scholarly commentators on Revelation rush in with armfuls of rocks. Every slur leveled at dogs in the ancient world is scented out, tracked down, dug up, and uncritically brought to the pile. The boundaries between human "dogs" and nonhuman canines are thoroughly blurred in the process. Dogs are "despicable" and "despised" creatures, we learn,[29] because they behave like degenerate humans: they are "concerned only about their physical well-being" (Beale 1999, 1141); are "unclean because of their habits"—they are, indeed, possessed of "disgusting habits" (Aune, 1998b, 1223);[30] are "sexually immoral" (Witherington 2003, 282),[31] "impure and malicious" (Mounce 1998, 408), "cowards, unfaithful ... , abominable" (Kraft 1974, 280), and much else of this ilk. Humans who practice such depravities are loathsome because they behave like dogs. Dogs are loathsome because they behave like depraved humans. Conspicuously absent from the constricting circle of this encaging logic, exiled from it as from the heavenly city itself, are dogs that simply behave like dogs.

The heavenly city is a little empty without the dogs. This continent-sized, lightly landscaped megamall contains but one named animal, sole companion to the single stream and the lone tree. That animal is, of course, the Lamb (21:22–23; 22:1, 3). How does it stack up against the dogs? Might this solitary animal be regarded as an "animal worthy of the name living for something other than to become a means for [the human]," to re-invoke Derrida's poignant formulation (2008a, 102)? It might and might not. It is represented as having died for human beings and at the

---

28. See, e.g., Ford 1975, 345; Metzger 1993, 106; Thomas 1995, 507; Mounce 1998, 408; Aune 1998b, 1222–23.

29. "The despicability of the godless is emphasized, since dogs are regarded as despised creatures throughout Scripture" (Beale 1999, 1141). Cf. Boxall 2006, 317: The word "dogs" refers "to despised groups, including evildoers."

30. See also Smalley 2005, 574: "dogs are associated with habits which are unsavory."

31. See also Caird 1966, 285: "the dogs ... are here defined as those heathen who are indelibly marked with the qualities of the monster and the whore"

hand of human beings to confer eternal life on human beings (5:6–10; 7:13–14, 17; 12:11; 13:8; 14:1, 4; 21:27), and to that extent emblematizes anthropocentric animality. And yet this solitary, seven-horned, seven-eyed sheep might also be said, against all the odds, to be an animal worthy of the name to the extent that it lives—lives eternally—for something other than as a means for the human. It does not exist to serve human animals because human animals now exist to serve it: "the throne of God and of the Lamb will be in it [the heavenly city], and his slaves [*hoi douloi autou*] will worship him" (22:3).[32] Revelation's heavenly megalopolis is far from being an ecological paradise, but its token animal is also far from being the final victim, the last animal standing, of the ecocidal drama that enabled the establishment of the heavenly city in the first place—the incremental demolition and progressive leveling of "the first heaven and the first earth" (21:1; see esp. 8:6–12; 16:1–12). In a final, bizarre twist of this hypersurreal animal tale, the only animal worthy of the name ("Thou art worthy," 5:9; cf. 5:12) has become other than a means to human ends because, in the end, all humans have become its property.

The hierarchical relationship of human beings and domestic animals is thus radically inverted in Revelation. But is it deconstructed? Hardly. The heavenly city remains a domestic enclosure constructed, like any enclosure, through the systemic exclusion of its others—in this case, undomesticated animality, wild animality, animality altogether unbeholden to the human. The "dogs" are excluded, as we saw. They trouble the human/domestic animal opposition, because as ancient Mediterranean dogs, perched precariously on the jagged edges of human society, they are neither fully domesticated nor fully wild. Still more unequivocally outside the domestic enclosure that is the heavenly city is the *thērion*, the beast or wild beast. Earlier, this wild beast was "thrown alive into the lake of fire that burns with sulfur" (19:20), probably not its optimal habitat; yet there it will survive "forever and ever," although only to be "tormented day and night" (20:10). Indeed, the consignment of the wild beast to unending destruction was one of the prime preconditions for the emergence of the heavenly city.

Is Revelation's heavenly city founded, then, like all our terrestrial cities, on the perpetual sacrifice of any and every "animal worthy of the name, living for something other than to become a means for the human"? Yes and no, yet again, because the beast of Revelation is, at base, a figure

---

32. See 212 n. 30 in "Quadrupedal Christ" above.

for human despotism, for Roman imperialism. Even the beast that must be eternally destroyed so that the domestic bliss of the bride and the Lamb can ensue within the urban enclosure of Revelation's heaven is not as wild, as untethered to the human, as ecotherology might wish it to be. All of which is to say that Revelation does, after all, "end in a world without animals," without "any animal worthy of the name," whether inside or outside the city. And to that extent at least, Revelation may, after all, be an unveiling of "what is" and "what must soon take place" (1:1, 19; 4:1; 22:6)—an apocalyptic uncovering of the already present future of our catastrophically theriocidal cultures.

And yet.... Revelation's multihorned, multieyed, multifaceted sheep is sufficiently complex, elusive, and exasperating to merit a final "and yet." The sheer ecotheological significance of this singular if sinister animal should not be lost sight of amid the paradoxes, contradictions, and disappointments that encircle it. Jesus of Nazareth enters Revelation as a Son of Humanity (1:12–13), transforms into a Lamb (5:6), and trots through the main body of the text in that theriomorphic guise and hence on all fours, only assuming anthropomorphic form again sporadically in 14:14–16 and 19:11–21. But when the shape-shifting eventually ends and the heavenly city arrives and God's Messiah is enthroned with God in the city center as eternal object of incessant worship, it is not as anthropomorph but as theriomorph that he comes into final focus (21:22–23; 22:3; see also 19:7, 9; 21:9, 14). Indeed, it is only in animal guise that Jesus is worshiped anywhere in Revelation (see also 5:8–14; cf. 7:9–12, 15–17). Revelation evinces a high Christology, as has often been remarked.[33] What has not been remarked is that Revelation's Christology is highest when it is an animal Christology. The ultimate christological image in Revelation, then—and the image most deeply stamped with the mark of divinity to the extent that it is the one explicitly marked for worship[34]—is the image of a quadrupedal Christ. In an age of mass extinc-

---

33. Thomas Schreiner's assessment (2008, 430) is typical: "Even though Revelation is an apocalyptic work, the Christology is astonishingly explicit and high. Indeed, the Christology is analogous to that in the Gospel of John. Jesus as the Lamb is on the same plane as God and is worshiped as a divine being.... Just as God is the Alpha and Omega and the first and the last, so too Jesus is the Alpha and Omega and the first and the last."

34. And in a book that is notably fastidious about proper and improper objects of worship (see 19:10; 22:8–9; cf. 9:20; 13:4, 8, 12, 15; 14:9–11; 16:2; 19:20–21; 20:4).

tion this surely qualifies as an ecotheological image with legs, despite the contradictions that repeatedly trip it up.

Bibliography

Adams, Edward. 2007. *The Stars Will Fall from Heaven: "Cosmic Catastrophe" in the New Testament and Its World*. LNTS 347. New York: T&T Clark.

Agamben, Giorgio. 2004. *The Open: Man and Animal*. Translated by Kevin Attell. Meridian, Crossing Aesthetics. Stanford: Stanford University Press.

Ahmed, Sara. 2004. *The Cultural Politics of Emotion*. New York: Routledge.

———. 2010a. Happy Objects. Pages 29–51 in *The Affect Theory Reader*. Edited by Melissa Gregg and Gregory J. Seigworth. Durham: Duke University Press.

———. 2010b. *The Promise of Happiness*. Durham: Duke University Press.

Allen, Leslie C. 2008. *Jeremiah: A Commentary*. OTL. Louisville: Westminster John Knox.

Amarasingam, Amarnath. 2011. "Baracknophobia and the Paranoid Style: Visions of Obama as the Anti-Christ on the World Wide Web." Pages 96–123 in *Network Apocalypse: Visions of the End in an Age of Internet Media*. Edited by Robert Glenn Howard. The Bible in the Modern World 36. Sheffield: Sheffield Phoenix.

Anderson, Graham. 1993. *The Second Sophistic: A Cultural Phenomenon in the Roman Empire* New York: Routledge.

Anderson, Janice Capel, and Jeffrey L. Staley, eds. 1995. *Taking It Personally: Autobiographical Biblical Criticism*. Semeia 72. Atlanta: Scholars Press.

Ashcroft, Bill, Gareth Griffiths, and Helen Tiffin. 2001. *Post-Colonial Studies: The Key Concepts*. 2nd ed. New York: Routledge.

Aune, David E. 1981. "The Social Matrix of the Apocalypse of John." *Papers of the Chicago Society of Biblical Research* 26:16–32.

———. 1983. "The Influence of Roman Imperial Court Ceremonial on the Apocalypse of John." *Papers of the Chicago Society of Biblical Research* 28:5–29. Reprinted in pages 99–119 of David E. Aune, *Apocalypticism*,

Prophecy, and Magic in Early Christianity: Collected Essays. WUNT 199. Tübingen: Mohr Siebeck, 2006.

———. 1997. *Revelation 1–5*. WBC 52A. Dallas: Word.

———. 1998a. *Revelation 6–16*. WBC 52B. Nashville: Thomas Nelson.

———. 1998b. *Revelation 17–22*. WBC 52C. Nashville: Thomas Nelson.

Aydemir, Murat, ed. 2011. *Indiscretions: At the Intersection of Queer and Postcolonial Theory*. Thamyris/Intersecting: Place, Sex and Race 22. Amsterdam and New York: Rodopi.

Badmington, Neil, ed. 2007. *DerridAnimals*. Oxford Literary Review 29.1 (thematic issue).

Barker, Margaret. 1992. *The Great Angel: A Study of Israel's Second God*. London: SPCK.

Barr, David L. 1998. *Tales of the End: A Narrative Commentary on the Book of Revelation*. Santa Rosa, Calif.: Polebridge.

———. 2001. "Waiting for the End That Never Comes: The Narrative Logic of John's Story." Pages 101–12 in *Studies in the Book of Revelation*. Edited by Steve Moyise. New York: T&T Clark.

———. 2003. "The Story John Told: Reading Revelation for Its Plot." Pages 11–23 in *Reading the Book of Revelation: A Resource for Students*. Edited by David L. Barr. SBLRBS 44. Atlanta: Society of Biblical Literature.

Barton, Carlin A. 1994. "Savage Miracles: The Redemption of Lost Honor in Roman Society and the Sacrament of the Gladiator and the Martyr." *Representations* 45:41–71.

Bartsch, Shadi. 1994. *Actors in the Audience: Theatricality and Doublespeak from Nero to Hadrian*. Revealing Antiquity 6. Cambridge: Harvard University Press.

Bauckham, Richard. 1993a. *The Climax of Prophecy: Studies on the Book of Revelation*. Edinburgh: T&T Clark.

———. 1993b. *The Theology of the Book of Revelation*. New Testament Theology. Cambridge: Cambridge University Press.

———. 2010. *The Bible and Ecology: Rediscovering the Community of Creation*. Waco, Tex.: Baylor University Press.

———. 2011. *Living with Other Creatures: Green Exegesis and Theology*. Waco, Tex.: Baylor University Press.

Beale, G. K. 1997. "Solecisms in the Apocalypse as Signals for the Presence of Old Testament Allusions: A Selective Analysis of Revelation 1–22." Pages 421–46 in *Early Christian Interpretation of the Scriptures of Israel*. Edited by Craig A. Evans and James A. Sanders. JSNTSup 148; SEJC 3. Sheffield: Sheffield Academic.

———. 1999. *The Book of Revelation: A Commentary on the Greek Text*. NIGTC. Grand Rapids: Eerdmans.

Beard, Mary, John North, and Simon Price. 1998. *A History*. Vol. 1 of *Religions of Rome*. Cambridge: Cambridge University Press.

Beasley-Murray, George R. 1978. *Revelation*. NCB. Grand Rapids: Eerdmans.

Beauvery, Robert. 1983. "L'Apocalypse au risqué de la numismatique: Babylone, la grande Prostituée et le sixième roi Vespasien et la déesse Rome." *RB* 90:243–60.

Bennett, Jane. 2010. *Vibrant Matter: A Political Ecology of Things*. Durham: Duke University Press.

Bennington, Geoffrey, et al. 2009. "General Introduction to the French Edition." Pages ix–xi in vol. 1 of Jacques Derrida, *The Beast and the Sovereign*. Edited by Geoffrey Bennington and Peggy Kamuf. The Seminars of Jacques Derrida 1. Chicago: University of Chicago Press.

Bentham, Jeremy. 1879. *Principles of Morals and Legislation*. Oxford: Clarendon.

Bergen, Wesley J. 2005. *Reading Ritual: Leviticus in Postmodern Culture*. JSOTSup 417. New York: Continuum.

Berger, Anne Emmanuelle, and Marta Segarra, eds. 2011. *Demenageries: Thinking (of) Animals after Derrida*. Amsterdam and New York: Rodopi.

Berlant, Lauren. 2012. *Desire/Love*. New York: Dead Letter Office.

Bhabha, Homi K. 1990. "Introduction: Narrating the Nation." Pages 1–7 in *Nation and Narration*. Edited by Homi K. Bhabha. New York: Routledge.

———. 1991. "The Third Space: Interview with Homi K. Bhabha." Pages 207–21 in *Identity: Community, Culture, Difference*. Edited by Jonathan Rutherford. London: Lawrence & Wishart.

———. 1992. "Postcolonial Criticism." Pages 437–65 in *Redrawing the Boundaries: The Transformation of English and American Literary Studies*. Edited by Stephen Greenblatt and Giles Gunn. New York: Modern Language Association of America.

———. 1994a. "Articulating the Archaic: Cultural Difference and Colonial Nonsense." Pages 123–38 in Bhabha 1994c.

———. 1994b. "The Commitment to Theory." Pages 19–39 in Bhabha 1994c.

———. 1994c. *The Location of Culture*. New York: Routledge.

———. 1994d. "Of Mimicry and Man: The Ambivalence of Colonial Discourse." Pages 85–92 in Bhabha 1994c.

———. 1994e. "Signs Taken for Wonders: Questions of Ambivalence and Authority under a Tree outside Delhi, May 1817." Pages 102–22 in Bhabha 1994c.

———. 2002. "Speaking of Postcoloniality, in the Continuous Present: A Conversation." Pages 15–46 in *Relocating Postcolonialism*. Edited by David Theo Goldberg and Ato Quayson. Oxford: Blackwell.

Bible and Culture Collective. 1995. *The Postmodern Bible*. New Haven: Yale University Press.

Bitel, Lisa. 1997. *Land of Women: Tales of Sex and Gender from Early Ireland*. Ithaca, N.Y.: Cornell University Press.

Blackman, Lisa. 2012. *Immaterial Bodies: Affect, Embodiment, Mediation*. Thousand Oaks, Calif.: Sage.

Blanchot, Maurice. 1986. *The Writing of the Disaster*. Translated by Ann Smock. Lincoln: University of Nebraska Press.

Bloch, Ernst. 1986. *The Principle of Hope*. 3 vols. Translated by Neville Plaice et al. German original 1954–59. Cambridge: MIT Press.

Blount, Brian K. 2005. *Can I Get a Witness? Reading Revelation through African American Culture*. Louisville: Westminster John Knox.

———. 2007. "Revelation." Pages 523–57 in *True to Our Native Land: An African American New Testament Commentary*. Edited by Brian K. Blount, Cain Hope Felder, Clarice J. Martin, and Emerson B. Powery. Minneapolis: Fortress.

Boesak, Allan A. 1987. *Comfort and Protest: The Apocalypse from a South African Perspective*. Philadelphia: Westminster.

Boggs, Colleen Glenney. 2009. "Emily Dickinson's Animal Pedagogies." *PMLA* 124:533–41.

Bolin, Anne. 1992. "Vandalized Vanity: Feminine Physiques Betrayed and Portrayed." Pages 79–99 in *Tattoo, Torture, Mutilation, and Adornment: The Denaturalization of the Body in Culture and Text*. Edited by Frances E. Mascia-Lees and Patricia Sharpe. SUNY Series on the Body in Culture, History, and Religion. Albany: State University of New York Press.

Boring, M. Eugene. 1989. *Revelation*. IBC. Louisville: John Knox.

Bousset, Wilhelm. 1906. *Die Offenbarung Johannis*. 6th ed. KEK 16. Göttingen: Vandenhoeck & Ruprecht.

Bowra, C. M. 1957. "Melinno's Hymn to Rome." *JRS* 47:21–28.

Boxall, Ian. 2001. "The Many Faces of Babylon the Great: *Wirkungsgeschichte* and the Interpretation of Revelation 17." Pages 51–68 in *Studies in the Book of Revelation*. Edited by Steve Moyise. New York: Continuum.

———. 2006. *The Revelation of Saint John*. BNTC 18. London: Black.

Braidotti, Rosi. 2009. "Animals, Anomalies, and Inorganic Others." *PMLA* 124:526–32.

———. 2013. *The Posthuman*. Cambridge: Polity.

Brasher, Brenda E. 1998. "From Revelation to *The X-Files*: An Autopsy of Millenialism in American Popular Culture." *Semeia* 82:281–95.

Bredin, Mark. 2010. *The Ecology of the New Testament: Creation, Re-Creation, and the Environment*. Colorado Springs: Biblica.

Brennan, Teresa. 2004. *The Transmission of Affect*. Ithaca, N.Y.: Cornell University Press.

Brenner, Athalya. 1995. "On Prophetic Propaganda and the Politics of 'Love': The Case of Jeremiah." Pages 256–74 in *A Feminist Companion to the Latter Prophets*. The Feminist Companion to the Bible 8. Edited by Athalya Brenner. Sheffield: Sheffield Academic.

———. 1997. *The Intercourse of Knowledge: On Gendering Desire and "Sexuality" in the Hebrew Bible*. Biblical Interpretation Series 26. Leiden: Brill.

Brent, Allen. 1999. *The Imperial Cult and the Development of Church Order: Concepts and Images of Authority in Paganism and Early Christianity before the Age of Cyprian*. VCSup 45. Leiden: Brill.

Brütsch, Charles. 1970. *Die Offenbarung Jesu Christi: Johannes-Apokalypse*. 2nd ed. ZBK. Zürich: Zwingli.

Burrell, Barbara. 2004. *Neokoroi: Greek Cities and Roman Emperors*. Leiden: Brill.

Burrus, Virginia. 2008. *Saving Shame: Martyrs, Saints, and Other Abject Subjects*. Divinations: Rereading Late Ancient Religion. Philadelphia: University of Pennsylvania Press.

Butler, George. 1990. *Arnold Schwarzenegger: A Portrait*. New York: Simon & Schuster.

Butler, Judith. 1990. *Gender Trouble: Feminism and the Subversion of Identity*. Thinking Gender. New York: Routledge.

———. 1993. *Bodies That Matter: On the Discursive Limits of "Sex."* New York: Routledge.

———. 1995. "Melancholy Gender/Refused Identification." Pages 21–36 in *Constructing Masculinities*. Edited by Maurice Berger, Brian Wallis, and Simon Watson. New York: Routledge.

———. 1999. Preface to *Gender Trouble: Feminism and the Subversion of Identity*. 10th Anniversary ed. New York: Routledge.

———. 2004. *Undoing Gender*. New York: Routledge.

———. 2006. Afterword to *Bodily Citations: Religion and Judith Butler*. Edited by Ellen T. Armour and Susan M. St. Ville. Gender, Theory, and Religion. New York: Columbia University Press.

Butler, Judith, with Peter Osborne and Lynne Segal. 1994. "Gender as Performance: An Interview with Judith Butler." *Radical Philosophy* 67:32–37.

Caird, G. B. 1966. *The Revelation of Saint John*. BNTC. London: Black.

Calarco, Matthew. 2008. *Zoographies: The Question of the Animal from Heidegger to Derrida*. New York: Columbia University Press.

Calarco, Matthew, and Peter Atterton, eds. 2004. *Animal Philosophy: Essential Readings in Continental Thought*. New York: Continuum.

Callahan, Allen D. 1995. "The Language of the Apocalypse." *HTR* 88:453–70.

Campbell, Jan. 2000. *Arguing with the Phallus: Feminist, Queer and Postcolonial Theory. A Psychoanalytic Contribution*. New York: St. Martin's.

Caputo, John D. 2006. *The Weakness of God: A Theology of the Event*. Indiana Series in the Philosophy of Religion. Bloomington: Indiana University Press.

Carey, Greg. 2006. "Revelation and Empire: Symptoms and Resistance." Pages 169–80 in *The Reality of Apocalypse: Rhetoric and Politics in the Book of Revelation*. Edited by David L. Barr. SBLSymS 39. Atlanta: Society of Biblical Literature.

———. 2008. "The Book of Revelation as Counter-Imperial Script." Pages 157–76 in *In the Shadow of Empire: Reclaiming the Bible as a History of Faithful Resistance*. Edited by Richard A. Horsley. Louisville: Westminster John Knox.

Carrell, Peter R. 1997. *Jesus and the Angels: Angelology and the Christology of the Apocalypse of John*. SNTSMS 95. Cambridge: Cambridge University Press.

Carroll, Robert P. 1995. "Desire under the Terebinths: On Pornographic Representation in the Prophets—A Response." Pages 275–307 in *A Feminist Commentary to the Latter Prophets*. Edited by Athalya Brenner. Feminist Companion to the Bible 8. Sheffield: Sheffield Academic.

Carson, Marion. 2011. "The Harlot, the Beast and the Sex Trafficker: Reflections on Some Recent Feminist Interpretations of Revelation 17–18." *ExpTim* 122:218–27.
Carter, Warren. 2009. "Accommodating 'Jezebel' and Withdrawing John: Negotiating Empire in Revelation Then and Now." *Int* 63:32–47.
———. 2011. *What Does Revelation Reveal? Unlocking the Mystery*. Nashville: Abingdon.
Cassuto, Umberto. 1961–64. *A Commentary on the Book of Genesis*. 2 vols. Translated by Israel Abrahams. Jerusalem: Magnes.
Castelli, Elizabeth A. 1991. *Imitating Paul: A Discourse of Power*. Literary Currents in Biblical Interpretation. Louisville: Westminster John Knox.
Cate, James Jeffrey. 2010. "How Green Was John's World? Ecology and Revelation." Pages 145–55 in *Essays on Revelation: Appropriating Yesterday's Apocalypse in Today's World*. Edited by Gerald L. Stevens. Eugene, OR: Wipf & Stock.
Cavell, Stanley, Cora Diamond, John McDowell, Ian Hacking, and Cary Wolfe. 2008. *Philosophy and Animal Life*. New York: Columbia University Press.
Chapkis, Wendy. 1997. *Live Sex Acts: Women Performing Erotic Labor*. New York: Routledge.
Charles, R. H. 1920. *A Critical and Exegetical Commentary on the Revelation to St John*. 2 vols. ICC. Edinburgh: T&T Clark.
Chrulew, Matthew. 2008. "Feline Divinanimality: Derrida and the Discourse of Species in Genesis." *The Bible and Critical Theory* 2.2. Online: http://novaojs.newcastle.edu.au/ojsbct/index.php/bct/article/view/87/73.
Clanton, Dan W., Jr., ed. 2012. *The End Will Be Graphic: Apocalyptic in Comic Books and Graphic Novels*. The Bible in the Modern World 43. Sheffield: Sheffield Phoenix.
Clark, J. Michael. 1999. "Queering the Apocalypse: Reading Catherine Keller's *Apocalypse Now and Then*." *Journal of Men's Studies* 7:233–44.
Clarke, John R. 2003. *Roman Sex 100 BC–AD 250*. New York: Abrams.
Clines, David J. A. 1995. "David the Man: The Construction of Masculinity in the Hebrew Bible." Pages 212–43 in *Interested Parties: The Ideology of Writers and Readers of the Hebrew Bible*. Gender, Culture, Theory 1. Sheffield: Sheffield Academic.
Clough, Patricia Ticineto, with Jean Halley, eds. 2007. *The Affective Turn: Theorizing the Social*. Durham: Duke University Press.

Coetzee, J. M. 1997. *Boyhood: Scenes from Provincial Life*. New York: Viking.
———. 1999a. *Disgrace*. New York: Viking.
———. 1999b. *The Lives of Animals*. Edited by Amy Gutmann. University Center for Human Values Series. Princeton: Princeton University Press.
Cohen, Edward E. 2006. "Free and Unfree Sexual Work: An Economic Analysis of Athenian Prostitution." Pages 95–124 in *Prostitutes and Courtesans in the Ancient World*. Edited by Christopher A. Faraone and Laura K. McClure. Madison: University of Wisconsin Press.
Cohn, Norman. 1970. *The Pursuit of the Millennium: Revolutionary Millenarians and Mystical Anarchists of the Middle Ages*. 2nd ed. Oxford: Oxford University Press.
Collins, Adela Yarbro. 1976. *The Combat Myth in the Book of Revelation*. HDR 9. Missoula, Mont.: Scholars Press.
———. 1984. *Crisis and Catharsis: The Power of the Apocalypse*. Philadelphia: Westminster.
———. 1987. "Women's History and the Book of Revelation." Pages 80–91 in *SBLSP* 26. Edited by Kent Harold Richards. Atlanta: Scholars Press.
———. 1992. "The 'Son of Man' Tradition and the Book of Revelation." Pages 536–68 in *The Messiah*. Edited by James H. Charlesworth. Minneapolis: Fortress.
———. 1993a. "Feminine Symbolism in the Book of Revelation." *BibInt* 1:20–33.
———. 1993b. "The Influence of Daniel on the New Testament." Pages 90–112 in John J. Collins. *Daniel: A Commentary on the Book of Daniel*. Hermeneia. Minneapolis: Fortress.
Conway, Colleen M. 2008. *Behold the Man: Jesus and Greco-Roman Masculinity*. Oxford: Oxford University Press.
Cooke, Miriam. 1993. "Wo-man, Retelling the War Myth." Pages 169–90 in *Gendering War Talk*. Edited by Miriam Cooke and Angela Woollacott. Princeton: Princeton University Press.
Corner, Sean. 2011. "Bringing the Outside In: The *Andrōn* as Brothel and the Symposium's Civic Sexuality." Pages 60–85 in *Greek Prostitutes in the Ancient Mediterranean, 800 BCE–200 CE*. Edited by Allison Glazebrook and Madeleine M. Henry. Wisconsin Studies in Classics. Madison: University of Wisconsin Press.
Court, John M. 1979. *Myth and History in the Book of Revelation*. Atlanta: John Knox Press.

Crane, Susan. 2013. *Animal Encounters: Contacts and Concepts in Medieval Britain*. Philadelphia: University of Pennsylvania Press.
Crossan, John Dominic. 1994. *Jesus: A Revolutionary Biography*. San Francisco: HarperSanFrancisco.
Curran, John R. 2000. *Pagan City and Christian Capital: Rome in the Fourth Century*. Oxford Classical Monographs. Oxford: Oxford University Press.
Cuss, Dominique. 1974. *Imperial Cults and Honorary Terms in the New Testament*. Paradosis 23. Fribourg: Fribourg University Press.
D'Ambra, Eve. 2007. *Roman Women*. Cambridge Introduction to Roman Civilization. Cambridge: Cambridge University Press.
Darden, Lynne St. Clair. 2011. "'Some folks is born wid they feet on the sun and they kin seek out de inside meanin' of words': An African American Scripturalization of the Book of Revelation Signified through the (Postcolonial) Middle Passage." Ph.D. diss., Drew University, Madison, N.J.
Darwin, Charles. 1872. *On the Expression of the Emotions in Man and Animals*. London: John Murray.
Davidson, James. 1994. "Consuming Passions: Appetites, Addiction and Spending in Classical Athens." Ph.D. diss., Trinity College, Oxford University.
———. 1997. *Courtesans and Fishcakes: The Consuming Passions of Classical Athens*. New York: St. Martin's.
Deissmann, Adolf. 1900. "Das vierte Makkabäerbuch." Pages 149–77 in vol. 2 of *Die Apokryphen und Pseudepigraphen des Alten Testament*. Edited by Emil Kautzsch. Tübingen: Mohr Siebeck.
———. 1923. *Licht vom Osten: Das Neue Testament und die neuentdeckten Texte der hellenistisch-römischen Welt*. 4th ed. Tübingen: Mohr Siebeck.
DeKoven, Marianne. 2009. "Guest Column: Why Animals Now?" *PMLA* 124:361–69.
DeKoven, Marianne, and Michael Lundblad, eds. 2012. *Species Matters: Humane Advocacy and Cultural Theory*. New York: Columbia University Press.
Deleuze, Gilles. 1986. *Cinema 1: The Movement-Image*. Translated by Hugh Tomlinson and Barbara Habberjam. Minneapolis: University of Minnesota Press.
———. 1989. *Cinema 2: The Time-Image*. Translated by Hugh Tomlinson and Robert Galeta. Minneapolis: University of Minnesota Press.

———. 1990. *The Logic of Sense*. Edited by Constantin V. Boundas. Translated by Mark Lester with Charles Stivale. New York: Columbia University Press.

———. 1993. *The Fold: Leibnitz and the Baroque*. Translated by Tom Conley. Minneapolis: University of Minnesota Press.

———. 1995. *Negotiations, 1972–1990*. Translated by Martin Joughin. New York: Columbia University Press.

———. 1998. "Nietzsche and Saint Paul, Lawrence and John of Patmos." Pages 36–52 in *Essays Critical and Clinical*. Translated by Daniel W. Smith and Michael A. Greco. London: Verso.

———. 2003. *Francis Bacon: The Logic of Sensation*. Translated by Daniel W. Smith. Minneapolis: University of Minnesota Press.

Deleuze, Gilles, and Félix Guattari. 1983. *Anti-Oedipus: Capitalism and Schizophrenia*. Translated by Robert Hurley, Mark Seem, and Helen R. Lane. Minneapolis: University of Minnesota Press.

———. 1986. *Kafka: Toward a Minor Literature*. Translated by Dana Polan. Theory and History of Literature 30. Minneapolis: University of Minnesota Press.

———. 1987. *A Thousand Plateaus: Capitalism and Schizophrenia*. Translated by Brian Massumi. Minneapolis: University of Minnesota Press.

Dellamora, Richard, ed. 1995. *Postmodern Apocalypse: Theory and Cultural Practice at the End*. Philadelphia: University of Pennsylvania Press.

DeLoughrey, Elizabeth, and George B. Handley, eds. 2011. *Postcolonial Ecologies: Literatures of the Environment*. Oxford: Oxford University Press.

De Man, Paul. 1979. *Allegories of Reading: Figural Language in Rousseau, Nietzsche, Rilke, and Proust*. New Haven: Yale University Press.

Derrida, Jacques. 1976. *Of Grammatology*. Translated by Gayatri Chakravorty Spivak. Baltimore: Johns Hopkins University Press.

———. 1978. *Writing and Difference*. Translated by Alan Bass. Chicago: University of Chicago Press.

———. 1981. "Economimesis." *Diacritics* 11:3–25.

———. 1982a. *Margins of Philosophy*. Translated by Alan Bass. Chicago: University of Chicago Press.

———. 1982b. "On a Newly Arisen Apocalyptic Tone in Philosophy." Translated by John P. Leavey Jr. *Semeia* 23:63–97.

———. 1983. *D'un ton apocalyptique adopté naguère en philosophie*. Paris: Galilée.

———. 1986. *Glas*. Translated by John P. Leavey Jr. Lincoln: University of Nebraska Press.

———. 1987. "Le facteur de la vérite." Pages 411–96 in *The Post Card: From Socrates to Freud and Beyond*. Translated by Alan Bass. Chicago: University of Chicago Press.

———. 1989. *Of Spirit: Heidegger and the Question*. Translated by Geoffrey Bennington and Rachel Bowlby. Chicago: University of Chicago Press.

———. 1992a. *Given Time: I. Counterfeit Money*. Translated by Peggy Kamuf. Chicago: University of Chicago Press.

———. 1992b. "Of an Apocalyptic Tone Newly Adopted in Philosophy." Translated by John P. Leavey Jr. Pages 25–72 in *Derrida and Negative Theology*. Edited by Harold Coward and Toby Foshay. Albany: State University of New York Press.

———. 1993a. "Circumfession: Fifty-Nine Periods and Periphrases." Pages 3–315 in Geoffrey Bennington and Jacques Derrida, *Jacques Derrida*. Chicago: University of Chicago Press.

———. 1993b. *Memoirs of the Blind: The Self-Portrait and Other Ruins*. Translated by Pascale Anne-Brault and Michael Naas. Chicago: University of Chicago Press.

———. 1994a. *Aporias*. Translated by Thomas Dutoit. Meridian: Crossing Aesthetics. Stanford: Stanford University Press.

———. 1994b. *Specters of Marx: The State of the Debt, the Work of Mourning, and the New International*. Translated by Peggy Kamuf. New York: Routledge.

———. 1995a. "'Eating Well,' or the Calculation of the Subject." Pages 255–83 in Derrida 1995e.

———. 1995b. *The Gift of Death*. Translated by David Wills. Chicago: University of Chicago Press.

———. 1995c. *On the Name*. Edited by Thomas Dutoit. Translated by David Wood, John P. Leavey Jr., and Ian McLeod. Meridian: Crossing Aesthetics. Stanford: Stanford University Press.

———. 1995d. "Passages—from Traumatism to Promises." Pages 372–98 in Derrida 1995e.

———. 1995e. *Points ... Interviews, 1974–1994*. Edited by Elisabeth Weber. Translated by Peggy Kamuf et al. Meridian: Crossing Aesthetics. Stanford: Stanford University Press.

———. 1997a. *Politics of Friendship*. Translated by George Collins. London: Verso.

———. 1997b. "The Villanova Roundtable." Pages 1–28 in *Deconstruction in a Nutshell: A Conversation with Jacques Derrida*. Edited with a commentary by John D. Caputo. Perspectives in Continental Philosophy. New York: Fordham University Press.

———. 1999a. "L'animal que donc je suis (à suivre)." Pages 251–303 in *L'animal autobiographique*. Edited by Marie-Louise Mallet. Paris: Galilée.

———. 1999b. "Marx and Sons." Pages 213–69 in *Ghostly Demarcations: A Symposium on Jacques Derrida's Specters of Marx*. Edited by Michael Sprinker. Radical Thinkers 33. London: Verso.

———. 2000. *Of Hospitality: Anne Dufourmantelle Invites Jacques Derrida to Respond*. Translated by Rachel Bowlby. Cultural Memory in the Present. Stanford: Stanford University Press.

———. 2002a. *Acts of Religion*. Edited by Gil Anidjar. Translated by Mark Dooley and Michael Hughes. New York: Routledge.

———. 2002b. "The Animal That Therefore I Am (More to Follow)." Translated by David Wills. *Critical Inquiry* 28:369–418.

———. 2002c. "Faith and Knowledge: The Two Sources of 'Religion' at the Limits of Reason Alone." Pages 1–39 in Derrida 2002a.

———. 2004. "Violence against Animals." Pages 62–76 in Jacques Derrida and Elisabeth Roudinesco, *For What Tomorrow … : A Dialogue*. Translated by Jeff Fort. Cultural Memory in the Present. Stanford: Stanford University Press.

———. 2005. *Rogues: Two Essays on Reason*. Translated by Pascale-Anne Brault and Michael Naas. Meridian: Crossing Aesthetics. Stanford: Stanford University Press.

———. 2007. "No Apocalypse, Not Now: Full Speed Ahead, Seven Missiles, Seven Missives." Translated by Catherine Porter and Philip Lewis. Pages 387–409 in vol. 1 of Jacques Derrida. *Psyche: Inventions of the Other*. Edited by Peggy Kamuf and Elizabeth Rottenberg. Meridian: Crossing Aesthetics. Stanford: Stanford University Press.

———. 2008a. *The Animal That Therefore I Am*. Edited by Marie-Louise Mallet. Translated by David Wills. Perspectives in Continental Philosophy. New York: Fordham University Press.

———. 2008b. "Heidegger's Hand (*Geschlecht* II)." Translated by John P. Leavey Jr. Pages 27–62 in vol. 2 of Jacques Derrida. *Psyche: Inventions of the Other*. Edited by Peggy Kamuf and Elizabeth Rottenberg. Meridian: Crossing Aesthetics. Stanford: Stanford University Press.

———. 2009. *The Beast and the Sovereign*. Vol. 1. Translated by Geoffrey Bennington. The Seminars of Jacques Derrida 1. Chicago: University of Chicago Press.

———. 2011. *The Beast and the Sovereign*. Vol. 2. Translated by Geoffrey Bennington. The Seminars of Jacques Derrida 1. Chicago: University of Chicago Press.

Derrida, Jacques, and Maurizio Ferraris. 2001. *A Taste for the Secret*. Translated by Giacomo Donis. Cambridge: Polity.

Descartes, René. 2000. *Philosophical Essays and Correspondence*. Edited by Roger Ariew. Translated by Roger Ariew et al. Indianapolis: Hackett.

———. 2006. *A Discourse on the Method*. Translated by Ian Maclean. Oxford World's Classics. Oxford: Oxford University Press.

DeSilva, David A. 2009. *Seeing Things John's Way: The Rhetoric of the Book of Revelation*. Louisville: Westminster John Knox.

Diehl, Judy. 2013. "'Babylon': Then, Now and 'Not Yet': Anti-Roman Rhetoric in the Book of Revelation." *Currents in Biblical Research* 11:168–95.

Ditmore, Melissa. 2007. "In Calcutta, Sex Workers Organize." Pages 170–86 in *The Affective Turn: Theorizing the Social*. Edited by Patricia Ticineto Clough with Jean Halley. Durham: Duke University Press.

Dover, K. J. 1989. *Greek Homosexuality: Updated and with a New Postscript*. 2nd ed. Cambridge: Harvard University Press.

Downing, Gerald F. 1988. "Pliny's Prosecutions of Christians: Revelation and 1 Peter. *JSNT* 34:105–23.

Duff, Paul B. 2001. *Who Rides the Beast? Prophetic Rivalry and the Rhetoric of Crisis in the Churches of the Apocalypse*. Oxford: Oxford University Press.

Duling, Dennis C. 2005. "Empire: Theories, Methods, Models." Pages 49–74 in *The Gospel of Matthew in Its Roman Imperial Context*. Edited by John Riches and David C. Sim. JSNTSup 276. New York: T&T Clark.

Duncan, Anne. 2006. "Infamous Performers: Comic Actors and Female Prostitutes in Rome." Pages 252–73 in *Prostitutes and Courtesans in the Ancient World*. Edited by Christopher A. Faraone and Laura K. McClure. Madison: University of Wisconsin Press.

Dupont-Sommer, André. 1939. *Le Quatrième Livre des Machabées: Introduction, traduction et notes*. Bibliothèque de l'École des Hautes Études 274. Paris: Librairie Ancienne Honoré Champion.

Dutton, Kenneth R. 1995. *The Perfectible Body: The Western Ideal of Male Physical Development*. New York: Continuum.

Edwards, Catherine. 1997. "Unspeakable Professions: Public Performance and Prostitution in Ancient Rome." Pages 66–95 in *Roman Sexualities*. Edited by Judith P. Hallett and Marilyn B. Skinner. Princeton: Princeton University Press.

Edwards, Ruth Dudley. 1977. *Pearse: The Triumph of Failure*. London: Victor Gollancz.

Egger-Wenzel, Renate, and Jeremy Corley, eds. 2012. *Emotions from Ben Sira to Paul*. Deuterocanonical and Cognate Literature Yearbook. Berlin: de Gruyter.

Ellis, Bret Easton. 1991. *American Psycho*. New York: Vintage.

Erskine, Andrew. 1995. "Rome in the Greek World: The Significance of a Name." Pages 368–86 in *The Greek World*. Edited by Anton Powell. New York: Routledge.

Exum, J. Cheryl. 1996. *Plotted, Shot, and Painted: Cultural Representations of Biblical Women*. Gender, Culture, Theory 3. Sheffield: Sheffield Academic.

Fanon, Frantz. 1968. *The Wretched of the Earth*. Translated by Constance Farrington. New York: Grove.

Faraone, Christopher A., and Laura K. McClure, eds. 2006. *Prostitutes and Courtesans in the Ancient World*. Madison: University of Wisconsin Press.

Feuerbach, Ludwig. 1841. *Das Wesen des Christentums*. Leipzig: Otto Wigand.

Findon, Joanne. 1997. *A Woman's Words: Emer and Female Speech in the Ulster Cycle*. Toronto: University of Toronto Press.

Fish, Stanley E. 1980. *Is There a Text in This Class? The Authority of Interpretive Communities*. Cambridge: Harvard University Press.

Fishwick, Duncan. 1987–92. *The Imperial Cult in the Latin West: Studies in the Ruler Cult of the Western Provinces of the Roman Empire*. 2 vols. EPRO 108. Leiden: Brill.

Flemming, Rebecca. 1999. "*Quae Corpore Quaestum Facit*: The Sexual Economy of Female Prostitution in the Roman Empire." *JRS* 89:38–61.

Foerster, Werner. 1964. Βδελύσσομαι, κτλ. In *TDNT* 1:598–600.

Ford, J. Massyngberde. 1975. *Revelation*. AB 38. Garden City, N.Y.: Doubleday.

Forster, E. M. 1927. *Aspects of the Novel*. London: Edward Arnold.

Foucault, Michel. 1977. *Discipline and Punish: The Birth of the Prison*. Translated by Alan Sheridan. New York: Pantheon.

———. 1985. *The Use of Pleasure.* Vol. 2 of *The History of Sexuality.* Translated by Robert Hurley. New York: Vintage.
Frankfurter, David. 2001. "Jews or Not? Reconstructing the 'Other' in Rev 2:9 and 3:9." *HTR* 94:403–25.
Freedman, Diane P., Olivia Frey, and Frances Murphy Zauhar, eds. 1993. *The Intimate Critique: Autobiographical Literary Criticism.* Durham: Duke University Press.
Freud, Sigmund. 1910. *Leonardo da Vinci and a Memory of His Childhood.* Pages 63–137 in vol. 11 of Freud 1953–74.
———. 1925. *An Autobiographical Study.* Pages 1–74 in vol. 20 of Freud 1953–74.
———. 1927. *The Future of an Illusion.* Pages 5–56 in vol. 21 of Freud 1953–74.
———. 1933. *New Introductory Lectures on Psycho-Analysis.* Pages 1–182 in vol. 22 of Freud 1953–74.
———. 1939. *Moses and Monotheism.* Pages 1–137 in vol. 23 of Freud 1953–74.
———. 1953–74. *The Standard Edition of the Complete Psychological Works of Sigmund Freud.* Edited and translated by James Strachey. London: Hogarth Press and the Institute of Psycho-Analysis.
Friesen, Steven J. 1993. *Twice Neokoros: Ephesus, Asia and the Cult of the Flavian Imperial Family.* Religions in the Graeco-Roman World 116. Leiden: Brill.
———. 2001. *Imperial Cults and the Apocalypse of John: Reading Revelation in the Ruins.* Oxford: Oxford University Press.
Frilingos, Christopher A. 2004. *Spectacles of Empire: Monsters, Martyrs, and the Book of Revelation.* Divinations: Rereading Late Ancient Religion. Philadelphia: University of Pennsylvania Press.
Frykholm, Amy Johnson. 2007. *Rapture Culture: Left Behind in Evangelical America.* Oxford: Oxford University Press.
Fussell, Samuel Wilson. 1991. *Muscle: Confessions of an Unlikely Bodybuilder.* New York: Avon.
———. 1994. "Bodybuilder Americanus." Pages 43–60 in *The Male Body: Features, Destinies, Exposures.* Edited by Laurence Goldstein. Ann Arbor: University of Michigan Press.
Gaines, Charles, and George Butler. 1974. *Pumping Iron: The Art and Sport of Bodybuilding.* London: Sphere.
Galinsky, Karl. 1998. *Augustan Culture: An Interpretive Introduction.* Princeton: Princeton University Press.

Gambs, Deborah. 2007. "Myocellular Transduction: When My Cells Trained my Body-Mind." Pages 106–18 in *The Affective Turn: Theorizing the Social*. Edited by Patricia Ticineto Clough with Jean Halley. Durham: Duke University Press.

Garrett, Susan R. 1992. "Revelation." Pages 377–82 in *Women's Bible Commentary*. Edited by Carol A. Newsom and Sharon H. Ringe. Louisville: Westminster John Knox.

Gentry Jr., Kenneth L. 1989. *Before Jerusalem Fell: Dating the Book of Revelation. An Exegetical and Historical Argument for a Pre-ad 70 Composition*. Tyler, Tex.: Institute for Christian Economics.

George, A. Raymond. 1953. *Communion with God in the New Testament*. London: Epworth.

Gersht, Rivka. 2001. "Aquatic Figure Types from Caesarea Maritima." *Assaph* 6:63–90.

Gibson, Nigel C. 2003. *Fanon: The Postcolonial Imagination*. Oxford: Polity.

Gilhus, Ingvild Saelid. 2006. *Animals, Gods and Humans: Changing Attitudes to Animals in Greek, Roman and Early Christian Ideas*. New York: Routledge.

Ginsburg, Judith. 2006. *Representing Agrippina: Constructions of Female Power in the Early Roman Empire*. Oxford: Oxford University Press.

Glazebook, Allison. 2011. "*Porneion*: Prostitution in Athenian Civic Space." Pages 34–59 in Glazebrook and Henry 2011b.

Glazebrook, Allison, and Madeleine M. Henry. 2011a. "Introduction: Why Prostitutes? Why Greek? Why Now?" Pages 3–14 in Glazebrook and Henry 2011b.

Glazebrook, Allison, and Madeleine M. Henry, eds. 2011b. *Greek Prostitutes in the Ancient Mediterranean, 800 BCE–200 CE*. Wisconsin Studies in Classics. Madison: University of Wisconsin Press.

Gleason, Maud. 1995. *Making Men: Sophists and Self-Presentation in Ancient Rome*. Princeton: Princeton University Press.

González, Justo L. 1999. *For the Healing of the Nations: The Book of Revelation in an Age of Cultural Conflict*. Maryknoll, NY: Orbis.

Gramsci, Antonio. 1971. *Selections from the Prison Notebooks*. Edited and translated by Quintin Hoare and Geoffrey Nowell Smith. London: Lawrence & Wishart.

Grant, Robert M. 1999. *Early Christians and Animals*. New York: Routledge.

Gregg, Melissa, and Gregory J. Seigworth, eds. 2010. *The Affect Theory Reader*. Durham: Duke University Press.

Gregory, Augusta. 1902. *Cuchulain of Muirthemne: The Story of the Men of the Red Branch of Ulster Arranged and Put into English by Lady Gregory, with a Preface by W. B. Yeats*. London: Murray.

Gribben, Crawford, and Mark S. Sweetnam, eds. 2011. *Left Behind and the Evangelical Imagination*. The Bible in the Modern World 35. Sheffield: Sheffield Phoenix.

Gross, Aaron S. 2009. "The Question of the Creature: Animals, Theology and Levinas's Dog." Pages 121–37 in *Creaturely Theology: On God, Humans and Other Animals*. Edited by Celia Deane-Drummond and David Clough. London: SCM.

Gundry, Robert H. 2005. "The New Jerusalem: People as Place, Not Place for People." Pages 399–412 in *The Old Is Better: New Testament Essays in Support of Traditional Interpretations*. WUNT 178. Tübingen: Mohr Siebeck.

Hadas, Moses. 1953. *The Third and Fourth Books of Maccabees*. New York: Harper & Row.

Hallet, Judith P., and Thomas Van Nortwick, eds. 1997. *Compromising Traditions: The Personal Voice in Classical Scholarship*. New York: Routledge.

Halperin, David M. 2002. *How to Do the History of Sexuality*. Chicago: University of Chicago Press.

Haraway, Donna. 1990. *Primate Visions: Gender, Race, and Nature in the World of Modern Science*. New York: Routledge.

———. 2003. *The Companion Species Manifesto: Dogs, People, and Significant Otherness*. Chicago: Prickly Paradigm.

———. 2007. *When Species Meet*. Posthumanities 3. Minneapolis: University of Minnesota Press.

Hardt, Michael, and Antonio Negri. 2000. *Empire*. Cambridge: Harvard University Press.

Harrington, Wilfred J. 1993. *Revelation*. SP. Collegeville, Minn.: Liturgical Press.

Hauck, Friedrich, and Siegfried Schulz. 1964. "Πόρνη, κτλ." In *TDNT* 6:579–95.

Hawley, John C., ed. 2001a. *Postcolonial and Queer Theories: Intersections and Essays*. Contributions to the Study of World Literature. Westport, CT: Greenwood.

———, ed. 2001b. *Postcolonial, Queer: Theoretical Intersections*. Explorations in Postcolonial Studies. Albany: State University of New York Press.

Hawkin, David J. 2003. "The Critique of Ideology in the Book of Revelation and Its Implications for Ecology." *Ecotheology* 8:161–72.

Heidegger, Martin. 1968. *What Is Called Thinking?* Translated by J. Glenn Gray. New York: Harper & Row.

———. 1971. "The Thing." Pages 161–85 in *Poetry, Language, Thought*. Translated by Albert Hofstadter. New York: Harper & Row.

Hemer, Colin J. 1986. *The Letters to the Seven Churches of Asia in Their Local Setting*. JSNTSup 11. Sheffield: JSOT Press.

Heyman, George. 2007. *The Power of Sacrifice: Roman and Christian Discourses in Conflict*. Washington, D.C.: Catholic University of America Press.

Hoffmann, Matthias Reinhard. 2005. *The Destroyer and the Lamb: The Relationship between Angelomorphic and Lamb Christology in the Book of Revelation*. WUNT 203. Tübingen: Mohr Siebeck.

Hoffner, Harry A. 1966. "Symbols for Masculinity and Femininity: Their Use in Ancient Near Eastern Sympathetic Magic Rituals." *JBL* 85:327–32.

Honer, Anne. 1985. "Beschreibung einer Lebenswelt: Zur Empirie des Bodybuilding." *Zeitschrift für Soziologie* 14:131–39.

Hood, Renate Viveen. 2010. "Pure or Defiled? A Sociological Analysis of John's Apocalypse." Pages 87–99 in *Essays on Revelation: Appropriating Yesterday's Apocalypse in Today's World*. Edited by Gerald L. Stevens. Eugene, Ore.: Pickwick.

Hope, Valerie. 2000. "The City of Rome: Capital and Symbol." Pages 63–94 in *Experiencing Rome: Culture, Identity and Power in the Roman Empire*. Edited by Janet Huskinson. New York: Routledge.

Horrell, David G. 2010. *The Bible and the Environment: Towards a Critical Ecological Biblical Theology*. London: Equinox.

Howard, Robert Glenn, ed. 2011. *Network Apocalypse: Visions of the End in an Age of Internet Media*. The Bible in the Modern World 36. Sheffield: Sheffield Phoenix.

Howard-Brook, Wes, and Anthony Gwyther. 1999. *Unveiling Empire: Reading Revelation Then and Now*. Bible and Liberation. Maryknoll, N.Y.: Orbis.

Huber, Lynn R. 2007. *Like a Bride Adorned: Reading Metaphor in John's Apocalypse*. Emory Studies in Early Christianity. New York: T&T Clark.

———. 2008. "Sexually Explicit? Re-reading Revelation's 144,000 Virgins as a Response to Roman Discourses." *Journal of Men, Masculinities and Spirituality* 2:3–28.

———. 2011. "Gazing at the Whore: Reading Revelation Queerly." Pages 301–20 in *Bible Trouble: Queer Readings at the Boundaries of Biblical Scholarship* Edited by Teresa J. Hornsby and Ken Stone. SemeiaSt 67. Atlanta: Society of Biblical Literature.

———. 2013. *Thinking and Seeing with Women in Revelation.* LNTS. New York: T&T Clark.

Huddart, David. 2006. *Homi K. Bhabha.* Routledge Critical Thinkers. New York: Routledge.

Huggan, Graham, and Helen Tiffin. 2010. *Postcolonial Ecocriticism: Literature, Animals, Environment.* New York: Routledge.

Hunzinger, C. H. 1965. "Babylon als Deckname für Rom und die Datierung des I. Petrusbriefes." Pages 67–77 in *Gottes Wort und Gottes Land: Hans-Wilhelm Hertzberg zum 70. Geburtstag am 16. Januar 1965 dargebracht von Kollegen, Freunden und Schülern.* Edited by H. G. Reventlow. Göttingen: Vandenhoeck & Ruprecht.

Ipsen, Avaren. 2009. *Sex Working and the Bible.* BibleWorld. London: Equinox.

Isaac, Benjamin. 2006. *The Invention of Racism in Classical Antiquity.* Princeton: Princeton University Press.

Isasi-Díaz, Ada María. 2004. *En la Lucha/In the Struggle: Elaborating a Mujerista Theology.* 2nd ed. Minneapolis: Fortress.

James, Sharon L. 2006. "A Courtesan's Choreography: Female Liberty and Male Anxiety at the Roman Dinner Party." Pages 224–62 in *Prostitutes and Courtesans in the Ancient World.* Edited by Christopher A. Faraone and Laura K. McClure. Madison: University of Wisconsin Press.

Jameson, Fredric. 1991. *Postmodernism, or, The Cultural Logic of Late Capitalism.* Post-Contemporary Interventions. Durham: Duke University Press.

JanMohamed, Abdul R. 1983. *Manichean Aesthetics: The Politics of Literature in Colonial Africa.* Amherst: University of Massachusetts Press.

———. 1995. "The Economy of Manichean Allegory." Pages 18–23 in *The Post-Colonial Studies Reader.* Edited by Bill Ashcroft, Gareth Griffiths, and Helen Tiffin. New York: Routledge.

Janzen, Ernest P. 1994. "The Jesus of the Apocalypse Wears the Emperor's Clothes." Pages 637–61 in *SBLSP* 33. Edited by E. H. Lovering Jr. Atlanta: Scholars Press.

Jobling, David, and Tina Pippin, eds. 1992. *Ideological Criticism of Biblical Texts. Semeia* 59. Atlanta: Scholars Press.

Jobst, Werner. 1976–77. "Das 'öffentliche *Freudenhaus*' in Ephesos." *JÖAI* 51:61–84.
Johns, Loren L. 2003. *The Lamb Christology of the Apocalypse of John.* WUNT 167. Tübingen: Mohr Siebeck.
Johnston, Anna, and Alan Lawson. 2000. "Settler Colonies." Pages 360–76 in *A Companion to Postcolonial Studies.* Edited by Henry Schwarz and Sangeeta Ray. Oxford: Blackwell.
Jones, Brian W. 1992. *The Emperor Domitian.* New York: Routledge.
Jones, C. P. 1987. "*Stigma*: Tattooing and Branding in Graeco-Roman Antiquity." *JRS* 77:139–55.
Jones, Donald L. 1980. "Christianity and the Roman Imperial Cult." In *ANRW* 23.2:1023–54.
Joshel, Sandra R. 1997. "Female Desire and the Discourse of Empire: Tacitus's Messalina." Pages 221–54 in *Roman Sexualities.* Edited by Judith P. Hallett and Marilyn B. Skinner. Princeton: Princeton University Press.
Kang, Shinwook. 2007. "'Come, Gather Together for the Great Supper of God': Christian Identity-Making through Alimentary Images in Revelation in the Context of Roman Imperialism." Ph.D. diss., Drew University, Madison, N.J.
Kapparis, Konstantinos K. 2011. "The Terminology of Prostitution in the Ancient Greek World." Pages 222–55 in *Greek Prostitutes in the Ancient Mediterranean, 800 BCE–200 CE.* Edited by Allison Glazebrook and Madeleine M. Henry. Wisconsin Studies in Classics. Madison: University of Wisconsin Press.
Kaster, Robert A. 2005. *Emotion, Restraint, and Community in Ancient Rome.* Classical Culture and Society. Oxford: Oxford University Press.
Keller, Catherine. 1996. *Apocalypse Now and Then: A Feminist Guide to the End of the World.* Boston: Beacon.
———. 2001. "Eyeing the Apocalypse." Pages 253–77 in *Postmodern Interpretations of the Bible—A Reader.* Edited by A. K. M. Adam. St. Louis: Chalice.
———. 2005. *God and Power: Counter-Apocalyptic Journeys.* Minneapolis: Fortress.
Keller, Catherine, and Stephen D. Moore. 2004. "Derridapocalypse." Pages 189–207 in *Derrida and Religion: Other Testaments.* Edited by Yvonne Sherwood and Kevin Hart. New York: Routledge.
Kellum, Barbara. 1996. "The Phallus as Signifier: The Forum of Augustus and Rituals of Masculinity." Pages 170–83 in *Sexuality in Ancient Art:*

Near East, Egypt, Greece, and Italy. Edited by Natalie Boymel Kampen. Cambridge Studies in New Art History and Criticism. Cambridge: Cambridge University Press.

Kelly, Daniel. 2011. *Yuck! The Nature and Moral Significance of Disgust*. Boston: MIT Press.

Kiberd, Declan. 1995. *Inventing Ireland*. London: Jonathan Cape.

Kim, Jean K. 1999. "Uncovering Her Wickedness: An Inter(con)textual Reading of Revelation 17 from a Postcolonial Feminist Perspective." *JSNT* 21:61–81.

Kinsella, Thomas, ed. and trans. 1970. *The Táin: From the Irish Epic Táin Bó Cuailnge*. Mountrath, Ireland: Dolmen Press; Oxford: Oxford University Press.

Kirk, Kenneth E. 1931. *The Vision of God: The Christian Doctrine of the Summum Bonum*. London: Longmans, Green.

Kitzberger, Ingrid Rosa, ed. 1998. *The Personal Voice in Biblical Interpretation*. New York: Routledge.

Klausner, Theodor. 1974. "Aurum Coronarium." Pages 292–309 in *Gesammelte Arbeiten zur Liturgiegeschichte, Kirchengeschichte und christlichen Archäologie*. Edited by Ernst Dassmann. JAC 3. Münster: Aschendorff.

Klein, Alan M. 1993. *Little Big Men: Bodybuilding Subculture and Gender Construction*. SUNY Series on Sport, Culture, and Social Relations. Albany: State University of New York Press.

Knight, James. 2005. "Was Roma the Scarlet Harlot? The Worship of the Goddess Roma in Sardis and Smyrna." Pages 104–22 in *Religious Rivalries and the Struggle for Success in Sardis and Smyrna*. Edited by Richard S. Ascough. Studies in Christianity and Judaism/Études sur le christianisme et le judaïsme. Waterloo, Ont.: Wilfred Laurier University Press.

Knust, Jennifer Wright. 2001. *Abandoned to Lust: Sexual Slander and Ancient Christianity*. New York: Columbia University Press.

Koester, Craig R. 2001. *Revelation and the End of All Things*. Grand Rapids: Eerdmans.

Kolnai, Aurel. 2004. *On Disgust*. Edited by Barry Smith and Carolyn Korsmeyer. Chicago and La Salle, Ill.: Open Court.

Konstan, David. 2006. *The Emotions of the Ancient Greeks: Studies in Aristotle and Classical Literature*. Robson Classical Lectures. Toronto: University of Toronto Press.

Koosed, Jennifer L., ed. 2014. *The Bible and Posthumanism*. SemeiaSt 72. Atlanta: Society of Biblical Literature.

Koosed, Jennifer L., and Robert Paul Seesengood. 2014. "Daniel's Animal Apocalypse." Pages 182–95 in *Divinanimality: Animal Theory, Creaturely Theology*. Edited by Stephen D. Moore. Transdisciplinary Theological Colloquia. New York: Fordham University Press.

Koosed, Jennifer L., and Stephen D. Moore, eds. 2014. "Affect Theory and the Bible." *Biblical Interpretation* 22.4 (thematic issue).

Korsmeyer, Carolyn. 2011. *Savoring Disgust: The Foul and the Fair in Aesthetics*. Oxford: Oxford University Press.

Kotrosits, Maia. 2010. "The Rhetoric of Intimate Spaces: Affect and Performance in the Corinthian Correspondence." *USQR* 62.3–4:134–51.

———. 2012. "Romance and Danger at Nag Hammadi." *The Bible and Critical Theory* 8.1:39–52.

———. 2013. "Affect, Violence, and Belonging in Early Christianity." Ph.D. diss., Union Theological Seminary.

———. 2014. "Seeing Is Feeling: Revelation's Enthroned Lamb and Ancient Visual Affects." *Biblical Interpretation* 22.4:473–502.

Kotrosits, Maia, and Hal Taussig. 2013. *Re-reading the Gospel of Mark Amidst Loss and Trauma*. New York: Palgrave Macmillan.

Kuhn, Karl Allen. 2009. *The Heart of Biblical Narrative: Rediscovering Biblical Appeal to the Emotions*. Minneapolis: Fortress.

Kovacs, Judith, and Christopher Rowland. 2004. *Revelation*. Blackwell Bible Commentaries. Oxford: Blackwell.

Kraft, Heinrich. 1974. *Die Offenbarung des Johannes*. HNT 16a. Tübingen: Mohr Siebeck.

Kraemer, Ross S. 1994. "The Other as Woman: An Aspect of Polemic among Pagans, Jews, and Christians in the Greco-Roman World." Pages 121–44 in *The Other in Jewish Thought and History: Constructions of Jewish Cultural Identity*. Edited by Laurence J. Silberstein and Robert L. Cohn. New York: New York University Press.

Kraybill, J. Nelson. 1999. *Imperial Cult and Commerce in John's Apocalypse*. JSNTSup 132. Sheffield: Sheffield Academic.

———. 2010. *Apocalypse and Allegiance: Worship, Politics, and Devotion in the Book of Revelation*. Grand Rapids: Brazos.

Krell, David Farrell. 2013. *Derrida and Our Animal Others: Derrida's Final Seminar, the Beast and the Sovereign*. Studies in Continental Thought. Bloomington: Indiana University Press.

Kristeva, Julia. 1982. *Powers of Horror: An Essay on Abjection*. Translated by Leon S. Roudiez. New York: Columbia University Press.

Krodel, Gerhard. 1989. *Revelation*. ACNT. Minneapolis: Fortress.

Kurke, Leslie. 1999. *Coins, Bodies, Games, and Gold: The Politics of Meaning in Archaic Greece*. Princeton: Princeton University Press.
Lacan, Jacques. 1982a. "Desire and the Interpretation of Desire in *Hamlet*." Edited by Jacques-Alain Miller. Translated by James Hulbert. Pages 11–52 in *Literature and Psychoanalysis: The Question of Reading, Otherwise*. Edited by Shoshana Felman. Baltimore: Johns Hopkins University Press.
———. 1982b. *Feminine Sexuality: Jacques Lacan and the* école freudienne. Edited by Juliet Mitchell and Jacqueline Rose. Translated by Jacqueline Rose. New York: Norton.
———. 2006. *Écrits: The First Complete Edition in English*. Translated by Bruce Fink. New York: Norton.
Lakoff, George, and Mark Johnson. 1981. *Metaphors We Live By*. 2nd ed. Chicago: University of Chicago Press.
Laughlin, T. Cowden. 1902. *The Solecisms of the Apocalypse*. Princeton: Princeton University Press.
Lawlor, Leonard. 2007. *This Is Not Sufficient: An Essay on Animality and Human Nature in Derrida*. New York: Columbia University Press.
Leibowitz, Nehama. 1981. *Studies in Bereshit in the Context of Ancient and Modern Bible Commentary*. Edited and translated by Aryeh Newman. 4th ed. Jerusalem: World Zionist Organization Department for Torah Education and Culture.
Levick, Barbara. 1967. *Roman Colonies in Southern Asia Minor*. Oxford: Clarendon.
Levinas, Emmanuel. 1990. "The Name of a Dog, or Natural Rights." Pages 151–53 in *Difficult Freedom: Essays on Judaism*. Translated by Sean Hand. Baltimore: Johns Hopkins University Press.
Levine, Amy-Jill, ed. 2009. *A Feminist Companion to the Apocalypse of John*. Feminist Companion to the New Testament and Early Christian Writings 13. New York: T&T Clark.
Leys, Ruth. 2011. "The Turn to Affect: A Critique." *Critical Inquiry* 37:434–72.
Liew, Tat-siong Benny. 1999. *Politics of Parousia: Reading Mark Inter(con)textually*. Biblical Interpretation Series 42. Leiden: Brill.
Lingis, Alphonso. 1994. *Foreign Bodies*. New York: Routledge.
Lintott, Andrew. 1993. *Imperium Romanum: Politics and Administration*. New York: Routledge.
Lippit, Akira Mizuta. 1998. "Magnetic Animal: Derrida, Wildlife, Animetaphor." *Modern Language Notes* 113:1111–25.

———. 2000. *Electric Animal: Toward a Rhetoric of Wildlife*. Minneapolis: University of Minnesota Press.
Lohmeyer, Ernst. 1926. *Die Offenbarung des Johannes*. HNT 16. Tübingen: Mohr Siebeck.
Lohse, Eduard. 1979. *Die Offenbarung des Johannes*. 12th ed. NTD 11. Göttingen: Vandenhoeck & Ruprecht.
Lopez, Davina C. 2008. *Apostle to the Conquered: Reimagining Paul's Mission*. Paul in Critical Contexts. Minneapolis: Fortress.
Lundblad, Michael. 2009. "From Animal to Animality Studies." *PMLA* 124:496–52.
Lyons, William John, and Jorunn Økland, eds. 2009. *The Way the World Ends? The Apocalypse of John in Culture and Ideology*. The Bible in the Modern World 19. Sheffield: Sheffield Phoenix.
MacKenzie, Robert K. 1997. *The Author of the Apocalypse: A Review of the Prevailing Hypothesis of Jewish-Christian Authorship*. Mellen Biblical Press 51. Lewiston, N.Y.: Mellen.
Maier, Harry O. 2002a. *Apocalypse Recalled: The Book of Revelation after Christendom* Minneapolis: Fortress.
———. 2002b. "There's a New World Coming! Reading the Apocalypse in the Shadow of the Canadian Rockies." Pages 166–79 in *The Earth Story in the New Testament*. Edited by Norman C. Habel and Vicki Balabanski. Earth Bible 5. Cleveland: Pilgrim.
Malabou, Catherine, and Jacques Derrida. 2004. *Counterpath: Traveling with Jacques Derrida*. Translated by David Wills. Cultural Memory in the Present. Stanford: Stanford University Press.
Marshall, John W. 2001. *Parables of War: Reading John's Jewish Apocalypse*. Studies in Christianity and Judaism 10. Waterloo, Ont.: Wilfred Laurier University Press.
———. 2007. "John's Jewish (Christian?) Apocalypse." Pages 233–56 in *Jewish Christianity Reconsidered: Rethinking Ancient Groups and Texts*. Edited by Matt A. Jackson McCabe. Minneapolis: Fortress.
———. 2009. "Gender and Empire: Sexualized Violence in John's Anti-Imperial Apocalypse." Pages 17–32 in *A Feminist Companion to the Apocalypse of John*. Edited by Amy-Jill Levine with Maria Mayo Robbins. Feminist Companion to the New Testament and Early Christian Writings 13. New York: T&T Clark.
Martin, Clarice J. 2005. "Polishing the Unclouded Mirror: A Womanist Reading of Revelation 18:13." Pages 82–109 in *From Every People and*

Nation: The Book of Revelation in Intercultural Perspective. Edited by David Rhoads. Minneapolis: Fortress.

Martin, Dale B. 1995. *The Corinthian Body.* New Haven: Yale University Press.

Martin, Thomas W. 2009. "The City as Salvific Space: Heterotopic Place and Environmental Ethics in the New Jerusalem." *SBL Forum* 7.2. Online: http://www.sbl-site.org/publications/article.aspx?ArticleId=801.

Massumi, Brian. 1987. "Translator's Foreword: Pleasures of Philosophy." Pages ix–xvi in Gilles Deleuze and Félix Guattari, *A Thousand Plateaus: Capitalism and Schizophrenia.* Minneapolis: University of Minnesota Press.

———. 2002. *Parables for the Virtual: Movement, Affect, Sensation.* Postcontemporary Interventions. Durham: Duke University Press.

Mayo, Philip L. 2006. *"Those Who Call Themselves Jews": The Church and Judaism in the Apocalypse of John.* Princeton Theological Monograph Series 60. Eugene, Ore.: Wipf & Stock.

McClure, Laura K. 2003. *Courtesans at Table: Gender and Greek Literary Culture in Athenaeus.* New York: Routledge.

———. 2006. "Introduction." Pages 3–18 in *Prostitutes and Courtesans in the Ancient World.* Edited by Christopher A. Faraone and Laura K. McClure. Wisconsin Studies in Classics. Madison: University of Wisconsin Press.

McCourt, Frank. 1997. *Angela's Ashes: A Memoir of a Childhood.* London: Flamingo.

McDonald, P. M. 1996. "Lion as Slain Lamb: On Reading Revelation Recursively." *Horizons* 23:29–47.

McDonnell, Myles. 2006. *Roman Manliness*: Virtus *and the Roman Republic.* Cambridge: Cambridge University Press.

McGinn, Colin. 2011. *The Meaning of Disgust.* Oxford: Oxford University Press.

McGinn, Thomas A. J. 1998a. "Caligula's Brothel on the Palatine." *Echos du Monde Classique* n.s. 17:95–107.

———. 1998b. *The Economy of Prostitution in the Roman World: A Study of Social History and the Brothel.* Ann Arbor: University of Michigan Press.

———. 1998c. *Prostitution, Sexuality, and the Law in Ancient Rome.* Oxford: Oxford University Press.

———. 2006. "Zoning Shame in the Roman City." Pages 171–75 in *Prostitutes and Courtesans in the Ancient World.* Edited by Christopher A.

Faraone and Laura K. McClure. Wisconsin Studies in Classics. Madison: University of Wisconsin Press.

———. 2011. "Conclusion: Greek Brothels and More." Pages 256–68 in *Greek Prostitutes in the Ancient Mediterranean, 800 BCE–200 CE.* Edited by Allison Glazebrook and Madeleine M. Henry. Wisconsin Studies in Classics. Madison: University of Wisconsin Press.

McGough, Peter. 1994. Hard Times. *Flex.* January, 114.

———. 1995. "At the Court of King Dorian." *Flex.* September, 197–98.

McGuire, William, ed. 1974. *The Freud/Jung Letters: The Correspondence between Sigmund Freud and C. G. Jung.* Translated by Ralph Manheim and R. F. C. Hull. Princeton: Princeton University Press.

McHugh, Susan. 2011. *Animal Stories: Narrating across Species Lines.* Posthumanities 15. Minneapolis: University of Minnesota Press.

McKinley, Judith E. 2004. "Decolonizing with the Goddess: Following John of Patmos and Witi Ihimaera." Pages 137–60 in *Reframing Her: Biblical Women in Postcolonial Focus.* The Bible in the Modern World 1. Sheffield: Sheffield Phoenix.

McLean, B. H. 2012. *Biblical Interpretation and Philosophical Hermeneutics.* Cambridge: Cambridge University Press.

McLoughlin, Michael. 1996. *Great Irish Speeches of the Twentieth Century.* Dublin: Poolbeg.

Mellor, Ronald. 1975. *ΘΕΑ ΡΩΜΗ: The Worship of the Goddess Roma in the Greek World.* Hypomnemata: Untersuchungen zur Antike und zu ihrem Nachleben 42. Göttingen: Vandenhoeck & Ruprecht.

Memmi, Albert. 1957. *Portrait du colonisé précédé du portrait du colonisateur.* Paris: Editions Buchet/Chastel. ET: *The Colonizer and the Colonized.* Translated by Howard Greenfeld. Boston: Beacon, 2001.

Mendenhall, George E. 1973. *The Tenth Generation: The Origins of the Biblical Tradition.* Baltimore: Johns Hopkins University Press.

Menninghaus, Winfried. 2003. *Disgust: The Theory and History of a Strong Sensation.* Translated by Howard Eiland and Joel Golb. Albany: State University of New York Press.

Merchant, Carolyn. 2003. *Reinventing Eden: The Fate of Nature in Western Culture.* New York: Routledge.

Mettler, John J., Jr. 2003. *Basic Butchering of Livestock and Game.* 2nd ed. North Adams, Mass.: Storey.

Metzger, Bruce M. *Breaking the Code: Understanding the Book of Revelation.* Nashville: Abingdon, 1993.

Michaelis, Johann David. 1801. *Introduction to the New Testament.* Translated and augmented by Herbert Marsh. 4 vols. Cambridge: John Burges.

Míguez, Néstor O. 1995. "Apocalyptic and the Economy: A Reading of Revelation 18 from the Experience of Economic Exclusion." Pages 250–62 in *Social Location and Biblical Interpretation in Global Perspective.* Vol. 2 of *Reading from This Place.* Edited by Fernando F. Segovia and Mary Ann Tolbert. Minneapolis: Fortress.

———. 2002. "The Old Creation in the New, the New Creation in the Old." Keynote address for the Oxford Institute.

Millar, Fergus. 1977. *The Emperor in the Roman World (31 BC–AD 337).* Ithaca, N.Y.: Cornell University Press.

Miles, Rosalind. 1991. *The Rites of Man: Love, Sex and Death in the Making of the Male.* London: Grafton.

Miller, Nancy K. 1991. *Getting Personal: Feminist Occasions and Other Autobiographical Acts.* New York: Routledge.

Miller, William Ian. 1997. *The Anatomy of Disgust.* Cambridge: Harvard University Press.

Moltmann, Jürgen. 1967. *Theology of Hope: On the Ground and the Implications of a Christian Eschatology.* Translated by James W. Leitch. New York: Harper & Row.

———. 1996. *The Coming of God: Christian Eschatology.* Translated by Margaret Kohl. Minneapolis: Fortress.

Moore, Stephen D. 1992. *Mark and Luke in Poststructuralist Perspectives: Jesus Begins to Write.* New Haven: Yale University Press.

———. 1994. *Poststructuralism and the New Testament: Derrida and Foucault at the Foot of the Cross.* Minneapolis: Fortress.

———. 1995a. "The Beatific Vision as a Posing Exhibition: Revelation's Hypermasculine Deity." *JSNT* 60:27–55.

———. 1995b. "True Confessions and Weird Obsessions: Autobiographical Interventions in Literary and Biblical Studies." *Semeia* 72:19–50.

———. 1996. *God's Gym: Divine Male Bodies of the Bible.* New York: Routledge.

———. 1998. "Revolting Revelations." Pages 183–200 in *The Personal Voice in Biblical Interpretation.* Edited by Ingrid Rosa Kitzberger. New York: Routledge.

———. 1999. "War Making Men Making War: The Performance of Masculinity in the Revelation to John." Pages 84–94 in *The Apocalyptic*

Imagination: Aesthetics and Ethics at the End of the World. Edited by S. Brent Plate. Glasgow: Trinity St. Mungo Press.

———. 2001. *God's Beauty Parlor: And Other Queer Spaces in and around the Bible*. Contraversions: Jews and Other Differences. Stanford: Stanford University Press.

———. 2006. *Empire and Apocalypse: Postcolonialism and the New Testament*. The Bible in the Modern World 12. Sheffield: Sheffield Phoenix.

———. 2009. "Metonymies of Empire: Sexual Humiliation and Gender Masquerade in the Book of Revelation." Pages 71–97 in *Postcolonial Interventions: Essays in Honor of R. S. Sugirtharajah*. Edited by Tat-siong Benny Liew. The Bible in the Modern World 23. Sheffield: Sheffield Phoenix.

———. 2010. *The Bible in Theory: Critical and Postcritical Essays*. Resources for Biblical Study 57. Atlanta: Society of Biblical Literature.

———. 2011a. "Paul after Empire." Pages 9–23 in *The Colonized Apostle: Paul through Postcolonial Eyes*. Edited by Christopher D. Stanley. Paul in Critical Contexts. Minneapolis: Fortress.

———. 2011b. "Why There Are No Humans or Animals in the Gospel of Mark." Pages 71–94 in *Mark as Story: Retrospect and Prospect*. Edited by Kelly R. Iverson and Christopher W. Skinner. SBLRBS 65. Atlanta: Society of Biblical Literature.

———. 2013. "The Dog-Woman of Canaan, and Other Animal Tales in the Gospel of Matthew." Pages 57–71 in *Soundings in Cultural Criticism: Perspectives and Methods in Culture, Power, and Identity in New Testament Interpretation*. Edited by Francisco Lozada Jr. and Greg Carey. Minneapolis: Fortress.

Moore, Stephen D., ed. 2014. *Divinanimality: Animal Theory, Creaturely Theology*. Transdisciplinary Theological Colloquia. New York: Fordham University Press.

Moore, Stephen D., and Janice Capel Anderson. 1998. "Taking It Like a Man: Masculinity in 4 Maccabees." *JBL* 117:249–73.

Moore, Stephen D., and Yvonne Sherwood. 2011. *The Invention of the Biblical Scholar: A Critical Manifesto*. Minneapolis: Fortress.

Moore-Gilbert, Bart. 1997. *Postcolonial Theory: Contexts, Practices, Politics*. London: Verso.

Morris, Rosalind C. 1995. "All Made Up: Performance Theory and the New Anthropology of Sex and Gender." *Annual Review of Anthropology* 24:567–92.

Morton, Stephen. 2003. *Gayatri Chakravorty Spivak*. Routledge Critical Thinkers. New York: Routledge.
Mounce, Robert H. 1998. *The Book of Revelation*. 2nd ed. NICNT. Grand Rapids: Eerdmans.
Mowry, Lucetta. 1952. "Revelation 4–5 and Early Christian Liturgical Usage." *JBL* 71:75–84.
Müller, Ulrich B. 1984. *Die Offenbarung des Johannes*. ÖTK 19. Gütersloh: Gütersloher Verlagshaus; Würzburg: Echter.
Mulvey, Laura. 1999. "Visual Pleasure and Narrative Cinema." Pages 833–44 in *Film Theory and Criticism: Introductory Readings*. Edited by Leo Braudy and Marshall Cohen. Oxford: Oxford University Press. Originally published in *Screen* 16 (1975): 6–18.
Munteanu, Dana LaCourse, ed. 2011a. *Emotion, Genre, and Gender in Classical Antiquity*. New York: Bristol Academic Press.
———. 2011b. "Introduction." Pages 1–14 in Munteanu 2011a.
Mussies, Gerhard. 1971. *The Morphology of Koine Greek as Used in the Apocalypse of St. John: A Study in Bilingualism*. NovTSup 27. Leiden: Brill.
Naipaul, V. S. 1967. *The Mimic Men*. London: André Deutsch.
Nathanson, Donald L. 2008. "Prologue." Pages xv–xvi in vol. 1 of Silvan S. Tomkins. *Affect Imagery Consciousness: The Complete Edition*. New York: Springer.
Nelavala, Surekha. 2009. "'Babylon the Great, Mother of Whores' (Rev. 17:5): A Postcolonial Feminist Perspective." *ExpTim* 121:60–65.
Newton, Esther. 1971. *Mother Camp: Female Impersonators in America*. Chicago: University of Chicago Press.
Ngai, Sianne. 2005. *Ugly Feelings*. Cambridge: Harvard University Press.
Nigianni, Chrysanthi, and Merl Storr, eds. 2009. *Deleuze and Queer Theory*. Deleuze Connections. Edinburgh: Edinburgh University Press.
Nussbaum, Martha C. 2004. *Hiding from Humanity: Disgust, Shame, and the Law*. Princeton: Princeton University Press.
O'Grady, Standish. 1878–1880. *History of Ireland*. 2 vols. London: Sampson, Low, Searle, Marston and Rivington.
Økland, Jorunn. 2005. "Why Can't the Heavenly Miss Jerusalem Just Shut Up?" Pages 311–32 in *Her Master's Tools? Feminist and Postcolonial Engagements of Historical-Critical Discourse*. Edited by Caroline Vander Stichele and Todd Penner. Global Perspectives on Biblical Scholarship 9. Atlanta: Society of Biblical Literature.
Oliver, Kelly. 2009. *Animal Lessons: How They Teach Us to Be Human*. New York: Columbia University Press.

Olson, Kelly. 2006. "*Matrona* and Whore: Clothing and Definition in Roman Antiquity." Pages 186–206 in *Prostitutes and Courtesans in the Ancient World*. Edited by Christopher A. Faraone and Laura K. McClure. Wisconsin Studies in Classics. Madison: University of Wisconsin Press.

———. 2008. *Dress and the Roman Woman: Self-Presentation and Society*. New York: Routledge.

Ormand, Kirk. 2009. *Controlling Desires: Sexuality in Ancient Greece and Rome*. Praeger Series on the Ancient World. Westport, Conn.: Praeger.

Osborne, Grant R. 2002. *Revelation*. Baker Exegetical Commentary on the New Testament. Grand Rapids: Baker Academic.

Paglia, Camille. 1992. "Alice in Muscle Land." Pages 79–82 in *Sex, Art, and American Culture: Essays*. New York: Vintage.

Parry, Benita. 2004. *Postcolonial Studies: A Materialist Critique*. Postcolonial Literatures. New York: Routledge.

Partridge, Christopher, ed. 2012. *Anthems of Apocalypse: Popular Music and Apocalyptic Thought*. The Bible in the Modern World 42. Sheffield: Sheffield Phoenix.

Paul, G. M. 1982. "Urbs Capta: Sketch of an Ancient Literary Motif." *Phoenix* 36:144–55.

Pearse, Padraic H. 1924a. "The Coming Revolution." Pages 88–99 in *Collected Works of Padraic H. Pearse*. Unnumbered vol., *Political Writings and Speeches*. Dublin: Phoenix.

———. 1924b. "O'Donovan Rossa Graveside Panegyric." Pages 133–37 in *Collected Works of Padraic H. Pearse*. Unnumbered vol., *Political Writings and Speeches*. Dublin: Phoenix.

Penney, James. 2013. *After Queer Theory: The Limits of Sexual Politics*. London: Pluto.

Peterson, Erik. 1926. *Heis Theos: Epigraphische, formgeschichtliche, und religionsgeschichtliche Untersuchungen*. Forschungen zur Religion und Literatur des Alten und Neuen Testaments 41. Göttingen: Vandenhoeck & Ruprecht.

———. 1964. *The Angels and the Liturgy*. Translated by Ronald Walls. New York: Herder & Herder.

Pilch, John J. 1999. *The Cultural Dictionary of the Bible*. Collegeville, Minn.: Liturgical Press.

Pippin, Tina. 1992a. *Death and Desire: The Rhetoric of Gender in the Apocalypse of John*. Literary Currents in Biblical Interpretation. Louisville: Westminster John Knox.

———. 1992b. "Eros and the End: Reading for Gender in the Apocalypse of John." *Semeia* 59:193–210.

———. 1992c. "The Heroine and the Whore: Fantasy and the Female in the Apocalypse of John." *Semeia* 60:67–82

———. 1994a. "Peering into the Abyss: A Postmodern Reading of the Bottomless Pit." Pages 251–68 in *The New Literary Criticism and the New Testament*. Edited by Edgar V. McKnight and Elizabeth Struthers Malbon. Sheffield: JSOT Press.

———. 1994b. "The Revelation to John." Pages 109–30 in *A Feminist Commentary*. Vol. 2 of *Searching the Scriptures*. Edited by Elisabeth Schüssler Fiorenza. New York: Crossroad.

———. 1995. "'And I Will Strike Her Children Dead': Death and the Deconstruction of Social Location." Pages 191–98 in *Social Location and Biblical Interpretation in the United States*. Vol. 1 of *Reading from This Place*. Edited by Fernando F. Segovia and Mary Ann Tolbert. Minneapolis: Fortress.

———. 1999. *Apocalyptic Bodies: The Biblical End of the World in Text and Image*. Biblical Limits. New York: Routledge.

———. 2005. "The Heroine and the Whore: The Apocalypse of John in Feminist Perspective." Pages 127–45 in *From Every People and Nation: The Book of Revelation in Intercultural Perspective*. Edited by David Rhoads. Minneapolis: Fortress.

———. 2012. "Revelation/Apocalypse of John." Pages 627–32 in *Women's Bible Commentary*. 3rd ed. Edited by Carol A. Newsom, Sharon H. Ringe, and Jacqueline E. Lapsley. Louisville: Westminster John Knox.

Pippin, Tina, and J. Michael Clark. 2006. "Revelation/Apocalypse." Pages 753–68 in *The Queer Bible Commentary*. Edited by Deryn Guest, Robert E. Goss, Mona West, and Thomas Bohache. London: SCM.

Porter, Stanley E. 1989. "The Language of the Apocalypse in Recent Discussion." *NTS* 35:582–603.

Price, Robert M. 1998. "Saint John's Apothecary: Différance, Textuality, and the Advent of Meaning." *BibInt* 6:105–12.

Price, S. R. F. 1984. *Rituals and Power: The Roman Imperial Cult in Asia Minor*. Cambridge: Cambridge University Press.

Punt, Jeremy. 2008. "Intersections in Queer Theory and Postcolonial Theory, and Hermeneutical Spin-Offs." *The Bible and Critical Theory* 4.2. Online: http://novaojs.newcastle.edu.au/ojsbct/index.php/bct/article/view/197/181.

Quinby, Lee. 1994. *Anti-Apocalypse: Exercises in Genealogical Criticism*. Minneapolis: University of Minnesota Press.

Rad, Gerhard von. 1972. *Genesis: A Commentary*. Translated by J. H. Marks. OTL. Philadelphia: Westminster.

Rainbow, Jesse. 2007. "Male *mastoi* in Revelation 1.13." *JSNT* 30:249–53.

Rajak, Tessa. 1997. "Dying for the Law: The Martyr's Portrait in Jewish-Greek Literature." Pages 39–68 in *Portraits: Biographical Representation in the Greek and Latin Literature of the Roman Period*. Edited by M. J. Edwards and Simon Swain. Oxford: Clarendon.

Rauh, Nicholas K. 2011. "Prostitutes, Pimps, and Political Conspiracies during the Late Roman Republic." Pages 197–221 in *Greek Prostitutes in the Ancient Mediterranean, 800 BCE–200 CE*. Edited by Allison Glazebrook and Madeleine M. Henry. Wisconsin Studies in Classics. Madison: University of Wisconsin Press.

Reid, Duncan. 2000. "Setting Aside the Ladder to Heaven: Revelation 21.1–22.5 from the Perspective of the Earth." Pages 232–45 in *Readings from the Perspective of Earth*. Edited by Norman C. Habel. Earth Bible 1. Sheffield: Sheffield Academic.

Resseguie, James L. 1998. *Revelation Unsealed: A Narrative Critical Approach to John's Apocalypse*. Biblical Interpretation Series 32. Leiden: Brill.

———. 2005. *Narrative Criticism of the New Testament: An Introduction*. Grand Rapids: Baker Academic.

———. 2009. *The Revelation of John: A Narrative Commentary*. Grand Rapids: Baker Academic.

Richard, Pablo. 1995. *Apocalypse: A People's Commentary on the Book of Revelation*. Translated by Phillip Berryman. The Bible and Liberation. Maryknoll, N.Y.: Orbis.

Rissi, Mathias. 1995. *Die Hure Babylon und die Verführung der Heiligen: Eine Studie zur Apokalypse des Johannes*. BWANT 136. Stuttgart: Kohlhammer.

Robinson, J. A. T. 1976. *Redating the New Testament*. Philadelphia: Westminster.

Roloff, Jürgen. 1993. *The Revelation of John*. Translated by John E. Alsup. Continental Commentaries. Minneapolis: Fortress.

Romanow, Rebecca Fine. 2006. *The Postcolonial Body in Queer Space and Time*. Newcastle-upon-Tyne, U.K.: Cambridge Scholars.

Rossing, Barbara R. 1999a. *The Choice between Two Cities: Whore, Bride, and Empire in the Apocalypse.* HTS 48. Harrisburg, Pa.: Trinity Press International.

———. 1999b. "River of Life in God's New Jerusalem: An Ecological Vision for Earth's Future." Pages 205–24 in *Christianity and Ecology.* Edited by Rosemary Radford Ruether and Dieter Hassel. Religions of the World and Ecology 3. Cambridge: Harvard Center for World Religions.

———. 2002. "Alas for Earth! Lament and Resistance in Revelation 12." Pages 180–92 in *The Earth Story in the New Testament.* Edited by Norman C. Habel and Vicki Balabanski. Earth Bible 5. Cleveland: Pilgrim.

———. 2005a. "For the Healing of the World: Reading Revelation Ecologically." Pages 165–82 in *From Every People and Nation: The Book of Revelation in Intercultural Perspective.* Edited by David Rhoads. Minneapolis: Fortress.

———. 2005b. *The Rapture Exposed: The Message of Hope in the Book of Revelation.* New York: Basic.

———. 2008. "'Hastening the Day' When the Earth Will Burn: Global Warming, 2 Peter, and a Nature Hike through the Book of Revelation." Pages 25–38 in *Reading the Signs of the Times: Taking the Bible into the Public Square.* Edited by Cynthia Briggs Kittredge, Ellen Aitken, and Jonathan Draper. Minneapolis: Fortress.

Rowland, Christopher. 1980. "The Vision of the Risen Christ in Rev. i.13ff.: The Debt of an Early Christology to an Aspect of Jewish Angelology." *JTS* n.s. 31:1–11.

———. 1982. *The Open Heaven: A Study of Apocalyptic in Judaism and Early Christianity.* New York: Crossroad.

———. 2004. "Revelation." Pages 559–70 in *Global Bible Commentary.* Edited by Daniel Patte. Nashville: Abingdon.

Rowland, Ingrid. 2013. "Roma Aeterna." Pages 558–74 in *The Cambridge Companion to Ancient Rome.* Edited by Paul Erdkamp. Cambridge: Cambridge University Press.

Royalty, Robert M., Jr. 1998. *The Streets of Heaven: The Ideology of Wealth in the Apocalypse of John.* Macon, Ga.: Mercer University Press.

———. 2004. "Don't Touch *This* Book: Rev. 22:18-19 and the Rhetoric of Reading (in) the Apocalypse of John." *BibInt* 12:282–300.

Ruffolo, David. 2009. *Post-Queer Politics.* Queer Interventions. Burlington, Vt.: Ashgate.

Ruiz, Jean-Pierre. 2003. "Taking a Stand on the Sand of the Seashore: A Postcolonial Exploration of Revelation 13." Pages 119–36 in *Reading the Book of Revelation: A Resource for Students.* Edited by David L.

Barr. Resources for Biblical Study 44. Atlanta: Society of Biblical Literature.

Runions, Erin. 1998. "Zion Is Burning: 'Gender Fuck' in Micah." *Semeia* 82:225–46.

———. 2002. *Changing Subjects: Gender, Nation and Future in Micah*. Playing the Texts 7. Sheffield: Sheffield Academic.

———. 2008a. "Queering the Beast: The Antichrist's Gay Wedding." Pages 97–110 in *Queering the Non-Human*. Edited by Noreen Giffney and Myra Hird. Farnham, U.K.: Ashgate.

———. 2008b. "From Disgust to Humor: Rahab's Queer Affect." *Postscripts* 4:41–69. Reprinted in pages 45–74 of *Bible Trouble: Queer Reading at the Boundaries of Biblical Scholarship*. Edited by Teresa J. Hornsby and Ken Stone. SemeiaSt 67. Atlanta: Society of Biblical Literature, 2011.

———. 2014. *The Babylon Complex: Theopolitical Fantasies of War, Sex, and Sovereignty*. New York: Fordham University Press.

Said, Edward W. 1978. *Orientalism: Western Conceptions of the Orient*. New York: Pantheon.

———. 1993. *Culture and Imperialism*. New York: Vintage.

Salisbury, Joyce E. 1994. *The Beast Within: Animals in the Middle Ages*. New York: Routledge.

Samuel, Simon. 2002. "The Beginning of Mark: A Colonial/Postcolonial Conundrum." *BibInt* 10:405–19.

Samuelsson, Maria Jansdotter. 2012. "The Final Apocalypse of Phallocentrism: Irigarayan Openings to the Matrix of Male Desire and Correction of the Non-male Subject in the Book of Revelation." *Feminist Theology* 21:101–15.

Sánchez, David A. 2008. *From Patmos to the Barrio: The Subversion of Imperial Myths from the Book of Revelation to the Present*. Minneapolis: Fortress.

Sarna, Nahum M. 1989. *Genesis: The traditional Hebrew Text with New JPS Translation*. JPS Torah Commentary. Philadelphia: Jewish Publication Society.

Sauter, Franz. 1934. *Der römische Kaiserkult bei Martial und Statius*. Tübinger Beiträge zum Altertumswissenschaft 21. Stuttgart: Kohlhammer.

Schaberg, Jane. 1998. "Luke." Pages 275–92 in *Women's Bible Commentary*. 2nd ed. Edited by Carol A. Newsom and Sharon H. Ringe. Louisville: Westminster John Knox.

Scherrer, Steven J. 1984. "Signs and Wonders in the Imperial Cult: A New

Look at a Roman Religious Institution in Light of Rev. 13:13–15." *JBL* 103:599–610.
Schreiner, Thomas R. 2008. *New Testament Theology: Magnifying God in Christ*. Grand Rapids: Baker Academic.
Schüssler Fiorenza, Elisabeth. 1981. *Invitation to the Book of Revelation: A Commentary on the Apocalypse with Complete Text from the Jerusalem Bible*. Garden City, N.Y.: Image.
———. 1985. *The Book of Revelation: Justice and Judgment*. Philadelphia: Fortress.
———. 1991. *Revelation: Vision of a Just World*. Proclamation Commentaries. Minneapolis: Fortress.
———. 1998. *The Book of Revelation: Justice and Judgment*. 2nd ed. Minneapolis: Fortress.
———. 2007. *The Power of the Word: Scripture and the Rhetoric of Empire*. Minneapolis: Fortress.
Schwarzenegger, Arnold, and Bill Dobbins. 1985. *Encyclopedia of Modern Bodybuilding*. New York: Simon & Schuster.
Scott, Kenneth. 1975. *The Imperial Cult under the Flavians*. New York: Arno.
Scott, R. B. Y. 1928. *The Original Language of the Apocalypse*. Toronto: University of Toronto Press.
Sedgwick, Eve Kosofsky. 2003. *Touching Feeling: Affect, Pedagogy, Performativity*. Series Q. Durham: Duke University Press.
Seesengood, Robert Paul. 2006. *Competing Identities: The Athlete and the Gladiator in Early Christianity*. LNTS 346. New York: T&T Clark.
Seigworth, Gregory J., and Melissa Gregg. 2010. "An Inventory of Shimmers." Pages 1–25 in *The Affect Theory Reader*. Edited by Melissa Gregg and Gregory J. Seigworth. Durham: Duke University Press.
Selvidge, Marla J. 1996. "Reflections on Violence and Pornography: Misogyny in the Apocalypse and Ancient Hebrew Prophecy." Pages 274–85 in *A Feminist Companion to the Hebrew Bible in the New Testament*. Edited by Athalya Brenner. Feminist Companion to the Bible 10. Sheffield: Sheffield Academic.
Setel, T. Drorah. 1985. "Prophets and Pornography: Female Sexual Imagery in Hosea." Pages 86–95 in *Feminist Interpretation of the Bible*. Edited by Letty M. Russell. Philadelphia: Westminster.
Shannon, Laurie. 2009. "The Eight Animals in Shakespeare; or, Before the Human." *PMLA* 124:472–79.

———. 2013. *The Accommodated Animal: Cosmopolity in Shakespearean Locales*. Chicago: University of Chicago Press.

Shaw, Brent D. 1996. "Body/Power/Identity: Passions of the Martyrs." *JECS* 4:269–312.

Shepherd, M. H. 1960. *The Paschal Liturgy and the Apocalypse*. London: Lutterworth.

Simpson, Mark. 1994. *Male Impersonators: Men Performing Masculinity*. New York: Routledge.

Sintado, Carlos Alberto. 2010. "Social Ecology, Ecojustice, and the New Testament." Ph.D. diss., Drew University, Madison, N.J.

Sissa, Giulia. 2008. *Sex and Sensuality in the Ancient World*. New Haven: Yale University Press.

Skinner, Marilyn B. 2005. *Sexuality in Greek and Roman Culture*. Ancient Cultures. Oxford: Blackwell. 2nd ed. 2013.

Slater, Thomas B. 1999. *Christ and Community: A Socio-Historical Study of the Christology of Revelation*. JSNTSup 178. Sheffield: Sheffield Academic.

"Slaughtered Lamb Head Teacher Resigns from Kent School." 2010. *BBC News*. February 10. Online: http://news.bbc.co.uk/2/hi/uk_news/england/kent/8508975.stm.

Smalley, Stephen S. 2005. *The Revelation to John: A Commentary on the Greek Text of the Apocalypse*. Downers Grove, Ill.: IVP Academic.

Smith, Shanell T. 2012. "Empire, Gender, and Ambi*veilence*: Toward a Postcolonial Womanist Interpretation of the Woman Babylon in the Book of Revelation." Ph.D. diss., Drew University, Madison, N.J.

———. 2014. "The Construction of Gender in Revelation." Pages 340–41 in *Global Perspectives on the Bible*. Edited by Mark Roncace and Joe Weaver. Upper Saddle River, N.J.: Pearson Prentice Hall.

Spaeth, Barbette Stanley. 1996. *The Roman Goddess Ceres*. Austin: University of Texas Press.

Spittler, Janet E. 2008. *Animals in the Apocryphal Acts of the Apostles: The Wild Kingdom of Early Christian Literature*. WUNT 2/247. Tübingen: Mohr Siebeck.

Spivak, Gayatri Chakravorty. 1987a. *In Other Worlds: Essays in Cultural Politics*. New York: Methuen.

———. 1987b. "A Literary Representation of the Subaltern: A Woman's Text from the Third World." Pages 241–68 in Spivak 1987a.

———. 1987c. "Subaltern Studies: Deconstructing Historiography." Pages 270–304 in Spivak 1987a.

———. 1988. "Can the Subaltern Speak?" Pages 271–313 in *Marxism and the Interpretation of Culture*. Edited by Cary Nelson and Larry Grossberg. Urbana: University of Illinois Press.

———. 1990. *The Post-colonial Critic: Interviews, Strategies, Dialogues*. Edited by Sarah Harasym. New York: Routledge.

———. 1991. "Identity and Alterity: An Interview (with Nikos Papastergiadis)." *Arena* 97:65–76.

———. 1996. "More on Power/Knowledge." Pages 143–54 in *The Spivak Reader: Selected Works of Gayatri Chakravorty Spivak*. Edited by Donna Landry and Gerard Maclean. New York: Routledge.

———. 1999. *A Critique of Postcolonial Reason: Toward a History of the Vanishing Present*. Cambridge: Harvard University Press.

Stauffer, Ethelbert. 1955. *Christ and the Caesars*. Translated by K. Gregor Smith and R. Gregor Smith. London: SCM.

Staley, Jeffrey L. 1995a. "Narrative Structure (Self-Stricture) in Luke 4:14-9:62: The United States of Luke's Story World." *Semeia* 72:173–213.

———. 1995b. *Reading with a Passion: Rhetoric, Autobiography, and the American West in the Gospel of John*. New York: Continuum.

Steel, Karl. 2011. *How to Make a Human: Animals and Violence in the Middle Ages*. Columbus: Ohio State University Press.

Steiner, Gary. 2005. *Anthropocentrism and Its Discontents: Animals and Their Moral Status in the History of Western Philosophy*. Pittsburgh: University of Pittsburgh Press.

Stenström, Hanna. 2009. "'They Have Not Defiled Themselves with Women…': Christian Identity According to the Book of Revelation." Pages 33–54 in *A Feminist Companion to the Apocalypse of John*. Edited by Amy-Jill Levine with Maria Mayo Robbins. Feminist Companion to the New Testament and Early Christian Writings 13. New York: T&T Clark.

Stewart, Kathleen. 2007. *Ordinary Affects*. Durham: Duke University Press.

Stone, Ken. 2014. "The Dogs of Exodus and the Question of the Animal." Pages 36–50 in *Divinaminality: Animal Theory, Creaturely Theology*. Edited by Stephen D. Moore. Transdisciplinary Theological Colloquia. New York: Fordham University Press.

Strelan, Rick. 2003. "'Outside Are the Dogs and the Sorcerers…' (Revelation 22:15). *BTB* 33:148–57.

Stulman, Louis. 2005. *Jeremiah*. AOTC. Nashville: Abingdon.

Suetonius. 2000. *Lives of the Caesars*. Translated by Catherine Edwards. Oxford World's Classics. Oxford: Oxford University Press.

Sweet, J. P. M. 1979. *Revelation.* Westminster Pelican Commentaries. Philadelphia: Westminster.

Swete, Henry Barclay. 1906. *The Apocalypse of St. John.* London: Macmillan.

Tasker, Yvonne. 1993. *Spectacular Bodies: Gender, Genre and the Action Cinema.* New York: Routledge.

Taylor, Lily Ross. 1931. *The Divinity of the Roman Emperor.* Middletown, Conn.: American Philological Association.

Terada, Rei. 2001. *Feeling in Theory: Emotion after the "Death of the Subject."* Cambridge: Harvard University Press.

Thimmes, Pamela. 2009. "'Teaching and Beguiling My Servants': The Letter to Thyatira (Rev. 2.18-29)." Pages 69–87 in *A Feminist Companion to the Apocalypse of John.* Edited by Amy-Jill Levine with Maria Mayo Robbins. Feminist Companion to the New Testament and Early Christian Writings 13. New York: T&T Clark.

Thomas, Robert L. 1995. *Revelation 8–22: An Exegetical Commentary.* Wycliffe Exegetical Commentary. Chicago: Moody Press.

Thompson, Leonard L. 1990. *The Book of Revelation: Apocalypse and Empire.* Oxford: Oxford University Press.

Thompson, Steven. 1985. *The Apocalypse and Semitic Syntax.* SNTSMS 52. Cambridge: Cambridge University Press.

Thompson, W. E., and J. H. Bair. 1982. "A Sociological Analysis of Pumping Iron." *Free Inquiry in Creative Sociology* 10:192–96.

Thrift, Nigel. 2010. "Understanding the Material Practices of Glamour." Pages 289–308 in *The Affect Theory Reader.* Edited by Melissa Gregg and Gregory J. Seigworth. Durham: Duke University Press.

Thurman, Eric. 2003. "Looking for a Few Good Men: Mark and Masculinity." Pages 137–65 in *New Testament Masculinities.* Edited by Stephen D. Moore and Janice Capel Anderson. SemeiaSt 45. Atlanta: Society of Biblical Literature.

Torrey, C. C. 1958. *The Apocalypse of John.* New Haven: Yale University Press.

Vander Stichele, Caroline. 2000a. "Apocalypse, Art and Abjection: Images of the Great Whore." Pages 124–38 in *Culture, Entertainment, and the Bible.* Edited by George Aichele. JSOTSup 309. Sheffield: Sheffield Academic.

———. 2000b. "Just a Whore: The Annihilation of Babylon According to Revelation 17:16." *Lectio difficilior* 1. Online: http://www.lectio.unibe.ch/00_1/j.htm.

———. 2009. "Re-membering the Whore: The Fate of Babylon According to Revelation 17.16." Pages 106–20 in *A Feminist Companion to the Apocalypse of John*. Edited by Amy-Jill Levine with Maria Mayo Robbins. Feminist Companion to the New Testament and Early Christian Writings 13. New York: T&T Clark.

Veeser, H. Aram, ed. 1996. *Confessions of the Critics*. New York: Routledge.

Vermeule, Cornelius C. 1959. *The Goddess Roma in the Art of the Roman Empire*. Cambridge: Spink & Son.

Voorwinde, Stephen. 2005. *Jesus' Emotions in the Fourth Gospel: Human or Divine?* LNTS 284. New York: T&T Clark.

———. 2011. *Jesus' Emotions in the Gospels*. New York: T&T Clark.

Wall, Robert W. 1991. *Revelation*. NIBCNT 18. Peabody, Mass.: Hendrickson.

Waller, Alexis G. 2014. "Violent Spectacles and Public Feelings: Trauma and Affect in the Gospel of Mark and *The Thunder: Perfect Mind*." *BibInt* 22.4:450–72.

Walliss, John, and Lee Quinby, eds. 2010. *Reel Revelations: Apocalypse and Film*. The Bible in the Modern World 31. Sheffield: Sheffield Phoenix.

Walters, Margaret. 1978. *The Nude Male: A New Perspective*. New York: Paddington.

Warner, William. 1992. "Spectacular Action: Rambo and the Popular Pleasures of Pain." Pages 672–88 in *Cultural Studies*. Edited by Lawrence Grossberg, Cary Nelson, and Paula A. Treichler. New York: Routledge.

Washington, Harold C. 1997. "Violence and the Construction of Gender in the Hebrew Bible: A New Historicist Approach." *BiBInt* 5:324–63.

Weil, Kari. 2012. *Thinking Animals: Why Animal Studies Now?* New York: Columbia University Press.

Weinfeld, Moshe. 1972. *Deuteronomy and the Deuteronomic School*. Oxford: Clarendon.

Weinrich, William C., ed. 2005. *Revelation*. ACCS NT 12. Downers Grove, Ill.: InterVarsity Press.

Weinstock, Stefan. 1971. *Divus Julius*. Oxford: Clarendon.

Westhelle, Vitor. 2005. "Revelation 13: Between the Colonial and the Postcolonial, a Reading from Brazil." Pages 183–99 in *From Every People and Nation: The Book of Revelation in Intercultural Perspective*. Edited by David Rhoads. Minneapolis: Fortress.

Williams, Craig A. 1999. *Roman Homosexuality: Ideologies of Masculinity in Classical Antiquity*. Oxford: Oxford University Press.

———. 2010. *Roman Homosexuality*. 2nd ed. Oxford: Oxford University Press.
Williams, Jeromie. 2011. "US Soldiers in Afghanistan Kill Sheep with Metal Baseball Bat." *Digital Journal*. December 10. Online: http://digitaljournal.com/article/315911.
Williams, Wesley. 2009. "A Body Unlike Bodies: Transcendent Anthropomorphism in Ancient Semitic Tradition and Early Islam." *JAOS* 129:19–44.
Wise, Michael, Martin Abegg, Jr., and Edward Cook, eds. and trans. 2005. *The Dead Sea Scrolls: A New Translation*. 2nd ed. New York: HarperCollins.
Witherington, Ben, III. 2003. *Revelation*. New Cambridge Bible Commentary. Cambridge: Cambridge University Press.
Wolfe, Cary. 2003. *Animal Rites: American Culture, the Discourse of Species, and Posthumanist Theory*. Chicago: University of Chicago Press.
———. 2009a. "Human, All Too Human: 'Animal Studies' and the Humanities." *PMLA* 124:564–75.
———. 2009b. *What Is Posthumanism?* Posthumanities 8. Minneapolis: University of Minnesota Press.
Wood, David. 2007. "Specters of Derrida: On the Way to Econstruction." Pages 264–89 in *Ecospirit: Religions and Philosophies for the Earth*. Edited by Laurel Kearns and Catherine Keller. Transdisciplinary Theological Colloquia. New York: Fordham University Press.
Young, Robin Darling. 1991. "The 'Woman with the Soul of Abraham': Traditions about the Mother of the Maccabean Martyrs." Pages 67–81 in *"Women Like This": New Perspectives on Jewish Women in the Greco-Roman World*. Edited by Amy-Jill Levine. SBLEJL 1. Atlanta: Scholars Press.
Zane, Frank. 1994. "Train with Zane: Bodybuilding in Future-World." *Muscle & Fitness*. June, 247.

Index of Modern Authors

Abegg, Martin, Jr.   236
Adams, Edward   4 n. 12
Agamben, Giorgio   206 n. 16
Ahmed, Sara   8, 158, 162–78
Allen, Leslie C.   127 n. 3
Amarasingam, Amarnath   229 n. 7
Anderson, Janice Capel   39, 64, 69 n. 28, 137 n. 29
Ashcroft, Bill   22 n. 7
Atterton, Peter   206 n. 16
Aune, David E.   4, 32 n. 21, 67 n. 5, 69 n. 26, 84–86, 104 n. 3, 113 n. 23, 129 n. 13, 185 n. 8, 211 n. 28, 213 n. 33, 215 n. 41, 229 n. 9, 235 n. 19, 236–37, 240
Aydemir, Murat   145 n. 41

Badmington, Neil   204 n. 6
Barker, Margaret   68 n. 24
Barr, David L.   4 n. 10
Barth, Karl   198
Barton, Carlin   62
Bauckham, Richard   4 n. 12, 43, 47, 49, 51–53, 55–56, 67 n. 6, 67 n. 10, 68 n. 19, 68 n. 23, 71 n. 40, 77–80, 86–89, 202 n. 1, 235 n. 54
Beale, G. K.   4, 68 nn. 20–21, 69 n. 26, 71 n. 41, 105, 108 n. 12, 185 n. 8, 209 n. 25, 212 n. 30, 235 n. 19, 240
Beard, Mary   24 n. 11
Beasley-Murray, G. R.   95
Beauvery, Robert   129 n. 13
Bennett, Jane   161
Bennington, Geoffrey   226 n. 1
Bentham, Jeremy   214
Bergen, Wesley J.   221–22 n. 49

Berger, Anne Emmanuelle   204 n. 6
Berlant, Lauren   1–2
Bhabha, Homi K.   7, 13–16, 24–28, 30–31
Bible and Culture Collective   3 n. 2, 10
Bitel, Lisa   71 n. 38
Blackman, Lisa   155
Blanchot, Maurice   187
Bloch, Ernst   191–92 n. 17
Blount, Brian K.   3 n. 7, 4 n. 14
Boesak, Allan A.   3, 37
Boggs, Colleen Glenney   227–28
Bolin, Anne   81 n. 8
Boring, Eugene M.   84, 86–87
Bousset, Wilhelm   33 n. 23
Bowra, C. M.   133 n. 22
Boxall, Ian   128 n. 7, 240 n. 29
Braidotti, Rosi   204 n. 9, 228
Brasher, Brenda E.   4 n. 13
Bredin, Mark   4 n. 12, 202 n. 1, 235 n. 20
Brennan, Teresa   158 n. 5
Brenner, Athalya   67 n. 15, 127 n. 3
Brütsch, Charles   210
Burrell, Barbara   129 n. 10
Burrus, Virginia   166 n. 23
Butler, George   82 n. 15, 93 n. 35, 94 n. 38
Butler, Judith   8, 67 n. 7, 125–26, 144–45, 147–52

Caird, G. B.   82 n. 12, 240 n. 31
Calarco, Matthew   206 n. 16
Callahan, Allen D.   71 n. 41
Campbell, Jan   145 n. 41
Cantarella, Eve   135 n. 27

Caputo, John D. 37 n. 30, 198
Carey, Greg 4 n. 8, 22 n. 8
Carrell, Peter R. 68 n. 24, 208 n. 18
Carroll, Robert P. 67 n. 15
Carson, Marion 3 n. 3, 105 n. 7
Carter, Warren 4 n. 8
Cate, James Jeffrey 4 n. 12, 202 n. 1, 235 n. 20
Cavell, Stanley 206 n. 16
Césaire, Aimé 13
Chapkis, Wendy 170 n. 31
Charles, R. H. 67 n. 11, 68 n. 20, 79 n. 6, 113 n. 22, 148 n. 52
Chrulew, Matthew 4 n. 11, 204 n. 6
Clanton, Dan W., Jr. 4 n. 14
Clark, J. Michael 3 n. 6, 144 n. 41
Clarke, John R. 109 n. 14, 110–11
Clines, David J. A. 45–46, 67 n. 9
Clough, Patricia Ticineto 155
Coetzee, J. M. 202, 212, 216
Cohen, Edward E. 107, 110
Cohn, Norman 192 n. 17
Collins, Adela Yarbro 3, 29 n. 17, 68 n. 24, 69 n. 26, 90 n. 30, 129 n. 13
Conway, Colleen M. 148 n. 52
Cook, Edward 236
Cooke, Miriam 67 n. 8
Corley, Jeremy 156 n. 1
Corner, Sean 106, 107 n. 9
Court, John M. 112 n. 20
Crane, Susan 205
Curran, John R. 153
Cuss, Dominique 85

D'Ambra, Eve 113 n. 24
Darden, Lynne St. Clair 4 n. 8
Darwin, Charles 177 n. 38
Davidson, James 106–7
De Man, Paul 141 n. 34
DeKoven, Marianne 204 n. 8, 206, 207 n. 16
Deleuze, Gilles 159, 162 n. 12, 163–65, 168, 199
Dellamora, Richard 4 n. 13
DeLoughrey, Elizabeth 206 n. 16

Derrida, Jacques 4 n. 11, 8, 26 n. 14, 37 n. 30, 78, 159 n. 6, 160 n. 9, 164, 167, 172, 179–204, 206–7, 209–20, 226, 228 n. 3, 229–32, 237–40
Descartes, René 158–59, 204–7, 210–13, 218 n. 45, 219, 231 n. 14, 238–39
DeSilva, David A. 104 n. 3, 144 n. 40, 157 n. 1, 157 n. 3
Diehl, Judy 4 n. 8
Ditmore, Melissa 170
Dobbins, Bill 95
Dover, K. J. 107 n. 9, 135 n. 27
Duff, Paul B. 32, 148, 173 n. 34
Duncan, Anne 113 n. 24, 118 n. 36
Dutton, Kenneth R. 83, 93

Edwards, Catherine 116 n. 30, 118 n. 36
Egger-Wenzel, Renate 156 n. 1
Ellis, Brett Easton 63–64, 69 n. 27, 69–70 n. 32
Erskine, Andrew 130–31, 133 n. 22
Exum, J. Cheryl 67 n. 15

Fanon, Frantz 13, 35
Ferraris, Maurizio 195, 197, 199
Feuerbach, Ludwig 78
Findon, Joanne 71 n. 38
Fish, Stanley E. 9
Flemming, Rebecca 109 n. 15, 110, 111 nn. 18–19, 120 n. 40
Foerster, Werner 112 n. 21
Ford, Joan Massyngberde 81–82, 113 n. 22, 240 n. 28
Forster, E. M. 169 n. 30
Foucault, Michel 88–89, 135 n. 27, 136, 150
Frankfurter, David 164 n. 20, 165 n. 22
Freud, Sigmund 48, 78
Friesen, Steven J. 4–5, 21, 129 n. 10, 130
Frilingos, Christopher A. 3 n. 4, 4–5, 148 n. 52, 152
Frykholm, Amy Johnson 4 n. 14
Fussell, Samuel Wilson 81 n. 10, 93, 94 n. 39, 95, 97

INDEX OF MODERN AUTHORS

Gaines, Charles   93 n. 35, 94 n. 38
Galinsky, Karl   138
Gambs, Deborah   161
Garrett, Susan R.   3 n. 3
Gersht, Rivka   132 n. 21
Gibson, Nigel C.   35
Gilhus, Ingvild Saelid   203 n. 3, 204 n. 10, 209 n. 22, 220 n. 48, 222 n. 50, 227 n. 2
Ginsburg, Judith   137 n. 31
Glancy, Jennifer A.   7, 103–23
Glazebrook, Allison   107 n. 9, 110 n. 17
Gleason, Maud   135 n. 27, 141
González, Justo L.   3 n. 7
Gramsci, Antonio   17–18, 20, 33
Gregg, Melissa   155, 159
Gregory, Lady Augusta   71 n. 39
Gribben, Crawford   4 n. 14
Griffiths, Gareth   22 n. 7
Gross, Aaron S.   219 n. 47
Guattari, Félix   159, 162 n. 12, 165, 168
Gundry, Robert H.   234 n. 16
Gwyther, Anthony   3 n. 8, 15 n. 2, 22

Hallett, Judith P.   135 n. 27
Halley, Jean   155
Halperin, David M.   135 n. 27
Handley, George B.   206 n. 16
Haraway, Donna J.   205 n. 12, 206 n. 16
Hardt, Michael   18 n. 3, 194
Hart, Kevin   179
Hauck, Friedrich   113 n. 22
Hawkin, David J.   4 n. 12, 202 n. 1
Hawley, John C.   145 n. 41
Heidegger, Martin   190 n. 15, 198, 207, 213–14, 217–19, 231 n. 14
Henry, Madeleine M.   107 n. 9
Heyman, George   215 n. 41
Hobbes, Thomas   229 n. 8, 231 n. 14
Hoffmann, Matthias Reinhard 210 n. 25
Hoffner, Harry A.   48, 67 n. 13
Hood, Renate Viveen   164 n. 20
Horrell, David   4 n. 12, 202 n. 1, 203 n. 4, 236 n. 21
Howard-Brook, Wes   3 n. 8, 15 n. 2, 22

Howard, Robert Glenn   4 n. 14
Huber, Lynn R.   3 nn. 3–4, 3 n. 6, 105 n. 6, 144 n. 41, 209 n. 22
Huggan, Graham   206 n. 16, 228

Ipsen, Avaren   105 n. 7, 170 n. 31
Isaac, Benjamin   142 n. 37
Isasi-Díaz, Ada María   35 n. 28

James, Sharon L.   114
Jameson, Fredric   238 n. 25
JanMohamed, Abdul R.   35
Janzen, Ernest P.   86 n. 24
Jobling, David   3
Jobst, Werner   110 n. 16
Johns, Loren L.   208 n. 19, 208 n. 21, 211 n. 28, 215 n. 39, 215 n. 41
Johnson, Mark   208–9 n. 22
Johnston, Anna   17
Jones, Brian W.   91–92
Jones, C. P.   112–13
Joshel, Sandra R.   117–19, 121–22

Kang, Shinwook   4 n. 8
Kant, Immanuel   181, 207, 213, 231 n. 14
Kapparis, Konstantinos K. 104 n. 2
Kaster, Robert A.   156–57, 168
Keller, Catherine   3 n. 3, 4 n. 9, 4 nn. 11–12, 8, 67 n. 15, 69 n. 31, 144 n. 41, 179–81, 189–200, 203 n. 3, 210 n. 26
Kellum, Barbara   67 n. 14
Kelly, Daniel   164 n. 19
Kiberd, Declan   56–57, 71–72 n. 44
Kierkegaard, Søren   193 n. 20, 197–98
Kim, Jean K.   4 n. 9, 67 n. 15, 105 n. 7
Kinsella, Thomas   48–49, 52–53, 70 n. 36, 71 nn. 38–39
Kirk, Kenneth E.   76–77, 98
Klein, Alan M.   81, 83, 93 nn. 36–37
Knight, James   129 n. 13, 130 n. 15
Knust, Jennifer Wright   116 n. 29
Koester, Craig R.   104 n. 3
Kolnai, Aurel   168
Konstan, David   156, 158

Koosed, Jennifer L.   156, 201, 204 n. 6, 207 n. 17
Korsmeyer, Carolyn   173
Kotrosits, Maia   14, 156, 216 n. 41
Kovacs, Judith   208 n. 20
Kraemer, Ross S.   69 n. 30
Kraft, Heinrich   240
Kraybill, J. Nelson   134 n. 24
Krell, David Farrell   204 n. 6
Kristeva, Julia   160 n. 9, 170–72
Krodel, Gerhard A.   81 n. 11, 89 n. 27, 212 n. 30
Kuhn, Karl Allen   156 n. 1
Kurke, Leslie   105 n. 4, 107, 110 n. 17

Lacan, Jacques   160 n. 9, 184–85, 211 n. 27, 213, 231 n. 14, 232 n. 15
Ladd, George Eldon   212 n. 30
Lakoff, George   208–9 n. 22
Laughlin, T. Cowden   71 n. 41
Lawlor, Leonard   204 n. 6
Lawson, Alan   17
Levinas, Emmanuel   213, 218–19
Levine, Amy-Jill   3 n. 3, 75
Leys, Ruth   156
Liew, Tat-siong Benny   14 n. 1, 125
Lintott, Andrew   15
Lippit, Akira Mizuta   228 n. 4
Lohmeyer, Ernst   68 n. 17
Lopez, Davina C.   138 n. 32
Lundblad, Michael   204 n. 8, 207 n. 16
Lyons, William John   4 n. 14

MacKenzie, Robert K.   71 n. 41
Maier, Harry O.   4 n. 12, 24 n. 12, 202 n. 1, 235 n. 20, 239 n. 26
Marion, Jean-Luc   198
Marshall, John W.   4 n. 9, 148–49 n. 52, 164 n. 20, 165 n. 21
Martin, Clarice J.   3 n. 5, 4 n. 12
Martin, Thomas   4 n. 12, 203 n. 4, 236 n. 21
Massumi, Brian   159, 160 nn. 9–10, 161, 162 n. 11, 163
Mayo, Philip L.   164 n. 20

McClure, Laura K.   104–7, 109 n. 14, 110 n. 17, 114
McCourt, Frank   54–55, 61–62
McDonnell, Myles   136–37, 140–41 n. 33
McGinn, Colin   167 n. 24
McGinn, Thomas A. J.   108–11, 113, 115, 116 n. 28, 116 n. 31, 117 nn. 32–33, 120 n. 39, 121–22
McGough, Peter   82–83
McGuire, William   78 n. 5
McHugh, Susan   233
McKinley, Judith E.   4 n. 9
McLean, B. H.   162 n. 12
Mellor, Ronald   129 n. 13, 130, 131 n. 20, 133 n. 22, 153
Memmi, Albert   31
Menninghaus, Winfried   175
Merchant, Carolyn   237–38
Mettler, John J., Jr.   221
Metzger, Bruce M.   240 n. 28
Michaelis, Johann David   2
Míguez, Néstor O.   3 n. 7, 194 n. 22
Miles, Rosalind   93–94
Moltmann, Jürgen   192, 198
Moore-Gilbert, Bart   27
Morris, Leon   212 n. 30
Morton, Stephen   22 n. 7
Mounce, Robert H.   89 n. 29, 212 n. 30, 240
Mulvey, Laura   98
Munteanu, Dana LaCourse   156, 157 n. 2, 158 n. 4
Mussies, Gerhard   71 n. 41

Naipaul, V. S.   14, 26 n. 13
Negri, Antonio   18 n. 3, 194
Nelavala, Surekha   4 n. 9
Newton, Esther   145
Ngai, Sianne   176
Nietzsche, Friedrich   198
Nigianni, Chrysanthi   125
North, John   24 n. 11
Nussbaum, Martha C.   135 n. 26, 170

INDEX OF MODERN AUTHORS

O'Grady, Standish 56
Økland, Jorunn 4 n. 14, 234 n. 18
Oliver, Kelly 206 n. 16
Olson, Kelly 111, 113–14
Ormand, Kirk 135 n. 27
Osborne, Grant R. 4, 104 n. 3
Osborne, Peter 145 n. 41

Parry, Benita 27
Partridge, Christopher 4 n. 14
Pearse, Padraic H. 43, 55–61, 71 n. 43, 72 n. 45, 72 n. 48
Penney, James 125
Peterson, Erik 84 n. 18
Pilch, John J. 158 n. 4
Pippin, Tina 3, 5, 32 n. 21, 50, 67 n. 15, 122 n. 42, 144 n. 41
Porter, Stanley E. 71 n. 41
Price, Robert M. 4 n. 11
Price, S. R. F. 19 n. 4, 24 n. 11, 92, 94 n. 41, 129 n. 12, 130
Punt, Jeremy 145 n. 41

Quinby, Lee 4 n. 11, 4 n. 14

Rainbow, Jesse 150 n. 54
Rajak, Tessa 60–61, 69 n. 29
Reid, Duncan 4 n. 12, 202 n. 1, 235 n. 20
Resseguie, James L. 4 n. 10, 203 n. 3
Rhoads, David 3 n. 7
Richard, Pablo 3, 37
Richlin, Amy 135 n. 27
Roloff, Jürgen 77 n. 4
Romanow, Rebecca Fine 145 n. 41
Rossing, Barbara R. 3 n. 3, 4 n. 12, 4 n. 14, 5, 108 n. 12, 114 n. 26, 115, 122 n. 42, 202 n. 1, 203 n. 4, 235 n. 20, 239 n. 26
Rowland, Christopher 3 n. 7, 68 n. 24, 208 n. 20
Rowland, Ingrid 153
Royalty, Robert M., Jr. 4 n. 11, 28
Ruffolo, David 125
Ruiz, Jean-Pierre 3 n. 8, 37

Runions, Erin 3 n. 6, 4 n. 14, 14 n. 1, 144 n. 41, 156
Russell, George William 56–57

Said, Edward W. 13–15, 25, 142 n. 37
Salisbury, Joyce E. 205 n. 13
Samuel, Simon 14 n. 1
Samuelsson, Maria Jansdotter 3 n. 3, 4 n. 11
Sánchez, David A. 4 n. 8
Schaberg, Jane 55, 68 n. 22
Schreiner, Thomas R. 242 n. 33
Schulz, Siegfried 113 n. 22
Schüssler Fiorenza, Elisabeth 3, 4 n. 9, 5, 105 n. 6, 122 n. 42, 129 n. 13
Schwarzenegger, Arnold 95
Scott, Kenneth 91–92, 95, 96 n. 44
Scott, R. B. Y. 71 n. 41
Sedgwick, Eve Kosofsky 160 n. 9, 162–63 n. 14, 173, 176
Seesengood, Robert Paul 4 n. 8, 30 n. 18, 207 n. 17
Segal, Lynne 145 n. 41
Segarra, Marta 204 n. 6
Segovia, Fernando F. 13
Seigworth, Gregory J. 155, 159
Selvidge, Marla J. 67 n. 15, 127 n. 6
Setel, T. Drorah 67 n. 15
Shannon, Laurie 204–6
Shaw, Brent D. 59–60
Sherwood, Yvonne 179
Simpson, Mark 82, 93–94, 97
Sintado, Carlos Alberto 4 n. 12
Sissa, Giulia 135 n. 27
Skinner, Marilyn B. 135 n. 27, 141, 143 n. 38, 147 n. 48
Slater, Thomas B. 212 n. 30
Smalley, Stephen S. 240 n. 30
Smith, Shanell T. 3 n. 5, 4 n. 9
Spaeth, Barbette Stanley 135 n. 25
Spittler, Janet E. 204 n. 10
Spivak, Gayatri Chakravorty 13–14, 22, 25, 34 n. 24, 37, 190
Staley, Jeffrey L. 39, 67 n. 4
Steel, Karl 205 n. 13

Steiner, Gary 204 n. 10, 222 n. 50
Stenström, Hanna 164 n. 20, 165
Stewart, Kathleen 162 n. 13
Stone, Ken 219 n. 47
Storr, Merl 125
Strelan, Rick 239 n. 27
Stulman, Louis 127 n. 3
Sugirtharajah, R. S. 13
Sweet, J. P. M. 212 n. 30
Sweetnam, Mark S. 4 n. 14
Swete, Henry Barclay 33 n. 23, 67 n. 11, 148 n. 52

Taussig, Hal 156
Terada, Rei 159 n. 6
Thimmes, Pamela 149 n. 52
Thomas, Robert L. 240 n. 28
Thompson, Leonard L. 32 n. 21, 69 n. 26
Thompson, Steven 71 n. 41
Thrift, Nigel 161
Thurman, Eric 14 n. 1
Tiffin, Helen 22 n. 7, 206 n. 16, 228
Tomkins, Silvan 166 n. 23, 177 n. 37
Torrey, C. C. 71 n. 41
Tracy, David 198

Vander Stichele, Caroline 3 n. 3, 4 n. 13, 5, 105 n. 7, 122 n. 42, 127 n. 6
Vermeule, Cornelius C. 133, 136
Veyne, Paul 135 n. 27
Voorwinde, Stephen 156 n. 1

Waller, Alexis 156
Walliss, John 4 n. 14
Washington, Harold C. 45, 48–50, 67 n. 8, 67 nn. 12–14
Weil, Kari 206 n. 15
Weinrich, William C. 185 n. 8, 209 n. 24
Westhelle, Vitor 3 n. 8
Whitehead, Alfred North 198–99
Williams, Craig A. 135–37, 140–42, 147 n. 48
Williams, Jeromie 222–23
Williams, Wesley 236 n. 23

Winkler, John J. 135 n. 27
Wise, Michael 236
Witherington, Ben, III 113 n. 22, 240
Wolfe, Cary 204 n. 9, 206 n. 16, 216 n. 42

Young, Robin Darling 69 n. 28

Zane, Frank 93

www.ingramcontent.com/pod-product-compliance
Lightning Source LLC
Chambersburg PA
CBHW020642300426
44112CB00007B/211